Structural Mechanics

Analytical and Numerical Approaches for Structural Analysis

Lingyi Lu†
School of Civil Engineering
Southeast University, Nanjing, China

Junbo Jia
Aker Solutions, Bergen, Norway

Zhuo Tang
National Wind Institute
Texas Tech University, Lubbock, Texas, USA

CRC Press
Taylor & Francis Group
Boca Raton London New York

CRC Press is an imprint of the
Taylor & Francis Group, an **informa** business
A SCIENCE PUBLISHERS BOOK

First edition published 2022
by CRC Press
6000 Broken Sound Parkway NW, Suite 300, Boca Raton, FL 33487-2742

and by CRC Press
2 Park Square, Milton Park, Abingdon, Oxon, OX14 4RN

CRC Press is an imprint of Taylor & Francis Group, LLC

Library of Congress Cataloging-in-Publication Data (applied for)

ISBN: 978-0-367-55912-0 (hbk)
ISBN: 978-0-367-55915-1 (pbk)
ISBN: 978-1-003-09569-9 (ebk)

DOI: 10.1201/9781003095699

Typeset in Times New Roman
by Radiant Productions

In Memory of Professor Lingyi Lu

Preface

On 13th of July, 2019, I was informed about sad news of the passing away of Professor Lingyi Lu, after battling pancreatic cancer for more than half a year.

Just two days before this, I visited him at a hospital in Shanghai, and he informed me about the progress of this book and asked me to finalize it. Together with efforts of Dr. Zhuo Tang, a former PhD student of Prof. Lu, the book has finally been finished with a maximum preservation of Professor Lu's academic thoughts.

This book covers essential topics on structural mechanics and dynamics. It is mainly based on the contents of an undergraduate course in structural mechanics taught by Professor Lu at Southeast University, from the 1990s in the last century to 2019, and is an important contribution of Professor Lu's lifetime academic and pedagogic activities that he has been deeply engaged in, reflecting the forefront education status of structural mechanics in China during that period.

Modern structural mechanics originated from Galileo Galilei, Isaac Newton, and Robert Hooke. It is the computation of deformations, deflections, and internal forces or stresses within structures, either for design or for performance evaluation of existing structures. With the objective of providing principles of structural mechanics, topics covered in this book include both basic and advanced ones. Basic topics include geometric stability, internal forces and deflections of statically determinate structures, force and displacement method, and influence lines. Advanced topics include matrix displacement method for structural analysis, dynamics of structures, and limit load analysis.

The book serves as a classroom textbook in structural mechanics. It is written in such a way that it can be followed by anyone with a basic knowledge of classical and material mechanics.

While the book does not seek to promote any specific "school of thought," it inevitably reflects the authors' best practices and working habits. This is particularly apparent in the topics selected and the level of detail devoted to each of them, the choice of mathematical treatments and symbolic notations. It should not deter readers from seeking to find their own best practices and working habits.

Many individuals have provided valuable supports to finalize the book. Those contributions come from the family members of Professor Lu: his wife, Professor Hong Zhuang, and his daughter, Hongling Lu, and from his former students: Hongsheng Wang, Jie Zhang, Yuanzhi Liu, Min Chen, and Wensong Hu.

Professor Gang Wu from Southeast University has inquired about the book status several times and expressed an inspiring encouragement on behalf of Southeast University.

When I visited Professor Lu in the hospital in his last stage, he had requested me to write a preface for this book, as both his friend and former undergraduate student, I came up with this preface with emotion, and with a deep memory of my undergraduate study at Southeast University, that has had a strong influence on my life and career after I left the University. Hope this book will be useful for both students and academic professionals.

Junbo Jia

Contents

Chapter 1
Geometric Stability and Types of Structures

1.1 Classifications of Structural Members and Connections

A ***bar structure*** is an assemblage of connected bars used to support loads. In general, a complex structure can be decomposed into two types of components: ***structural elements*** and ***connections***. For example, Figure 1.1a shows a structure composed of 3 connected bars, *AB*, *BC* and *CD*. The deformation in the structure, shown in Figure 1.1b, caused by the concentrated and distributed loads, *P* and *q*, indicates that the members, *AB* and *BC*, of the structure have undergone bending. Usually, a bar undergoing bending has three kinds of internal forces: axial, shear and bending moments. Such a member is often referred to as a ***beam element***. This implies that the structural elements, *AB* and *BC*, of the structure shown in Figure 1.1a, are beam elements. However, there are some differences between the members *BC* and *AB*. The member *BC* primarily resists load *q* applied laterally to its axis. Its deflection is dominated by bending. Such a member undergoing bending primarily is termed a ***beam***. On the other hand, members, like element *AB* which are generally vertical and resist axial compressive loads, are called ***beam-columns***. However, since there is only an axial internal force in member *CD*, it is called a ***column***.

In addition to beam elements, there is another kind of structural member subjected to forces (no moments) acting at only two locations. This type of structural member

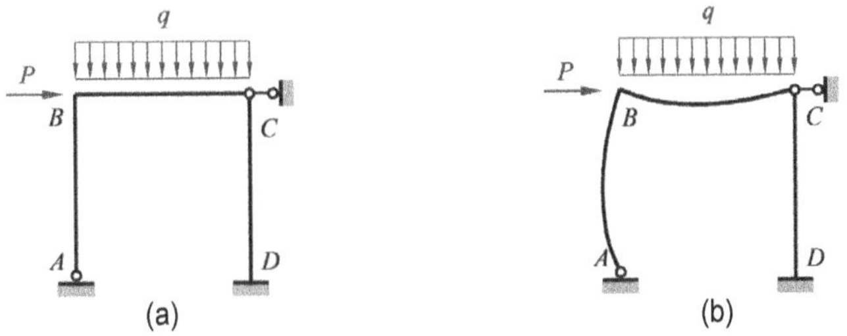

Figure 1.1. (a) A structure composed of a beam, a beam-column and a column which are pin-connected at nodes *A* and *C*, and fixed-connected at nodes *B* and *D*. (b) The deformation of the structure caused by the concentrated and distributed loads *P* and *q*.

is called a ***two-force member***, which is frequently used in structural engineering. Furthermore, for it to be in static equilibrium, the net force at each location must be equal, opposite, and collinear. Figure (1.2a) shows a structure whose members are connected together by pin-connected joints, and where the external loads are considered to act only at the nodes and result in forces in the members that are either tensile or compressive. Thus, it is a structure consisting of two-force members only. A structure may comprise of beam elements and two-force members. For example, the structure shown in Figure (1.2b) is composed of a two-force member *BD*, and four beam elements *AB*, *BC*, *CD*, *DE*.

Structural members are connected together by various ***joints*** depending on the designer's intent. The three types of simple joints most often specified are the ***pin-connected joint*** (or ***hinged joint***), the ***fixed-connected joint*** (or ***rigid joint***), and the ***sliding-connected joint***. Figure (1.3) shows the symbols and the internal forces of the three joints.

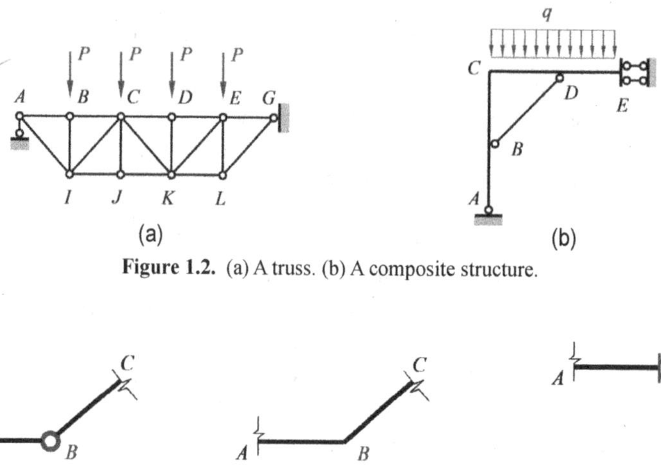

Figure 1.2. (a) A truss. (b) A composite structure.

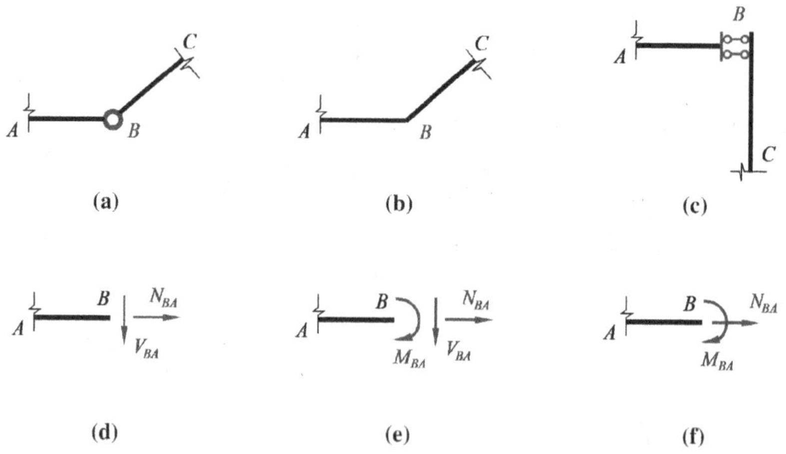

Figure 1.3. The three types of simple joints frequently used in structural engineering. (a) The idealized symbol of a pin. (b) The idealized symbol of a fixed joint. (c) The idealized symbol of a sliding joint. (d) The action at section *BA* on the bar from pin *B*. (e) The action at section *BA* on the bar from the fixed joint. (f) The action at section AB on the bar from the sliding joint *B*.

It should be noticed that there are no end moments at a pin, and that end shears vanish at a sliding-connected joint. In addition, a pin joint couples linear end displacements of structural members. For example, the pin shown in Figure (1.3a) couples the linear displacements at node B, that is:

$$u_{BA} = u_{BC} = u_B, \text{ and } v_{BA} = v_{BC} = v_B \tag{1.1}$$

where u_{BA}, u_{BC} are the horizontal displacements of members BA and BC, and v_{BA}, v_{BC} are the vertical displacements of members BA and BC. Thus, the linear end displacements at pin B can be denoted by u_B and v_B where the subscript B indicates the pin joint. A rigid joint couples not only the linear displacements, but also the angular displacements at the joint. For example, the fixed joint shown in Figure (1.3b), yields

$$\begin{cases} u_{BA} = u_{BC} = u_B \\ v_{BA} = v_{BC} = v_B \\ \theta_{BA} = \theta_{BC} = \theta_B \end{cases} \tag{1.2}$$

where θ_{BA} and θ_{BC} are the angular displacements at members BA and BC respectively. This implies that the rotations at a rigid joint can be denoted by a symbol θ_B in which the subscript B indicates the joint.

A joint connecting only two structural members is usually called a ***simple joint***. However, a joint may be used to connect more than two structural members. Such a joint is referred to as a ***multiple-connected joint*** (or simply ***multiple joint***). Figure (1.4) shows the four types of multiple joints frequently used in structural engineering practice. Figures (1.4a) and (1.4b) show a multiple pin and a multiple rigid joint, respectively. The multiple joint D shown in Figure (1.4c) combines a rigid joint between the bars DA and DB with the pin between DC and ADB. Such a multiple joint is termed a ***composite pin***. Similarly, the multiple joint D shown in Figure (1.4d) is the combination of the rigid joint between DB and DC and the sliding device between AD and BDC, which is referred to as a ***composite sliding joint***.

Figure (1.5) shows the actions on the structural members at the composite pin D, which couples the end linear displacements, and some of the end rotations at D, that is,

$$\begin{cases} u_{DA} = u_{DB} = u_{DC} = u_D \\ v_{DA} = v_{DB} = v_{DC} = v_D \\ \theta_{DA} = \theta_{DB} \end{cases} \tag{1.3}$$

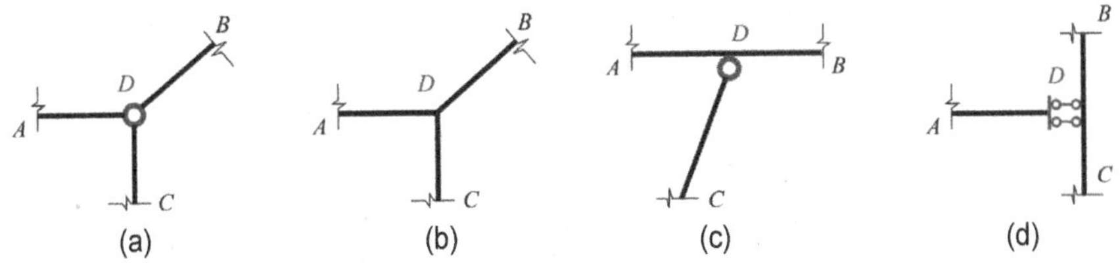

Figure 1.4. The four types of multiple-connected joints often used in structural engineering. (a) A multiple pin. (b) A multiple rigid joint. (c) A composite pin joint. (d) A composite sliding joint.

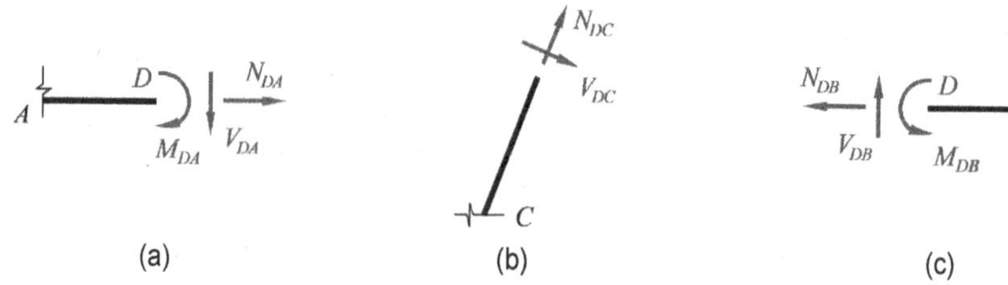

Figure 1.5. The end forces at the composite pin. (a) The action at member *DA* from the composite pin *D*. (b) The action at member *DC* from the composition pin. (c) The action at member *DB* from the composite pin.

The end rotation θ_{DC} is independent of θ_{DA} and θ_{DB} because they are not coupled by the composite pin.

Figure (1.6) shows the actions on the structural members from the composite slider *D*, which couples the end rotations, and some of the end linear displacements at *D*. It gives:

$$\begin{cases} u_{DA} = u_{DB} = u_{DC} = u_D \\ v_{DB} = v_{DC} \\ \theta_{DA} = \theta_{DB} = \theta_{DC} = \theta_D \end{cases} \tag{1.4}$$

The vertical displacement v_{DA} is independent of v_{DB} and v_{DC} because the sliding device does not couple them.

A ***support*** is a connection between a structural member (or members) and the *foundation* (or, more commonly *base*) of a structure. It corresponds to the locally inhibited displacement and rotation capacities (degrees of freedom) of structures. As shown in Figure 1.1a, the support at *A* is a ***pin support*** while the support at *D* is a ***fixed support***. In addition, the support at *E* in Figure (1.2b) is a ***sliding support***. All the displacements at a fixed support vanish due to the rigid coupling. Similarly, all the linear displacements at a pinned support are zero. However, the end rotations of the members at a pin support are independent and unknown. The actions on the structural members from a support are similar to those from the corresponding joint.

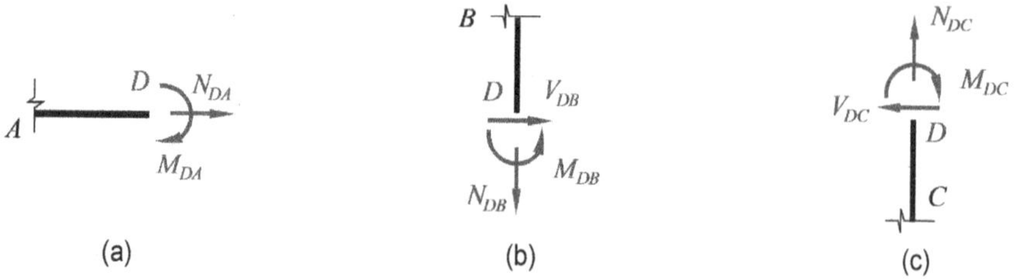

Figure 1.6. The end forces at the composite sliding joint. (a) The action at section *DA* from the composite slider *D*. (b) The action at section *DB* from the composite joint *D*. (c) The action at section *DC* from the joint.

1.2 Introduction to Geometrically Stable and Unstable Systems

In general, engineering structures include various types of structures such as buildings, bridges, offshore structures (fixed bottom and floating ones), cranes, and many other types of infrastructures. These are all examples of load-bearing structures. Members of a structure are arranged and joined together by various structural connections to resist loads. The magnitude of loads that a structure can withstand may depend on the geometric and physical properties of the structure. However, whether a system can support loads depends primarily on the geometric arrangement of the structural elements in the system.

If a system collapses when a tiny lateral perturbation is applied on it, it is said to be ***geometrically unstable***. In general, a bar system is said to be geometrically unstable if it moves continuously or collapses without accountable deformations of its members. Consider the two systems shown in Figure (1.7), composed of four bars. The system shown in Figure (1.7a) is geometrically unstable. On the other hand, the system shown in Figure (1.7b) cannot move if its members are assumed to be rigid. Such a system can sustain coplanar loads as long as they are not too large. A system is called ***geometrically stable*** if it cannot move without deforming any of its members. The system shown in Figure (1.7b) is geometrically stable.

In addition to geometrically stable and unstable systems, there exists another special system as shown in Figure 1.8a, in its initial state with three collinear hinges. In this case, hinge C is unconstrained in the vertical direction and movable.

However, after an infinitesimal displacement, the system becomes geometrically stable (see Figure 1.8b). In general, a system is said to be ***instantaneously unstable*** if it becomes stable after undergoing an infinitesimal displacement.

Usually, a geometrically unstable system, like the system shown in Figure (1.7a), cannot be used as a structure because it may collapse under a tiny load. The question

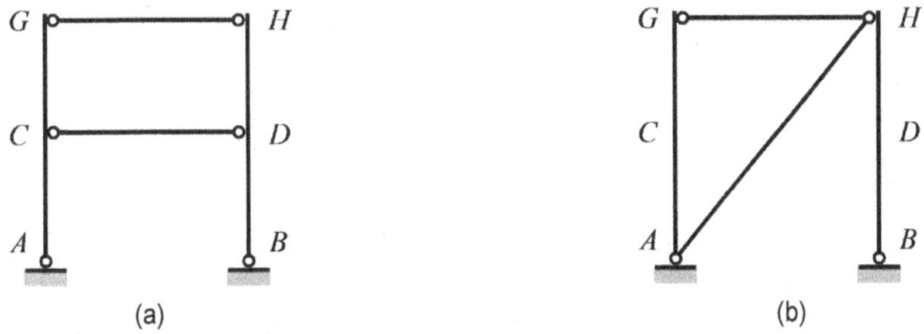

Figure 1.7. Two systems composed of four members. (a) A geometrically unstable system. (b) A geometrically stable system.

Figure 1.8. An instantaneously unstable system. (a) The initial state of the instantaneously unstable system. (b) The stable state of the system after an infinitesimal displacement.

now arises: can an instantaneously unstable structure be used as a structure? To answer this question, consider the two systems of Figure (1.9) with elastic modulus $E = 2.0 \times 10^{11}$ N/m^2, length of each member $l = 0.3$ m, cross-sectional area of each member $A = 1 \times 10^{-4}$ m^2. The system on the left side is instantaneously unstable, while that on the right side is geometrically stable.

Table 1.1 lists the responses of the two systems under a vertical load of 5 N. The responses of the instantaneously unstable system are one hundred times larger than the corresponding responses of the geometrically stable system. All this shows that instantaneously unstable systems are not suitable either.

In summary, systems consisting of one-dimensional members can be classified as *geometrically stable*, *geometrically unstable*, or *instantaneously unstable*. Usually, only the geometrically stable systems are suitable structural designs. This conclusion immediately raises a question: "What are the rules for constructing geometrically stable systems?", addressed in the rest of this chapter.

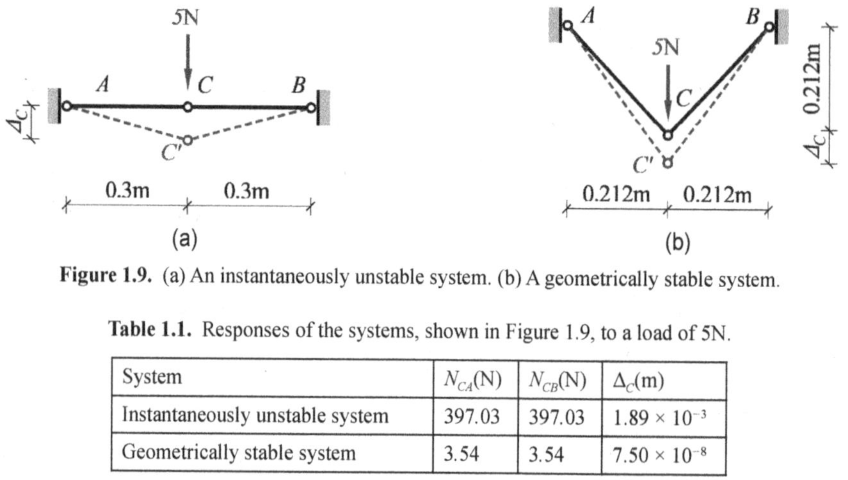

Figure 1.9. (a) An instantaneously unstable system. (b) A geometrically stable system.

Table 1.1. Responses of the systems, shown in Figure 1.9, to a load of 5N.

System	N_{CA}(N)	N_{CB}(N)	Δ_C(m)
Instantaneously unstable system	397.03	397.03	1.89×10^{-3}
Geometrically stable system	3.54	3.54	7.50×10^{-8}

† The properties of the systems are as follows: elastic modulus $E = 2.0 \times 10^{11}$ N/m^2, length of each member $l = 0.3$ m, cross-sectional area of each member $A = 1 \times 10^{-4}$ m^2.

1.3 Rigid Degrees of Freedom and Constraints

Before developing a formal procedure for constructing a geometrically stable system, it is necessary to establish some preliminary definitions and concepts. A ***rigid degree of freedom*** (or ***degree of freedom, DOF***) of a system is an independent parameter that is required to define its configuration at any time (neglecting the deformations of its members).

For example, a free particle (or mass point) in three-dimensional space has three degrees of freedom. Consider a particle shown in Figure (1.10a). In order to locate the particle at time t, a Cartesian coordinate system needs to be established. The position of the particle at any time can then be defined by three independent coordinates (x_P, y_P, z_P). This shows that the particle has three DOFs. Similarly, a free particle in two-dimensional space has two DOFs.

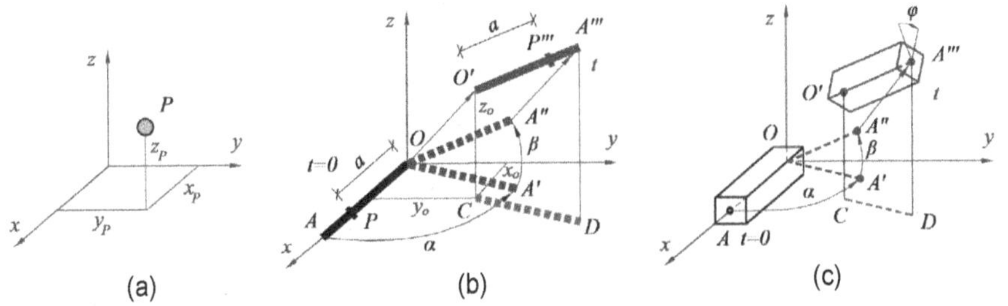

Figure 1.10. (a) A particle in three-dimensional space has 3 DOFs. (b) A rigid bar in three-dimensional space has 5 DOFs. (c) A rigid body has 6 DOFs.

A rigid bar in three-dimensional space has five degrees of freedom. Consider an unconstrained bar OA of length l, shown in Figure (1.10b), which moves to $O'A'''$ at time t from the initial position OA. We can decompose the total displacement into two rotations, α, β, and a translation, OO'. First, OA is rotated about the z-axis by α to the position OA', which is parallel to the projection, CD, of $O'A'''$ onto the xy-plane. Then, OA' is rotated about the axis perpendicular to the plane $A'Oz$ by β to OA'' which is parallel to $O'A'''$. Finally, the bar moves to $O'A'''$ by translation OO'. Thus, the position of the bar at time t can be defined by α, β and (x_O, y_O, z_O), which are the 5 DOFs of the rigid bar. Similarly, a rigid body in a three-dimensional space has 6 DOFs. As shown in Figure (1.10c), the five DOFs, α, β and (x_O, y_O, z_O) are used to define the position of an axis OA on the body, and the additional parameter, ϕ, is used to define the rotation of the body about the axis. Similarly, a rigid body in a plane has three degrees of freedom.

Next, consider two constrained systems shown in Figure (1.11). The system on the left side has a single degree of freedom (SDOF) because its configuration at any time can be defined by the rotation angle θ. The system on the right side has zero degree of freedom because it is properly constrained on both ends of member AB and has a time-invariant configuration.

The opposite of freedom is constraint. In the analysis of geometric stability, a ***constraint*** is defined as a restriction on a specific movement. The prototype of a constraint is a ***link*** which is defined as a bar hinged between two objects. For example, the system shown in Figure (1.11b) has two links, a long link and a short link. A link restrains the two points from moving away from each other. Point B in Figure (1.11a) can only rotate around point A because it is connected to the ground by a link AB. Since a link provides one and only one constraint, a link is said to be equivalent to a constraint. A pin offers two constraints and is said to be equivalent to two constraints as it prevents both vertical and horizontal displacements.

Figure 1.11. A rigid bar AB connected to the ground in a different manner. (a) A single-degree-of-freedom system. (b) A geometrically stable system.

The question now arises: is a pin equivalent to two links? The answer is *yes*. A rigid body in Figure (1.12a) is connected to the ground by two links, 1 and 2, intersecting at the real point *A*. The rigid body can only rotate around the intersection *A*. Thus, links, 1 and 2, are equivalent to a pin at *A*. Consider the rigid body in Figure (1.12b) connected to the ground by two links that do not intersect. At that instant the rigid body can only rotate about the intersection *A* of the extensions of the links 1 and 2 because their upper ends can only move in directions perpendicular to links 1 and 2, respectively. In this case *A* is an instantaneous center of the rigid body and is referred to as a ***virtual pin*** (or ***virtual hinge***). What happens if two links are parallel? The rigid body shown in Figure (1.12c) translates only in a direction perpendicular to the two links. It seems that the rigid body rotates about a point that is infinitely distant. In this case, the two links form an ***infinite pin*** (or ***infinite hinge***). In addition, a fixed support or fixed joint provides three constraints because it prevents relative rotation and translations in the horizontal and vertical directions.

The preceding discussions show that a constraint prevents a specific movement of an object. Does this mean that a constraint must reduce the number of DOFs of an object? Consider a system in Figure (1.13a), in which a particle is connected to the ground by three links, 1, 2 and 3. In fact, the particle constrained by the two links is properly constrained and has no degree of freedom. Any additional link added to the system will no longer reduce the DOFs and is called a ***redundant constraint*** (or ***redundant restraint***). The system in Figure (1.13b) has two redundant constraints because two extra links, 1 and 2, are added to a cantilever beam *AB* which is a natural expansion of the ground. It is noticed that an instantaneous unstable system has at least one redundant constraint. The proof of this proposition is left to the reader.

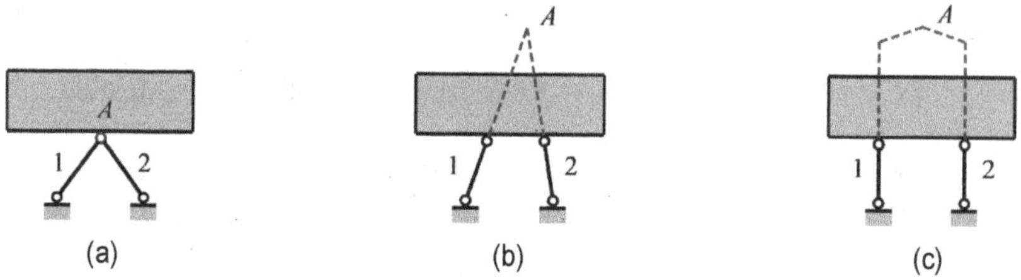

Figure 1.12. A pin is equivalent to two constraints. (a) A (real) pin. (b) A virtual pin. (c) An infinite pin.

Figure 1.13. Systems with redundant constraints. (a) A system with one redundant constraint. (b) A system having two redundant constraints.

1.4 Rules for Constructing Geometrically Stable Systems

Structures are usually designed to be geometrically stable so that they can resist loads in any direction. For a simple system, one may tend to judge whether it is geometrically stable or unstable using intuition. However, intuition offers little information on the stability of a complex system having a lot of bars. This section will develop a formal procedure for constructing geometrically stable systems.

1.4.1 Two-Body Rule

As shown in Figure (1.14a), two rigid bodies, *I* and *II*, are connected together by a hinge and a link. When rigid bodies *I* and *II* are connected by pin *A*, the assemblage has one and only one relative DOF—the relative rotation. As long as link *I* is non-collinear with the pin, it shall eliminate the relative rotation. This implies that the rigid-body assemblage has no relative DOF. Since a link can offer one and only one constraint, the assemblage has no redundant constraint. The discussion above leads to the following theorem.

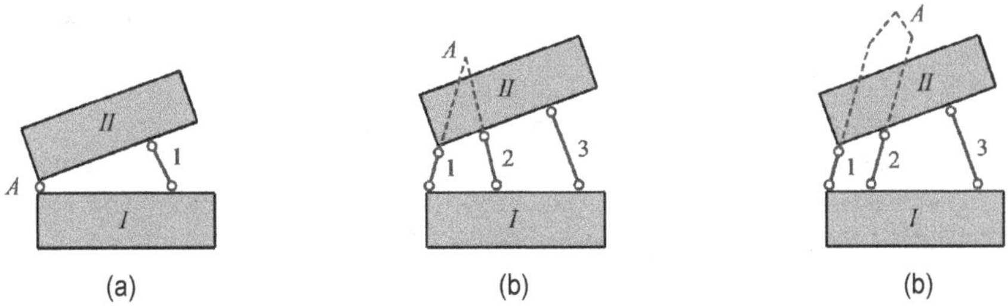

Figure 1.14. Two-body rule. (a) Two rigid bodies connected by a pin and a link. (b) Two rigid bodies connected by a virtual hinge and a link. (c) Two rigid bodies connected by an infinite pin and a link.

Theorem 1.1 (Two-Body Rule) If two rigid bodies in a plane are connected by a pin and a link that is non-collinearly placed with the pin, the rigid-body assemblage is then geometrically stable and has no redundant constraints.

The real pin connecting *I* and *II* may be replaced by a virtual hinge or an infinite hinge. See Figure (1.14b). The three types of systems shown in Figure 1.15 are excluded by the non-collinearity of the pin and the link.

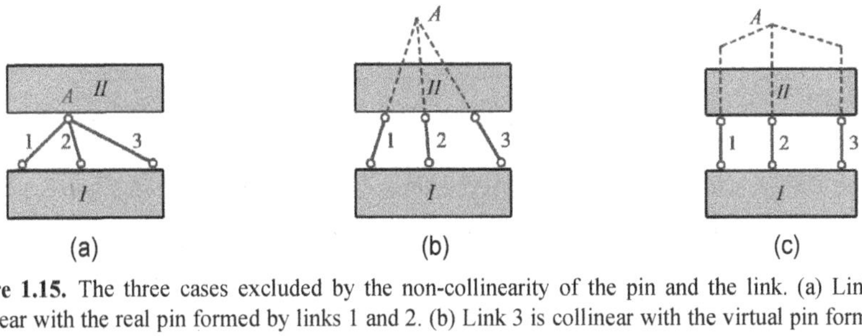

Figure 1.15. The three cases excluded by the non-collinearity of the pin and the link. (a) Link 3 is collinear with the real pin formed by links 1 and 2. (b) Link 3 is collinear with the virtual pin formed by links 1 and 2. (c) Link 3 is collinear with the infinite pin formed by links 1 and 2.

The two-body rule, Theorem 1.1, enables us to analyze the geometrical stability of a complex bar system, which will be illustrated by the following examples.

EXAMPLE 1.1

Analyze the geometric stability of the multi-span beams shown in Figure 1.16.

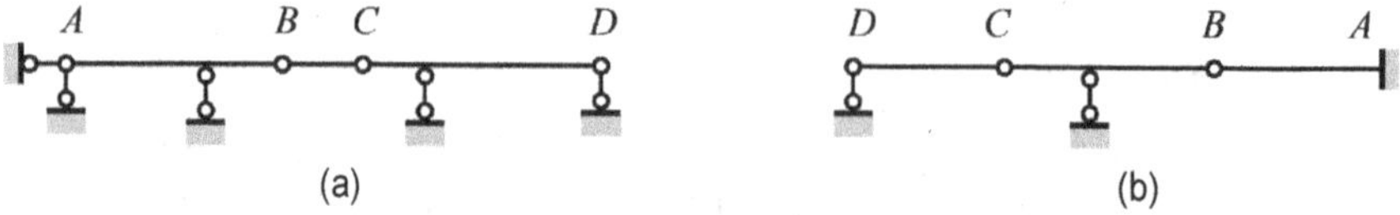

Figure 1.16. Systems of Example 1.1. (a) A three-span beam. (b) A two-span beam.

***Solution*:** First, let us analyze the geometric stability of the three-span beam shown in Figure (1.16a). As shown in Figure (1.17a), connecting the ground and the rigid bar *AB* by pin *A* and link 1 forms a new rigid body *I* enclosed by the envelope. It is a geometrically stable system without redundant constraints according to Theorem 1.1. Again, the rigid body *I* and the rigid bar *CD* are connected together by link 4 and pin $O_{I,II}$ formed by links 2 and 3, and these form a geometrically stable system without redundant constraints according to Theorem 1.1. Therefore, the system shown in Figure (1.16a) is geometrically stable and has no redundant constraints.

Next, consider the geometric construction of the two-span beam in Figure (1.16b). As shown in Figure (1.17b), the cantilever beam *AB* is a natural expansion of the ground, and forms a rigid body *I*. It is geometrically stable and has no redundant constraints. The rigid body *I* and the rigid bar *BC* are connected by hinge *B* and link 1 and form a rigid body *II*. It is geometrically stable and has no redundant constraints according to Theorem 1.1. Again, the rigid body *II* and the rigid body *CD* are connected by hinge *C* and link 2, and form a geometrically stable system without redundant constraints. Therefore, the two-span beam shown in Figure (1.16b) is geometrical stable and has no redundant constraints.

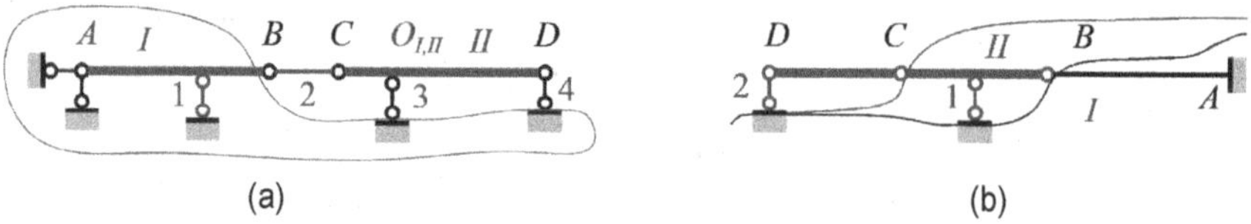

Figure 1.17. Analysis of the geometric stability for the systems in Example 1.1. (a) A three-span beam. (b) A two-span beam.

EXAMPLE 1.2

Analyze the geometric construction of the three-span beam shown in Figure (1.18).

***Solution*:** As shown in Figure (1.19a), the ground and the rigid bar *DE* are taken as the rigid bodies *I* and *II* respectively, while the rigid bar *BD* is viewed as a link connecting the rigid bodies *I* and *II*. Links 1 and 2 form a virtual hinge $O_{I,II}$ connecting *I* and *II*. Then, the rigid bodies *I* and *II* are connected by the virtual hinge $O_{I,II}$ and link 3, and form a rigid body *III enclosed* by the curve in Figure (1.19b).

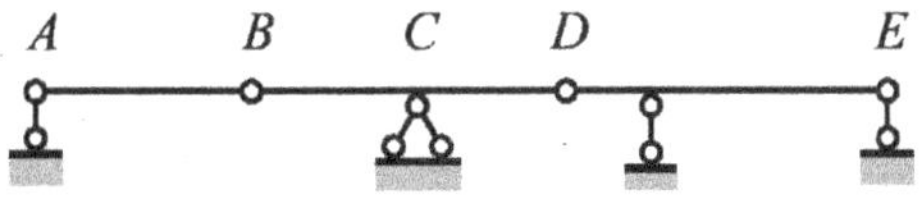

Figure 1.18. The three-span beam in Example 1.2.

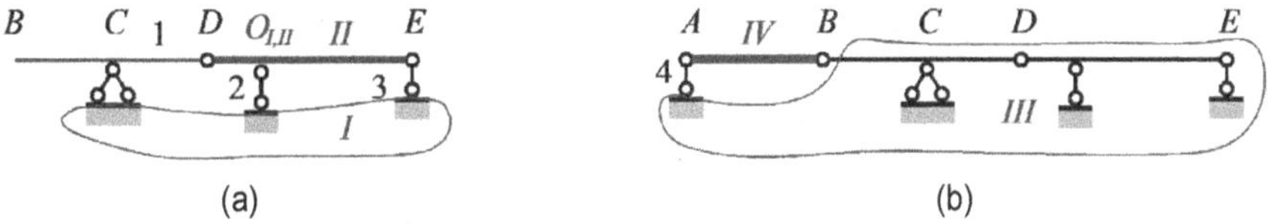

Figure 1.19. Analysis of the geometric stability of the three-span beam in Example 1.2. (a) The geometric construction of the rigid body *III*. (b) The geometric construction of the three-span beam.

It is a geometrically stable system without redundant constraints according to Theorem 1.1. The rigid bar *AB* viewed as the rigid body *IV* and the rigid body *III* are connected by pin *B* and link 4 and form a geometrically stable system without redundant constraints.

1.4.2 Three-Body Rule

This section will be devoted to developing another useful rule, called ***three-body rule***. As shown in Figure (1.20a), three rigid bodies, *I*, *II* and *III*, are mutually pinned together. The two-body rule immediately provides the following theorem by taking the rigid body *I* as a link connecting the rigid bodies *II* and *III*.

Figure 1.20. The three bodies pairwise pinned. (a) Three rigid bodies are pairwise connected together by three real pins. (b) Three rigid bodies are pairwise joined together by the infinite hinge *A*, a real hinge *B* and the virtual hinge *C*.

Theorem 1.2 (Three-Body Rule) A system consisting of three pairwise hinged rigid bodies without redundant constraints is geometrically stable and has no redundant constraints if the three hinges are non-collinear.

Here, the term *non-collinear* is used to exclude the three cases shown in Figure 1.21.

The following example illustrates the application of Theorem 1.2.

EXAMPLE 1.3

Analyze the geometric stability of the two systems shown in Figure (1.22).

Solution: (1) Analyze the geometric construction of the system shown in Figure (1.22a). As shown in Figure (1.23a), take the ground, the members *ACEF* and

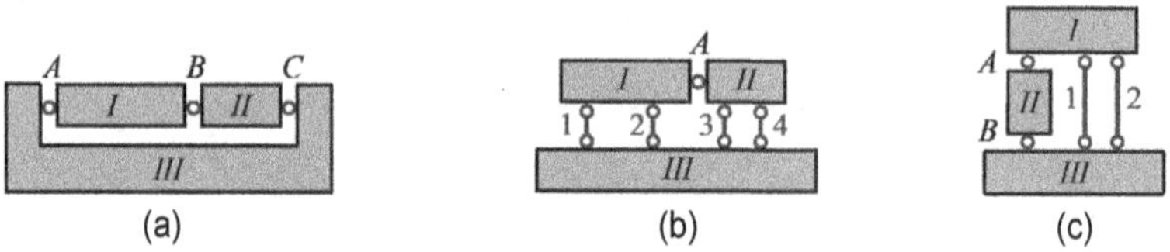

Figure 1.21. The cases excluded by Theorem 1.2. (a) Three real pins are collinear. (b) A real pin and two infinite pins are collinear. (c) An infinite pin and two real pins are collinear.

Figure 1.22. The two systems of Example 1.3. (a) A composite system consisting of flexural members and links. (b) A system consisting of links.

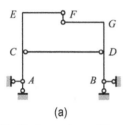

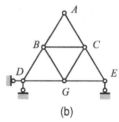

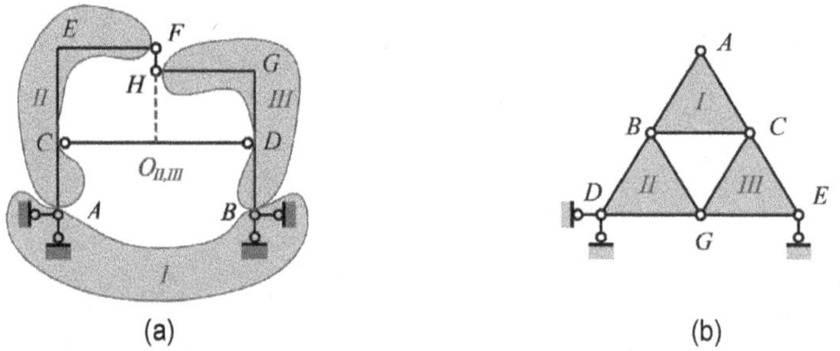

Figure 1.23. The geometric constructions of the systems in Example 1.3.

BDGH as the rigid bodies *I*, *II* and *III*. Links *FH* and *CD* form a virtual hinge, $O_{II,III}$, connecting the rigid bodies *II* and *III*. Because the three rigid bodies *I*, *II* and *III* are pairwise connected together by the hinges *A*, $O_{II,III}$ and *B*, the system is geometrically stable, and has no redundant constraints according to Theorem 1.2.

(2) Analyze the geometric construction of the system shown in Figure (1.22b). As shown in Figure (1.23b), note that Theorem 1.2 states that three rigid bars pairwise pinned form a geometrically stable system without redundant constraints, one can then take the triangles *ABC*, *BDG* and *CGE* as the rigid bodies *I*, *II* and *III*. Since the rigid bodies *I*, *II* and *III* are pairwise jointed by the non-collinear hinges *B*, *C* and *G*, they form a large rigid body *ADE* without redundant constraints. The large rigid body *ADE* and the ground are connected by pin *D* and link *E* and form a geometrically stable system without redundant constraints.

1.4.3 The Dual-Link Rule

In addition to the two- and three-body rules, structural engineers often use a rule, named ***dual-link rule***, to design structures. Figure (1.24a) shows two non-collinear

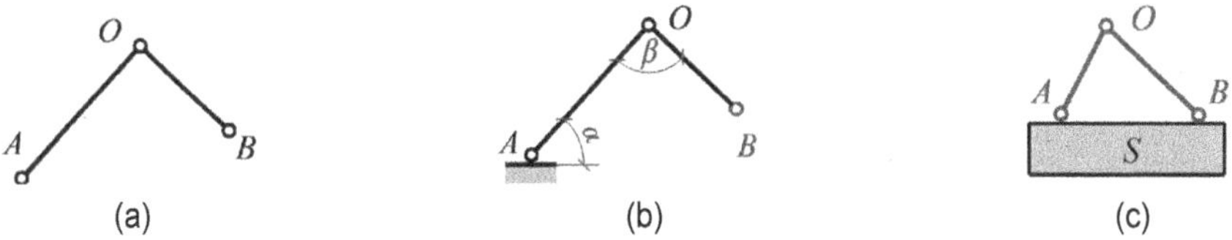

Figure 1.24. Dual-link rule. (a) Two non-collinear links hinged at O. (b) A dual-link object connecting the ground and a free point B. (c) An extended system by adding a dual-link object AOB to S.

links hinged at O, which are called a ***dual-link assembly*** denoted by AOB. In general, a dual-link assembly is defined as an assemblage in which a point is connected to two and only two non-collinear links.

Theorem 1.3 (Dual-Link Rule) The geometric stability of a system remains unchanged when a dual-link assembly is added to or dropped from the system.

***Solution*:** As shown in Figure (1.24b), consider a dual-link object connecting the ground and a free point B. Then, two independent parameters α and β are required to define the position of point B. Since a point in a plane has two degrees of freedom, this shows that a dual-link assembly is equivalent to no constraint between A and B. The geometric stability of a system S remains unchanged when no constraint is added or dropped from it.

The dual-link rule provides the following two propositions, which are frequently used in the analysis of geometric stability:

1. If a system S is geometrically stable, and has n redundant constraints, then the new system S^{I} obtained by adding (or dropping) a dual-link object to (or from) S is geometrically stable and has n redundant constraints.
2. If a system is geometrically unstable, and has n redundant constraints, then the new system S^{I} obtained by adding (or dropping) a dual-link object to (or from) S is geometrically unstable and has n redundant constraints.

For example, as shown in Figure (1.24c), if S is a geometrically stable system without redundant constraints, then the extended system by adding a dual-link object AOB to S is geometrically stable and has no constraint.

It should be noticed that the definition of a dual-link assembly excludes the three cases shown in Figure 1.25. In Figure (1.25a), AOB is not a dual-link assembly because OA and OB are flexural members rather than links. In fact, this system is geometrically stable, and has no redundant constraints because the three rigid bodies, the ground, OA and OB, are pairwise connected by the real pin O, and the two virtual

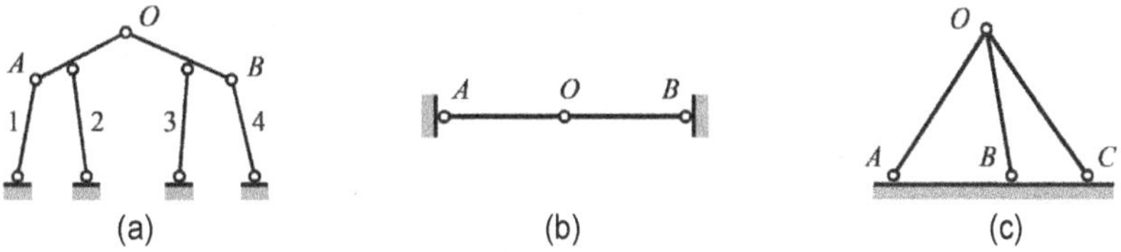

Figure 1.25. Three cases excluded by the definition of a dual-link assembly. (a) Point O is connected to two flexural members. (b) Point O is connected to two collinear links. (c) Point O is connected to three links.

pins formed by links 1 and 2, 3 and 4 respectively. *AOB* in Figure (1.25b) is not a dual-link object either because the two links *OA* and *OB* are collinear. *AOB* in Figure (1.25c) is not a dual-link object because *O* is also connected to *C*. However, if we remove link *OC* first, then the ground and the dual-link object *AOB* form a rigid body without redundant constraints, i.e., *OC* is a redundant constraint because it is a link connecting the rigid body to itself.

The following example illustrates the application of the dual-link rule.

EXAMPLE 1.4

Analyze the geometric stability of the two systems shown in Figure 1.26.

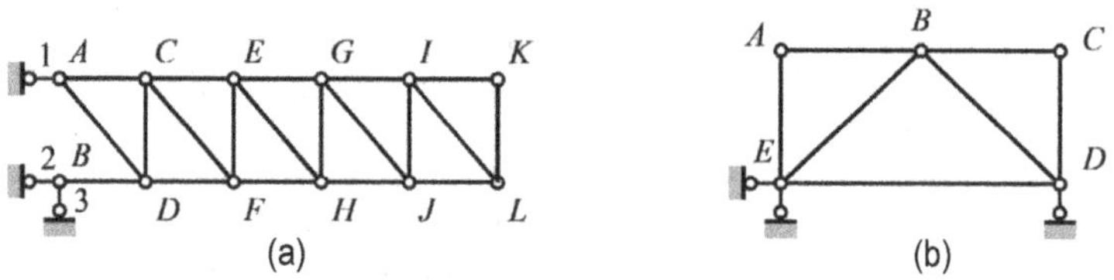

Figure 1.26. The two systems (a) and (b) for Example 1.4.

Solution: (1) Analyze the system shown in Figure (1.26a).

Theorem 1.3 (dual-link rule), shows that the dual-link object *IKL* can be dropped without changing the geometric stability of the system. If the dual-link object *IKL* is removed, *ILJ* immediately becomes a dual-link object, and can be dropped again. As shown in Figure (1.27a), after the dual-link objects 1, 2, ..., 11 are all removed sequentially, the rest of the system is link 1 having one degree of freedom. According to the dual-link rule, the original system is geometrically unstable, and has no redundant constraints.

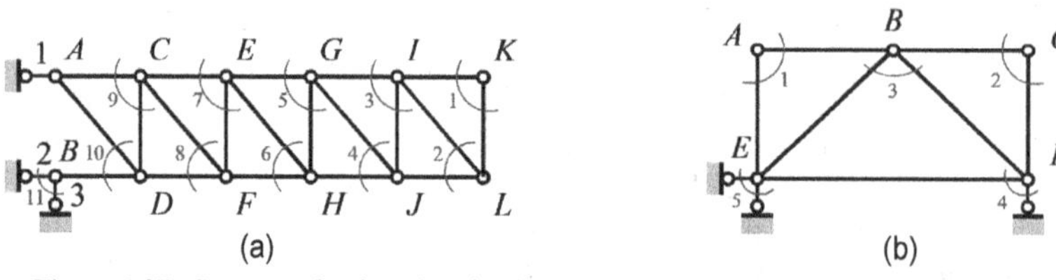

Figure 1.27. Sequence for dropping the dual-link for systems (a) and (b), respectively.

(2) Analyze the system shown in Figure (1.26b).

As shown in Figure (1.27b), after dropping all the dual-link objects 1,2,..., 5 sequentially, the remaining system is the ground. Thus, the original one is a geometrically stable system without redundant constraints.

1.4.4 The Simply-Supported Rule

If a system *S* is connected to the ground using a pin and a non-collinear link, then the combination of *S* and the ground is called a *simply-supported system*, and *S* is referred to as the *main part* of the simply-supported system. The prototype of

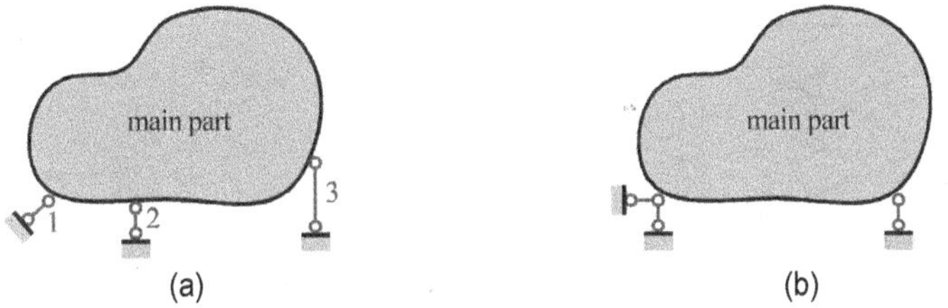

Figure 1.28. Simply-supported systems. (a) A simply-supported system consisting of a virtual pin and a link. (b) A simply-supported system consisting of link 3 and a real pin, or link 1 and the infinite pin.

simply-supported systems is the simply-supported beam. Figure (1.28a) shows a simply-supported system, whose main part is supported by a virtual pin and link. Figure (1.28b) shows a simply-supported system, whose main part is supported by link 3 and the real pin, or by link 1 and an infinite pin formed by links 2 and 3.

The following theorem shows that the geometric stability of a simply-supported system is dominated by its main part.

Theorem 1.4 (Simply-Supported Rule) A simply-supported system has the same geometric stability as its main part.

The following example demonstrates the application of Theorem 1.4—the simply-supported rule.

EXAMPLE 1.5

Analyze the geometric constructions of the two systems shown in Figure (1.29).

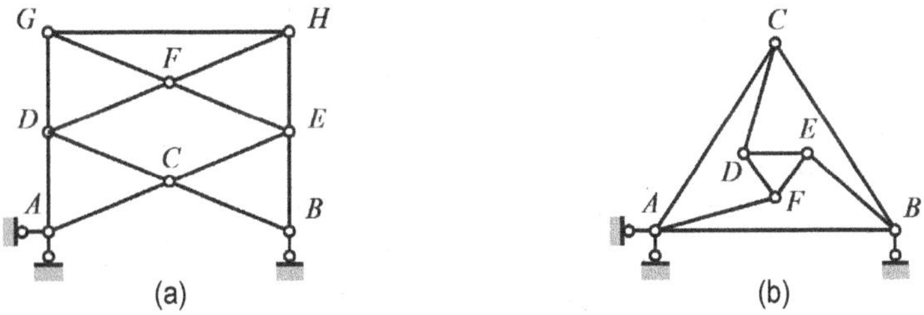

Figure 1.29. The original systems in Example 1.5.

***Solution*:** (1) Consider the system shown in Figure (1.29a).

This is a simply-supported system, whose main part is illustrated in Figure (1.30a). After dropping all the dual-link objects 1, 2, ...,6 sequentially, the rest of the main part is the rigid bar *GH*. Theorem 1.3, the dual-link rule, shows that the main part is a geometrically stable system without redundant constraints. Theorem 1.4, the simply-supported rule, shows that the simply-supported system shown in Figure (1.29a) is geometrically stable, and has no redundant constraints.

(2) Analyze the system shown in Figure (1.29b).

This is also a simply-supported system, whose main part is shown in Figure (1.30b). Theorem 1.2, the three-body rule, shows that *ABC* and *DEF* are two independent rigid bodies without redundant constraints. They are connected together by a virtual

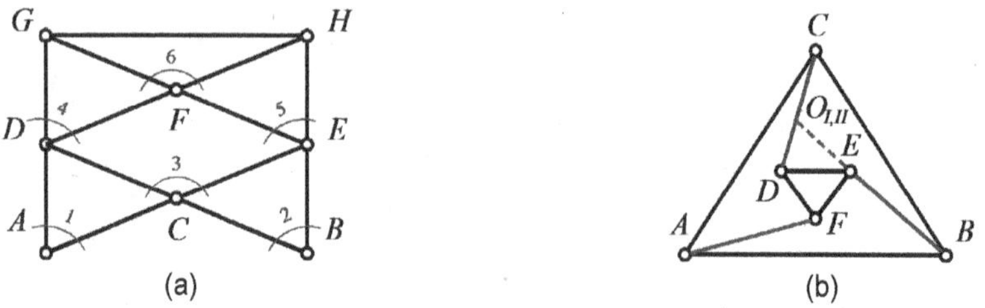

Figure 1.30. (a) The main part of the simply-supported system shown in Figure (1.29a). (b) The main part of the simply-supported system shown in Figure (1.29b).

hinge $O_{I,II}$ and link *AF*. Theorem 1.1, the two-body rule, shows that this main part is a geometrically stable system without redundant constraints. Finally, Theorem 1.4, the simply-supported rule, shows that the original system shown in Figure (1.29b) is geometrically stable and has no redundant constraints.

1.5 Examples of Geometric-Stability Analysis

This section will present applications of the four theorems presented in the previous section.

EXAMPLE 1.6

Analyze the geometric stability of the systems shown in Figure (1.31).

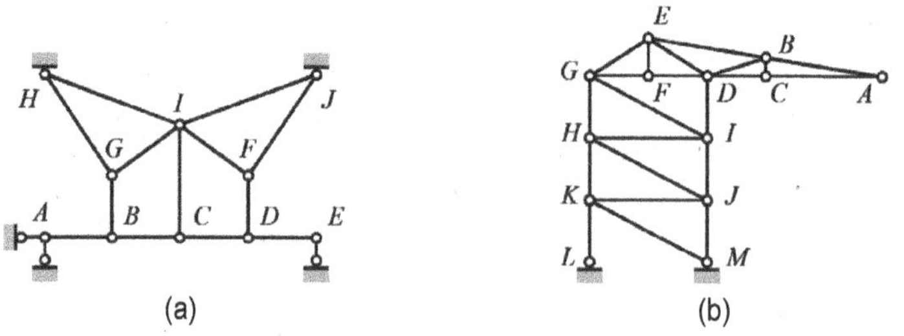

Figure 1.31. The original systems in Example 1.6.

Solution: (1) Analyze the geometric stability of the system shown in Figure (1.31a). Figure (1.32a) shows that the original system given in Figure (1.31a) can be constructed by sequentially adding the dual-line objects, 8, 7, 6, 5, 4, 3, 2, 1, to the ground. Theorem 1.3, the dual-link rule, shows that this is a geometrically stable system without redundant constraints.

(2) Consider the system shown in Figure (1.31b).

Figure (1.32b) shows the rest of the system obtained by sequentially dropping the dual-link objects, *BAC*, *BCD*, *EBD*, from the original system. The dual-link rule shows that extending the ground by sequentially adding the dual-line objects, *LKM*, *KJM*, *KHJ*, *HIJ*, *HGI*, leads to a rigid body *I* without redundant constraints. See Figure (1.32b). The rigid bodies *I* and *II* (the shaded regions in Figure 1.32b) are connected by the real hinge *G* and the non-collinear link *DI*, thus forming

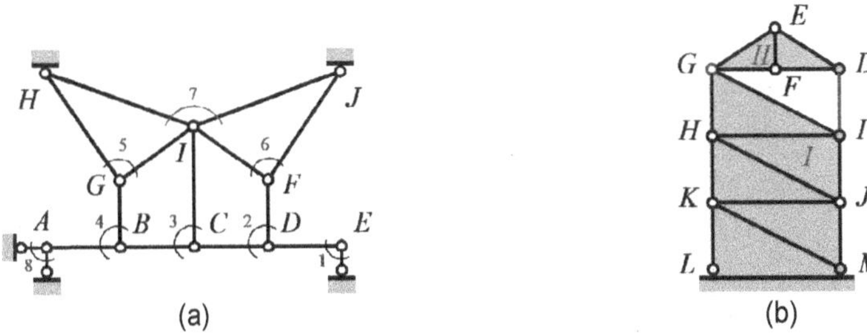

Figure 1.32. (a) The original system can be obtained by adding dual-link objects to the ground sequentially. (b) The rest of the system obtained by removing the dual-link objects sequentially.

a geometrically stable system without redundant constraints according to Theorem 1.1—the two-body rule. Finally, the dual-link rule shows that the original system is geometrically stable and has no redundant constraints.

EXAMPLE 1.7

Analyze the geometric stability of the systems shown in Figure (1.33).

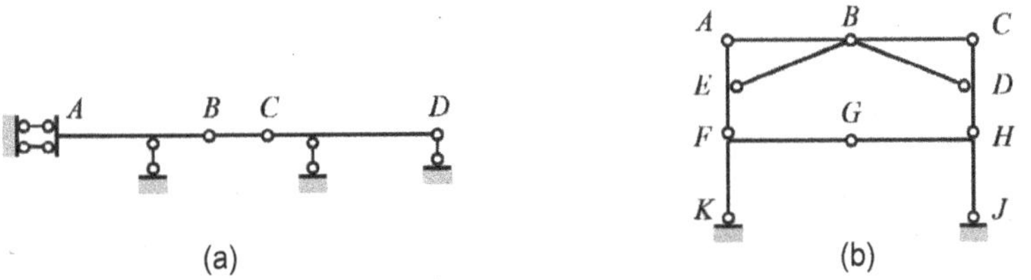

Figure 1.33. The original systems in Example 1.7.

***Solution*:** (1) Analyze the geometric stability of the systems shown in Figure (1.33a). See Figure (1.34a).

(2) Analyze the geometric stability of the systems shown in Figure (1.34a). See Figure (1.34b).

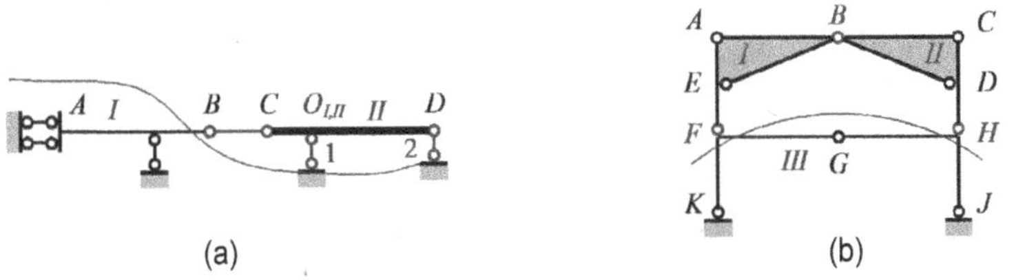

Figure 1.34. The geometric constructions of the systems in Example 1.7.

EXAMPLE 1.8

Analyze the geometric constructions of the system shown in Figure 1.35.

***Solution*:** (1) Analyze the geometric stability of the systems shown in Figure (1.35a). See Figure (1.36a).

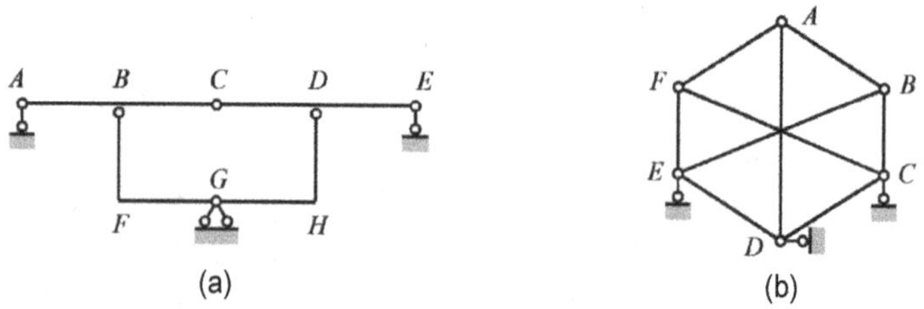

Figure 1.35. The original systems in Example 1.8.

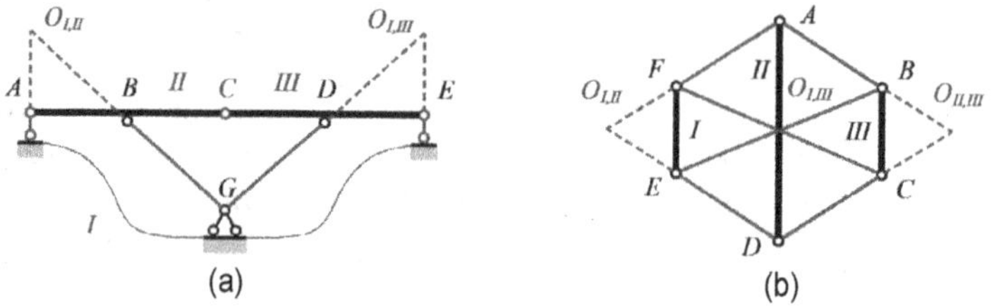

Figure 1.36. The geometric constructions of the systems in Example 1.8.

(2) Analyze the geometric stability of the system shown in Figure (1.35b). See Figure (1.36b).

All the previous examples show that the difficulty in analyzing the geometric stability is to properly select rigid bodies, links, and rules. Blindly attempting to select rigid bodies and links is not efficient. It is suggested that readers may follow the analysis procedure summarized in Table 1.2 when analyzing the geometric stability. This procedure will be implemented in the following examples.

Table 1.2. Analysis of the geometric stability of bar systems.

1. For a simply-supported system, separate the main part from the original system, and analyze the main part independently as illustrated in Examples 1.5 and 1.8b. It has the same geometric stability as the original system according to Theorem 1.4—the simply-supported rule.
2. Remove all the dual-link objects from the system. See Examples 1.4, 1.5a and 1.6. The rest object has the same geometric stability as the original one according to Theorem 1.3, the dual-link rule.
3. Try to extend the ground as illustrated in Example 1.7.
4. Apply the two- and three-body rules to analyze the geometric construction of the rest object obtained from steps 1, 2, 3 above.

EXAMPLE 1.9

Analyze the geometric stability of the systems shown in Figure (1.37).

Solution: (1) Analyze the geometric stability of the system shown in Figure (1.37a). See hints in Figure (1.38a).

(2) Analyze the geometric stability of the system shown in Figure (1.37b). See hints in Figures (1.38b) and (1.38c).

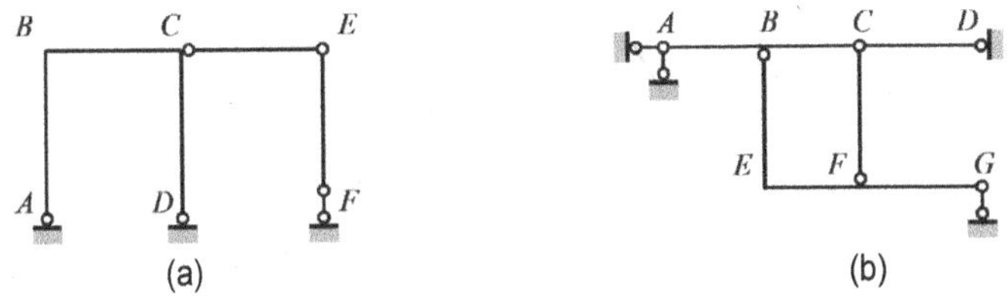

Figure 1.37. The original systems of Example 1.9.

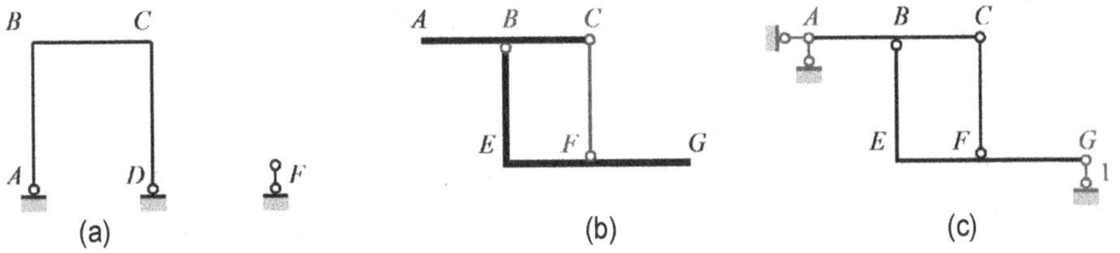

Figure 1.38. The constructions of the systems in Example 1.9.

EXAMPLE 1.10

Analyze the geometric stability of the systems shown in Figure 1.39.

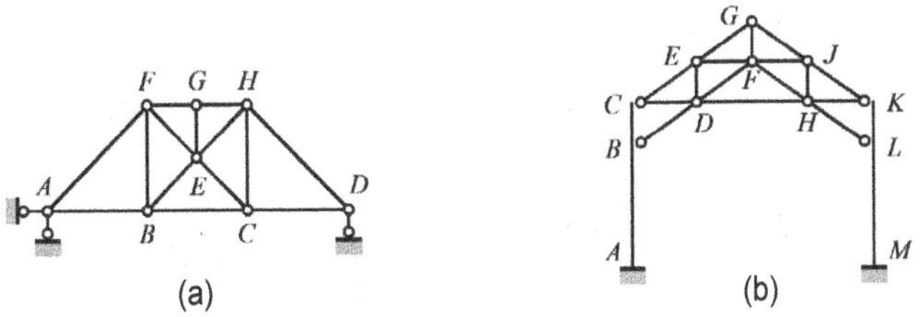

Figure 1.39. The original systems of Example 1.10.

Solution: (1) Analyze the geometric stability of the system shown in Figure (1.39a). See hints in Figure (1.40a).

(2) Analyze the geometric stability of the system shown in Figure (1.39b). See hints in Figure (1.40b).

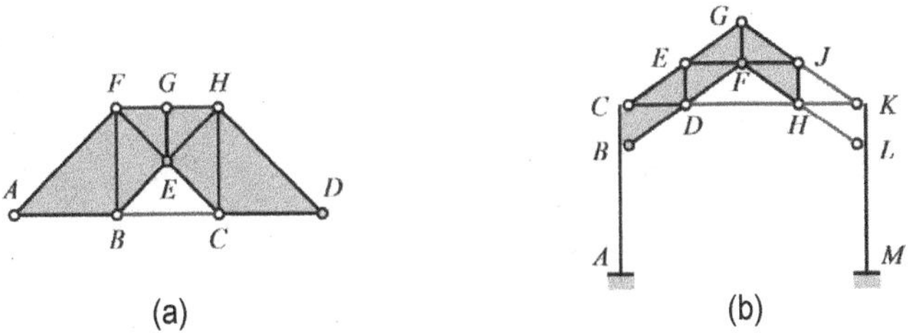

Figure 1.40. The geometric constructions of the systems in Example 1.10.

EXAMPLE 1.11

Solution: (1) Analyze the geometric stability of the system shown in Figure (1.41a). See hints in Figure (1.42a).

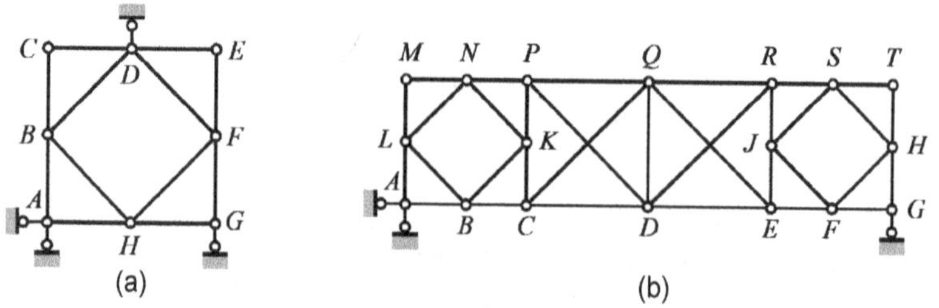

Figure 1.41. The original systems of Example 1.11.

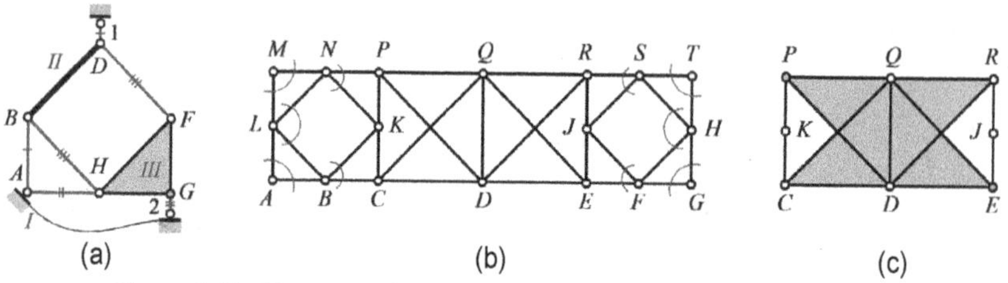

Figure 1.42. The geometric constructions of the systems in Example 1.11.

(2) Analyze the geometric stability of the system shown in Figure (1.41b). See hints in Figures (1.42b) and (1.42c).

EXAMPLE 1.12

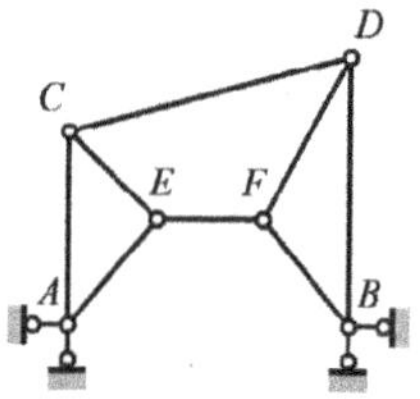

Figure 1.43. The original systems of Example 1.12.

Solution: Analyze the geometric stability of the system shown in Figure 1.44. See hints in Figures (1.44a) and (1.44b).

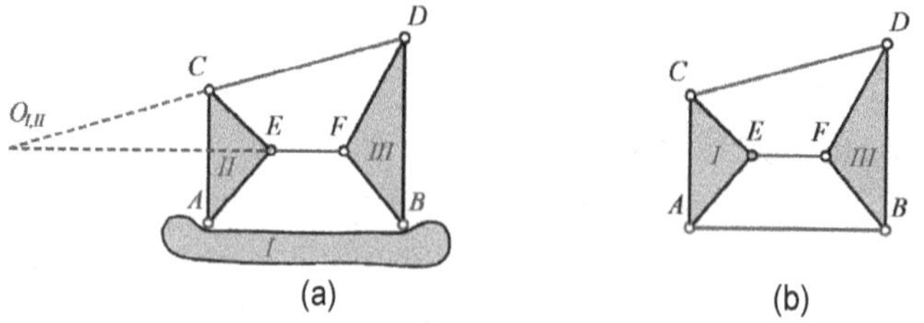

Figure 1.44. The geometric constructions of the system in Example 1.12.

PROBLEMS

Does there always exist a redundant constraint in an instantaneously-unstable system, and why?

EXERCISES

1.1 Analyze the geometric stability of the systems shown in Figure (1.45).

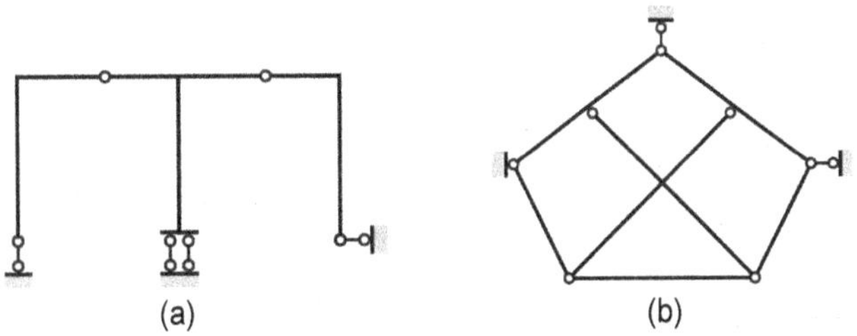

Figure 1.45. Exercise 1.1.

1.2 Analyze the geometric stability of the systems shown in Figure (1.46).

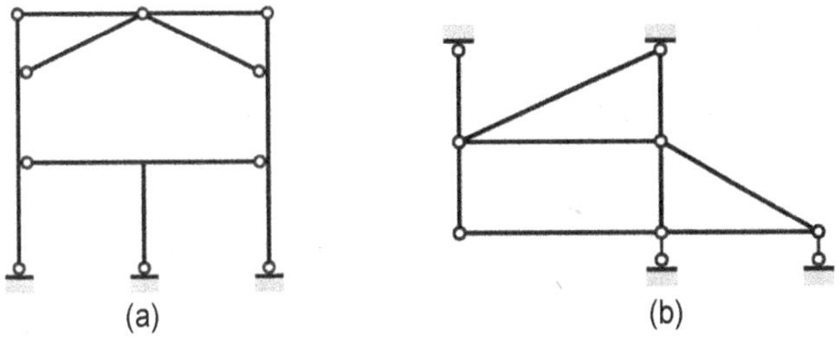

Figure 1.46. Exercise 1.2.

1.3 Analyze the geometric stability of the systems shown in Figure (1.47).

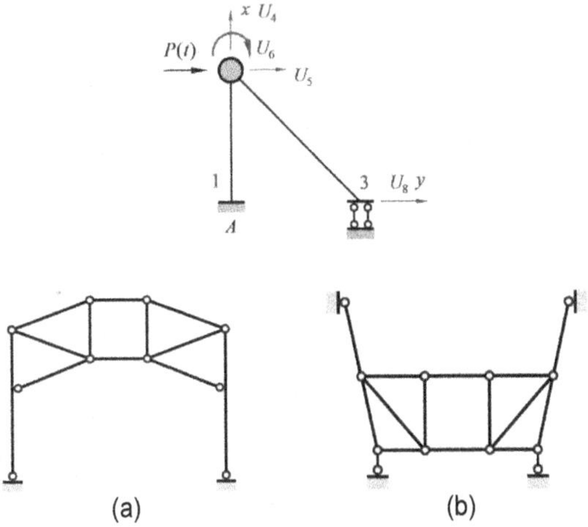

Figure 1.47. Exercise 1.3.

Chapter 2

Internal Forces in Statically Determinate Structures

A structure is said to be ***statically*** determinate if all its internal forces can be uniquely determined by the static equilibrium equations. The necessary and sufficient condition for a structure that is statically determinate is that it is a geometrically stable system without redundant constraints. In this chapter, we will focus on the calculation of the internal forces of statically determinate structures most often used in structural engineering. It is the foundation of structural analysis.

2.1 Analysis of Beams and Frames

Beams and frames may be the most popular engineering structures. The analysis of beams and frames is dominated by the constructions of their moment diagrams. In this section we begin by discussing the properties of the moment diagram of a straight-beam element, which is followed by presenting the superposition method for constructing a moment diagram. Furthermore, the method to determine the axial and shear forces from moment diagrams will be introduced. Finally, we will apply all these aforementioned techniques to analyze the internal forces of frames.

2.1.1 Properties of Moment Diagrams

Figure (2.1a) shows a straight-beam element subjected to distributed loads, q_x and q_y, along the beam, and concentrated loads, P_{yC} and M_C, at point C. The following two theorems are particularly important for constructing the moment diagrams of beams and frames.

Theorem 2.1 (Relation between Moments and Loads) *The moment diagram of a straight-beam segment shown in Figure (2.1a) has the following properties.*

1. *If* $q_y = 0$, $P_{yC} = 0$ *and* $M_C = 0$, *then the moment diagram of the beam segment AB is a straight line.*
2. *If* $M_C = 0$, *the moment diagram is continuous at point C.*

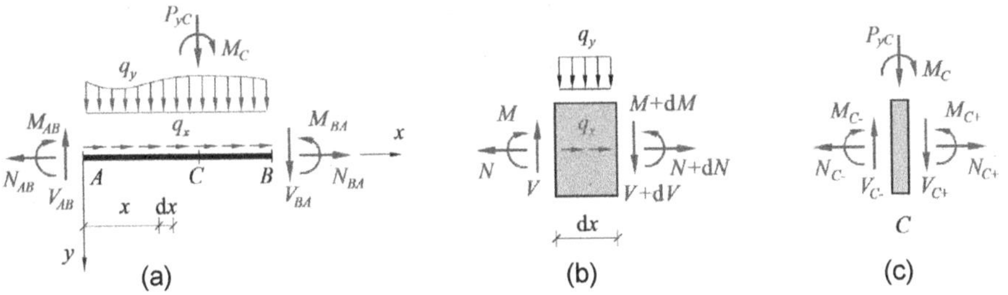

Figure 2.1. Theorem 2.1. (a) A beam element subjected to loads q_x, q_y, P_{yC}, M_C and end forces M, N, V. (b) The free-body diagram of an infinitesimal beam element. (c) The free-body diagram of point C.

3. *If $M_C \neq 0$, then, the moment diagram jumps down M_C at point C when M_C is clockwise, and jumps up M_C at C when M_C is anticlockwise.*
4. *If $P_{yC} \neq 0$, then the slopes of the moment at the left and the right of point C are different. Otherwise, they are the same.*

***Proof*:**

$$\frac{\mathrm{d}V}{\mathrm{d}x} = -q_y \tag{2.1}$$

$$\mathrm{d}M = V\mathrm{d}x + \frac{\mathrm{d}x}{2}\mathrm{d}V \tag{2.2}$$

$$\frac{\mathrm{d}M}{\mathrm{d}x} = V \tag{2.3}$$

$$\frac{\mathrm{d}^2 M}{\mathrm{d}x^2} = -q_y \tag{2.4}$$

$$\Delta M \equiv M_{C+} - M_{C-} = M_C \tag{2.5}$$

$$\Delta V \equiv V_{C+} - V_{C-} = P_{yC} \tag{2.6}$$

If $P_{yp} \neq 0$, then

$$\left.\frac{\mathrm{d}M}{\mathrm{d}x}\right|_{x_C^-} = \left.\frac{\mathrm{d}M}{\mathrm{d}x}\right|_{x_C^+} \tag{2.7}$$

Theorem 2.2 (Relation between Moments and Shears) The moment diagram of a straight beam segment without shear is a line parallel to the beam's neutral axis.

***Proof*:** This theorem follows immediately from Equation (2.3).

The following example illustrates the application of Theorems 2.1 and 2.2

EXAMPLE 2.1

Draw the moment diagrams of the beams shown in Figure 2.2.

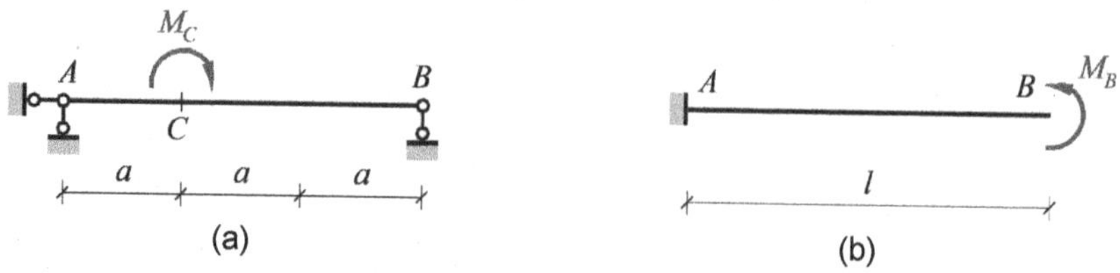

Figure 2.2. Example 2.1. (a) A simply-supported beam subjected to an external torque at point C. (b) A cantilever beam subjected to a torque M_B at the free end.

***Solution*:** (1) Draw the moment diagram of the simply-supported beam shown in Figure (2.2a).

Theorem 2.1 shows that the moment diagram jumps down M_C at point C because an external moment M_C is applied clockwise at point C, and that the moment diagrams on AC and CB are two parallel lines because no lateral load is applied on beam segments AC and CB. See Figure (2.3a). The similarity of triangles gives:

$$M_{CA} = \frac{M_C}{3} \quad \text{and} \quad M_{CB} = \frac{2M_C}{3} \tag{2.8}$$

(2) Draw the moment diagram of the cantilever beam shown in Figure 2.2b.

Theorem 2.2 shows that the moment diagram of the cantilever is parallel to the beam's neutral axis because the shear in the beam is zero. This leads to the moment diagram shown in Figure (2.3b).

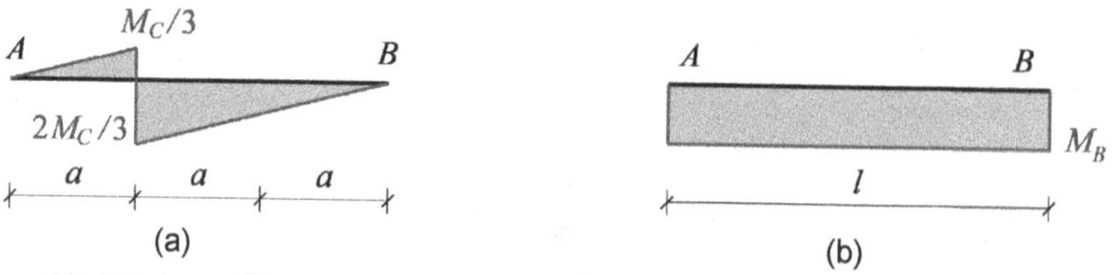

Figure 2.3. Solutions of Example 2.1. (a) The moment diagram of the simply-supported beam. (b) The moment diagram of the cantilever.

2.1.2 *Constructions of Moment Diagrams by the Superposition Method in Segments*

If the material of a structure is linear, i.e., the strain of the material increases or decreases proportionally to the increase or decrease in stress, and if the displacement responses of the structure due to external loading are small, then the structure is referred to as a *linear structure*. Since any response of a linear structure is governed by a system of linear equations, the total response of a linear structure to a linear combination of several loads can be expressed as the superposition of the responses of the structure to each individual load.

As an example to demonstrate how a moment diagram is constructed by the superposition method, Figure (2.4a), shows a simply-supported beam subjected to a uniform distributed load q and an external moment M_B at the right end B. These two external loads can be treated separately. The resultant moment diagram is obtained

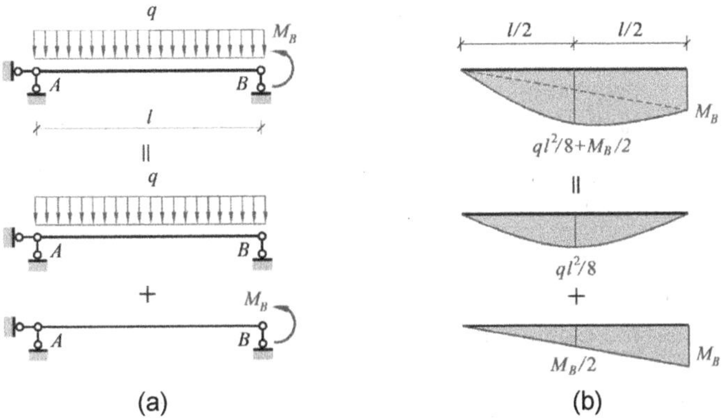

Figure 2.4. Constructing the moment diagram by the superposition method. (a) The superposition of external loads. (b) The superposition of moment diagrams.

by a superposition of the moment diagram of the beam to the distributed load q and that of the beam to the torque M_B, as given in the uppermost figure of Figure (2.4b). If one establishes a database of simple moment diagrams, the superposition method can then be conveniently used for the construction of a complex moment diagram. Table 2.1 lists some simple moment diagrams used frequently in structural analysis.

Table 2.1. Simple moment diagrams frequently used in structural analysis.

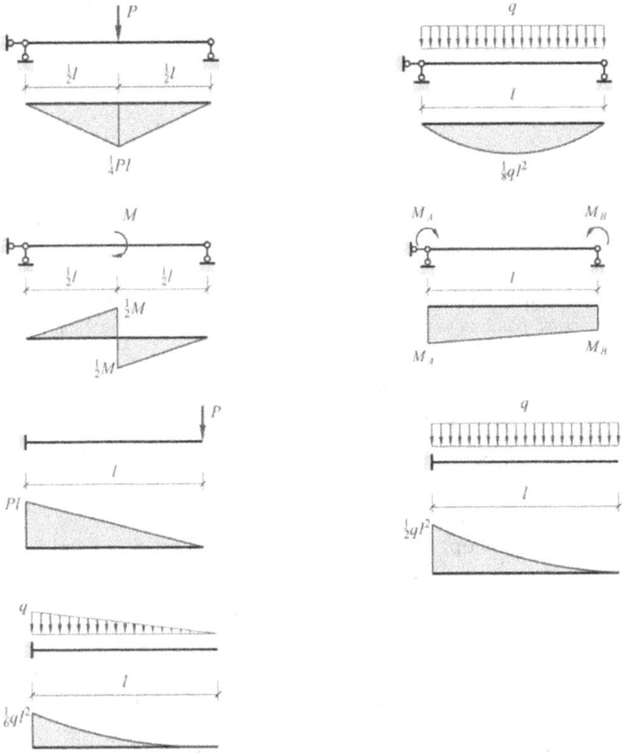

Next, let's introduce the ***superposition method in segments***. Consider a simply-supported beam shown in Figure (2.5). Its moment diagram cannot be constructed by the simple superposition method above because the required simple moment diagrams are typically not given in a handbook. As shown in Figure (2.6), let's draw the free-body diagrams of the simply-supported beam for each segment.

This shows that the moment diagram of the simply-supported beam shown in Figure (2.5) is the same as that of the system shown in Figure (2.6b), i.e., the moment diagram of the simply-supported beam of Figure (2.5) can be obtained by assembling the moment diagrams of each beam segment in the system shown in Figure (2.6b). It is noticed that the moment diagrams of each simple beam can be conveniently constructed by the simple superposition method as long as the internal moments M_{DA}, M_{ED}, M_{BE} of the simply-supported beam shown in Figure 2.5 have been determined.

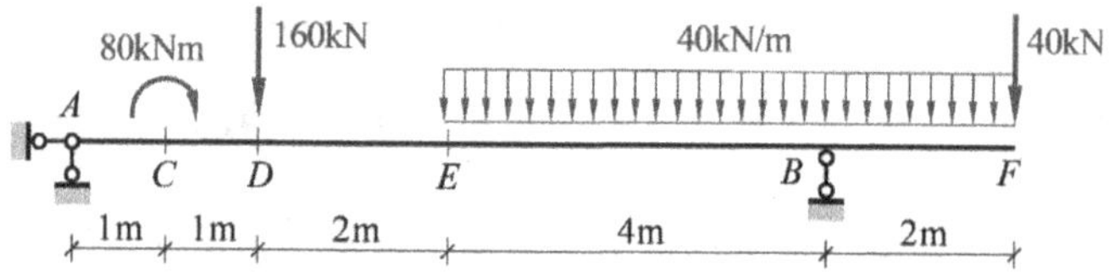

Figure 2.5. A simply-supported beam subjected to complex loads, whose moment diagram can be constructed by the superposition method in segments of the simple beam system shown in Figure 2.6.

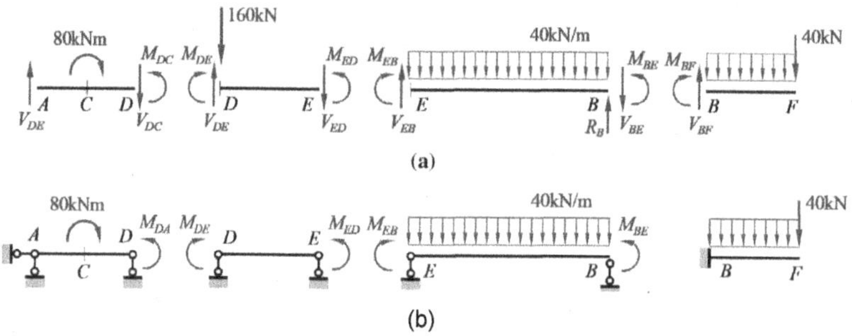

Figure 2.6. (a) The free-body diagrams of the beam shown in Figure 2.5. (b) A simple beam system corresponding to the free-body diagrams shown in Figure 2.6a.

In order to calculate the key moments M_D, M_E and M_B, we have to find the reaction, R_{yA}, from support A. The moment equilibrium about B of the free body of the entire simply-supported beam, shown in Figure 2.7, gives:

$$R_{yA} = 130\text{kN} \tag{2.9}$$

The moment equilibrium conditions of the free bodies of the beam segments AD, AE and AB lead to:

$$M_D = 340\text{kN}\cdot\text{m}, \quad M_E = 280\text{kN}\cdot\text{m}, \quad M_B = 160\text{kN}\cdot\text{m} \tag{2.10}$$

Based on the key moments M_{DA}, M_{ED} and M_{BE}, one can obtain the moment diagram of the system shown in Figure (2.6b) by the simple superposition method. Assembling them gives the moment diagram of the original simply-supported beam,

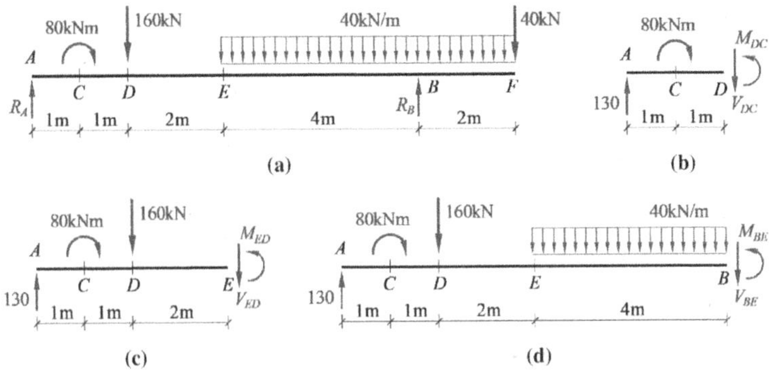

Figure 2.7. Free-body diagrams for determining the key moments. (a) The free-body diagram for calculating the support reaction R_A. (b) The free-body diagram for computing the moment M_{DC}. (c) The free-body diagram for calculating the moment M_{EB}. (d) The free-body diagram for the calculation of the moment M_{BE}.

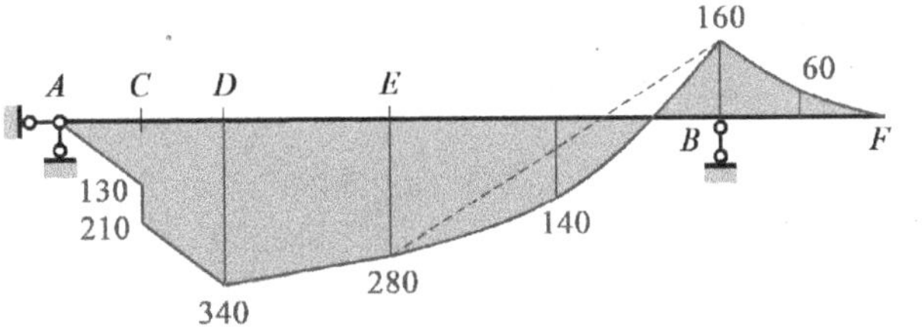

Figure 2.8. Assembling the key moments gives the moment diagram of the simply-supported beam.

as shown in Figure (2.8). By observing this figure, one may notice that the moment diagram is smooth at point E because no concentrated load is applied at E. In addition, the slope of the moment is not zero at point F because the shear force at F is not zero.

A procedure to construct the moment diagram of a structure by the superposition method in beam segments is summarized in Table 2.2.

Table 2.2. Construction of moment diagram by the superposition method in segments.

1. Select the key sections on the structure such that the moment diagrams of each beam element between adjacent key sections can be obtained by the simple superposition method.
2. Determine the key moments (the internal moments at key sections).
3. Use the superposition method in each beam segment to draw the moment diagrams of the structure.

The following examples illustrate the construction of the moment diagram of a two-span beam by the superposition method in a segment.

EXAMPLE 2.2

Draw the moment diagram of a two-span beam shown in Figure 2.9a.

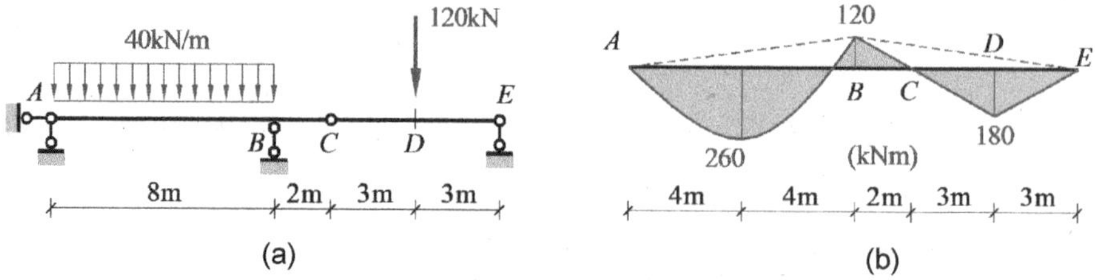

Figure 2.9. Example 2.2. (a) A two-span beam. (b) The moment diagram of the two-span beam shown in Figure 2.9a.

***Solution*:** Sections *AB*, *BC*, *CD* and *ED* are taken as the key sections for this problem. It is known that:

$$M_{AB} = 0, \quad M_{CD} = 0, \quad M_{ED} = 0$$

By using the simple superposition method, one can obtain the moment diagram of the beam segment *CD* as shown in Figure (2.9). Because none of the loads is applied to the beam segment *BD*, its moment diagram is a line according to Theorem 2.1. Then the similarity of triangles gives that

$$M_{BC} = \frac{2}{3} \times 180 = 120 \text{ kN} \cdot \text{m},$$

Again, by using the simple superposition method, the moment diagram of the beam segment *AB* can be obtained, as shown in Figure (2.9b).

Next, let us analyze a more complicated structure.

EXAMPLE 2.3

Draw the moment diagram of a four-span beam.

***Solution*:** Take sections *AB*, *CB*, *DC*, *FE*, *GF*, *IH*, *KJ* as the key sections. We are given that

$$M_{AB} = M_{DC} = M_{FE} = M_{KJ} = 0.$$

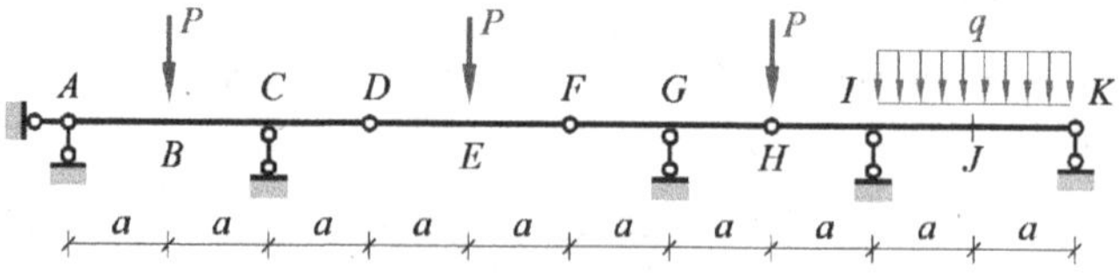

Figure 2.10. Example 2.3.

One can first obtain the moment diagram of the beam segment *DF* as shown in Figure (2.11a). Because the moment diagram of the beam segment *CE* is linear, the similarity of triangles gives

$$M_{CD} = M_{ED} = \frac{Pa}{2}.$$

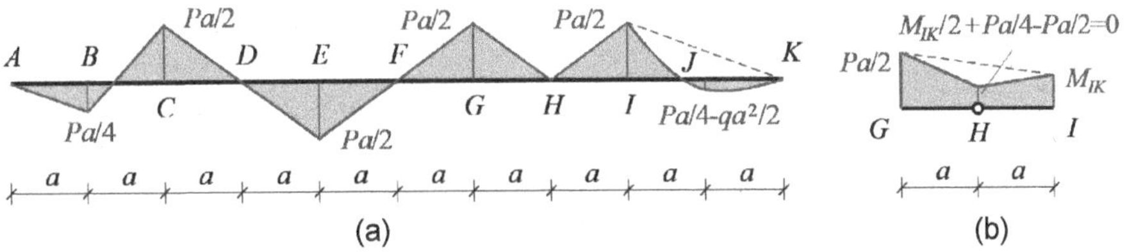

Figure 2.11. (a) The moment diagram of the four-span beam shown in Figure 3.10. (b) Calculation of the key moment M_{IH}.

Then, the moment diagram of segment AC can be obtained by using the superposition method as shown in Figure (2.11a). Similarly, one can also obtain $M_{GF} = \frac{Pa}{2}$.

As shown in Figure (2.11b), using the simple superposition method in segment GI gives

$$M_{HG} = \frac{M_{IK}}{2} + \frac{Pa}{4} - \frac{Pa}{2} \tag{2.11}$$

Since we are given that $M_{HG} = 0$, Equation (2.11) yields

$$\frac{M_{IK}}{2} + \frac{Pa}{4} - \frac{Pa}{2} = 0,$$

which leads to $M_{IK} = Pa/2$. Finally, the moment diagram of segment IK can be constructed by the superposition method as shown in Figure (2.11a).

Examples 2.2 and 2.3 show that we draw the moment diagram without computing any support reactions. Actually, for many problems, support reactions are not necessary for constructing moment diagrams.

2.1.3 Calculations of Internal Forces through Moment Diagrams

As soon as the moment diagram of a structure is obtained, all the internal forces—shears, axial force, support reactions—can conveniently be determined through the moment diagram. This section illustrates how to analyze beams based on their given moment diagram. This technique is particularly important for the advanced analysis of structures.

EXAMPLE 2.4

Compute the extreme moment of the beam segment AB in Example 2.2 based on the moment diagram shown in Figure (2.9b).

***Solution*:** The extreme moment of the beam segment AB may be computed by formulating its moment and differentiating it to find the stationary point. However, in structural analysis, the following procedure to determine the extreme point of the moment is preferred more.

The moment equilibrium of the free-body shown in Figure (2.12a) gives the end shears of the beam segment as follows.

$$V_{AB} = 145\text{kN}, \quad V_{BA} = 175\text{kN}$$

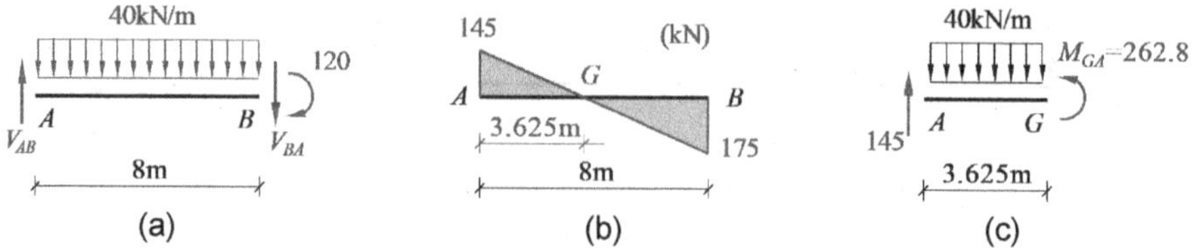

Figure 2.12. (a) The free-body diagram of the beam segment *AB* for the calculation of its end shears. (b) The shear diagram of the beam segment *AB*. (c) The free-body diagram for the calculation of extreme moments.

Draw the shear diagram of segment *AB* as shown in Figure (2.12b), where point *G* is the zero-shear point. The similarity of triangles gives $AG = 3.625$ m. Since, $dM/dx = V$, the extreme moment occurs at point *G*. The moment equilibrium of the free-body shown in Figure (2.12c) yields

$$M_{GA} = 262.8\text{kN} \cdot \text{m}$$

which is the extreme moment of segment *AB*.

EXAMPLE 2.5

Determine the support reaction at point *I* in Example 2.3 based on the moment diagram shown in Figure (2.11a).

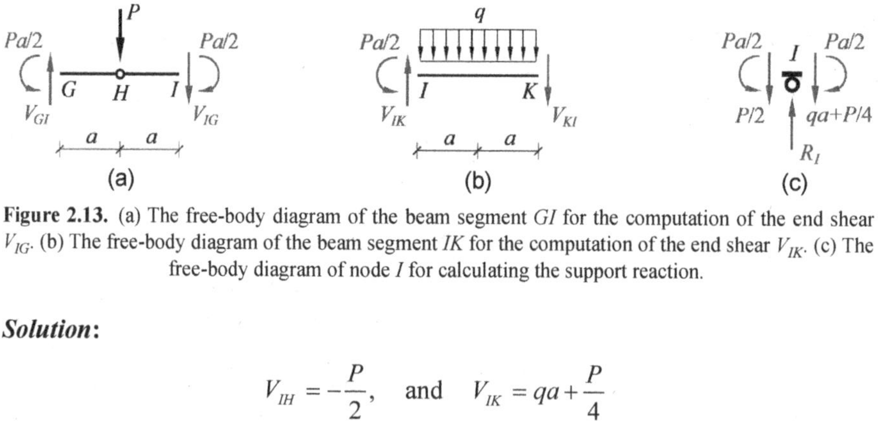

Figure 2.13. (a) The free-body diagram of the beam segment *GI* for the computation of the end shear V_{IG}. (b) The free-body diagram of the beam segment *IK* for the computation of the end shear V_{IK}. (c) The free-body diagram of node *I* for calculating the support reaction.

***Solution*:**

$$V_{IH} = -\frac{P}{2}, \quad \text{and} \quad V_{IK} = qa + \frac{P}{4}$$

$$R_I = qa + \frac{3P}{4} \tag{2.12}$$

2.1.4 Analysis of Frames

This section will demonstrate how to analyze the internal forces of statically determinate frames. Consider a frame shown in Figure (2.14a). It should be noticed that *the moments are drawn on the tension side of the member. Thus, it is not necessary to indicate if a moment is positive or negative.*

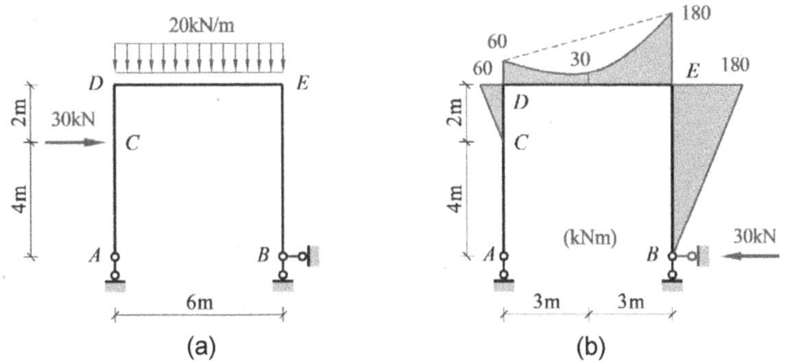

Figure 2.14. (a) A statically determinate frame. (b) The moment diagram of the frame shown in Figure 2.14a.

EXAMPLE 2.6

Analyze the internal forces of the three-hinged frame shown in Figure (2.15a).

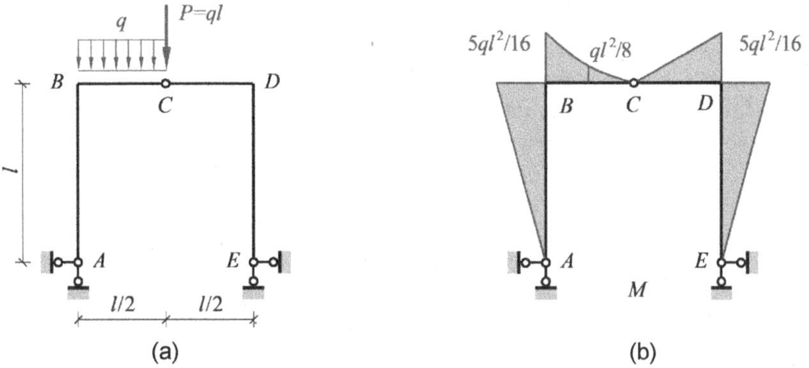

Figure 2.15. (a) A three-hinged frame. (b) The moment diagram of the three-hinged frame shown in Figure 2.15a.

Solution: For a three-hinged frame, usually, reactions from supports are required for drawing its moment diagram. To determine the support reactions, draw a free-body diagram of the entire frame as shown in Figure 2.16a. $M_A = 0$ and $M_E = 0$ lead to:

$$R_{vA} = \frac{7}{8}ql, \quad \text{and} \quad R_{vE} = \frac{5}{8}ql \tag{2.13}$$

where R_{vA} and R_{vE} are the reactions from the supports A and E in the vertical direction, respectively. In order to find the reaction, R_{hE}, of support E in the horizontal direction, we draw the free-body diagram of the right part of the frame as shown in Figure 2.16b. $M_C = 0$ gives

$$R_{hE} = \frac{5}{16}ql \tag{2.14}$$

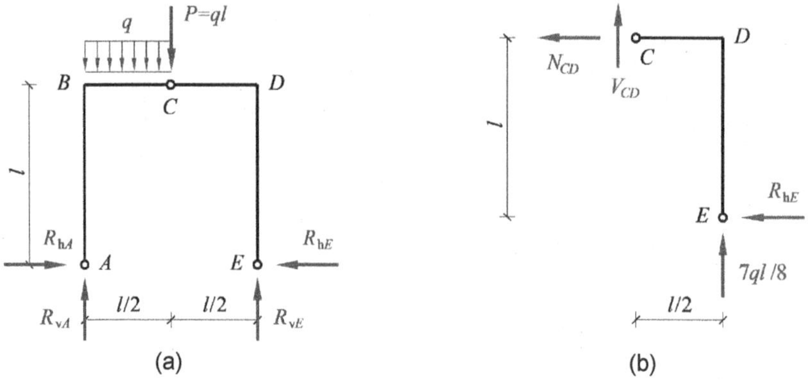

Figure 2.16. (a) Reaction forces of the frame. (b) Reaction forces of the right part of the frame.

The free-body equilibrium in the horizontal direction of the entire frame gives

$$R_{hA} = \frac{5}{16} ql \tag{2.15}$$

Draw the moment diagram as shown in Figure (2.15b).

Next, let's construct the shear diagram. It should be noticed that *it is necessary for a shear diagram to indicate if a shear is positive or negative.* Finally, draw the axial-force diagram as shown in Figure (2.17b). Similar to shear diagrams, *axial diagrams have to indicate if an axial force is positive or negative.*

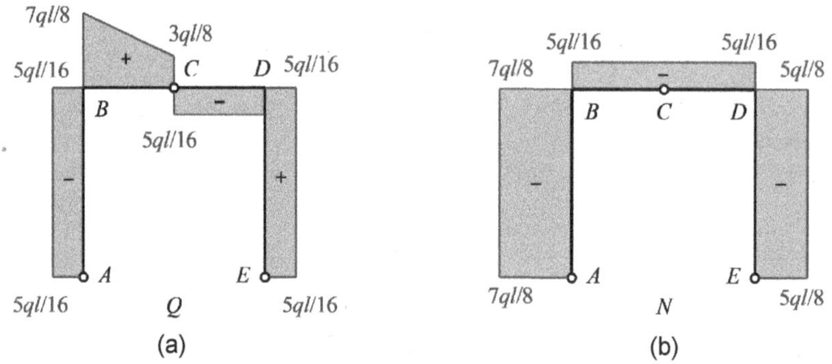

Figure 2.17. (a) The shear-force diagram. (b) The axial-force diagram.

EXAMPLE 2.7

Draw the moment diagrams of the frames shown in Figure (2.18).

Solution: (1) Consider the frame shown in Figure (2.18a). The moment distribution is shown in Figure (2.19c).

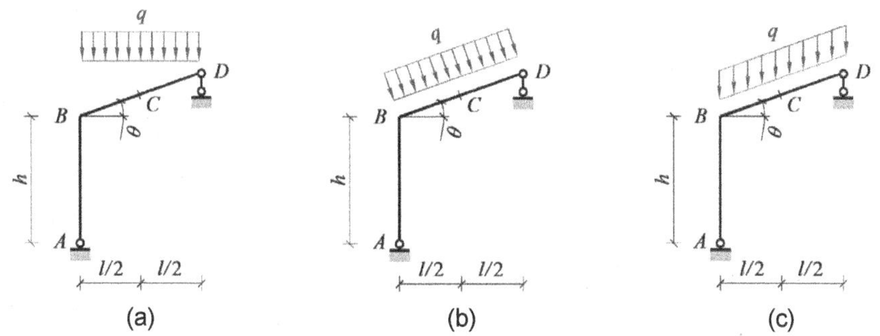

Figure 2.18. Three frames having an inclined beam subjected to distributed loads. (a) The vertical distributed load is defined on the projection of the inclined beam. (b) The distributed load is perpendicular to the inclined beam. (c) The vertical distributed load is defined on the axis of the inclined beam.

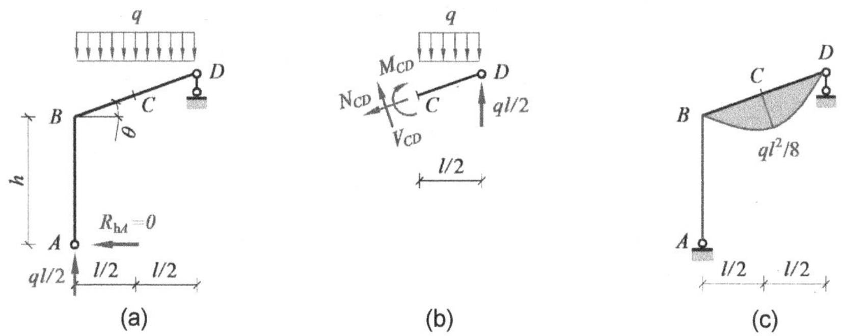

Figure 2.19. Forces and moments corresponding to Figure 2.18a: (a) Reaction forces of the frame. (b) Reaction forces in segment CD. (c) Moment diagram of BD.

(2) Consider the frame shown in Figure (2.18b). The free-body equilibrium in the horizontal direction of the entire frame shows that horizontal reaction, R_{hA}, from support A is:

$$R_{hA} = q \times \frac{l}{\cos\theta} \times \sin\theta = ql\tan\theta.$$

The moment distribution is shown in Figure (2.20c).

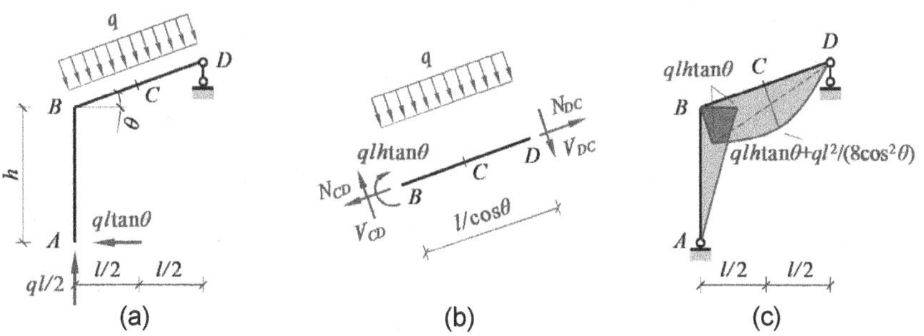

Figure 2.20. Forces and moments corresponding to Figure 2.18b: (a) Reaction forces of the frame. (b) Reaction forces in segment BD. (c) Moment diagram of the frame.

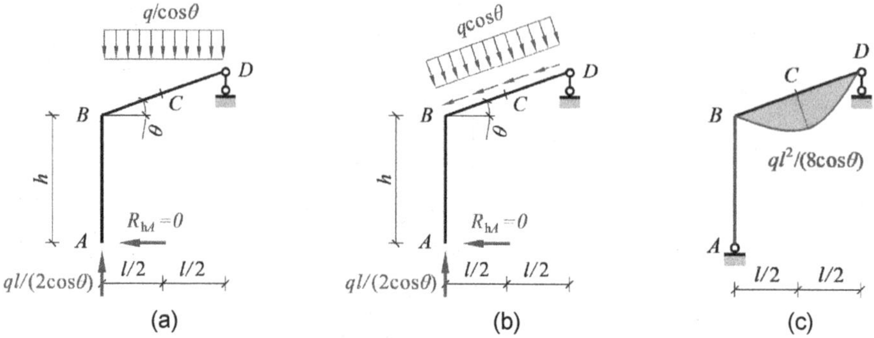

Figure 2.21. Forces and moments corresponding to Figure 2.18c: (a) Reaction forces of the frame. (b) Reaction forces in segment BD. (c) Moment diagram of the frame.

(3) Consider the frame shown in Figure (2.18c). The original virtual distributed load q is defined on the axis of the inclined beam. This original load is statically equivalent to the vertical distributed load, $q/\cos\,\theta$, defined on the horizontal projection of the inclined beam as shown in Figure (2.21a). The moment distribution is shown in Figure (2.21c).

EXAMPLE 2.8

Draw the moment diagram of the frame shown in Figure (2.22a).

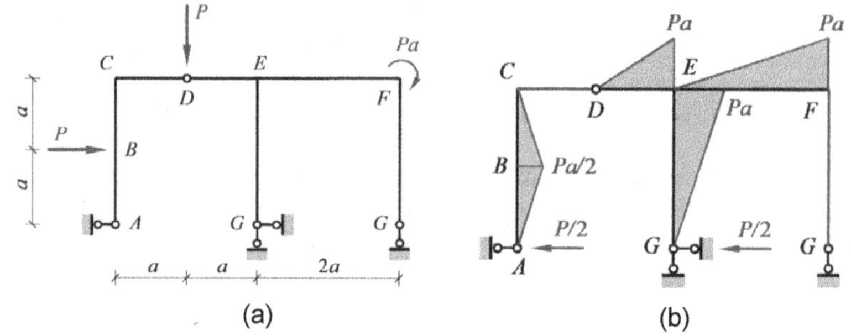

Figure 2.22. Example 2.8.

Solution: See Figure (2.23), the moment distribution is shown in Figure (2.22b).

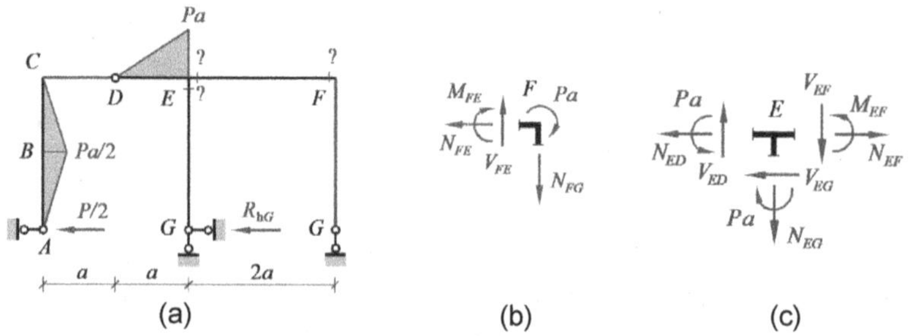

Figure 2.23. Example 2.8. (a) The moment diagram of the left part *ABCDE* of the frame. (b) The free-body diagram of joint *F*. (c) The free-body diagram of joint *E*.

EXAMPLE 2.9

Draw the moment diagrams of the three-hinged frames shown in Figure (2.24).

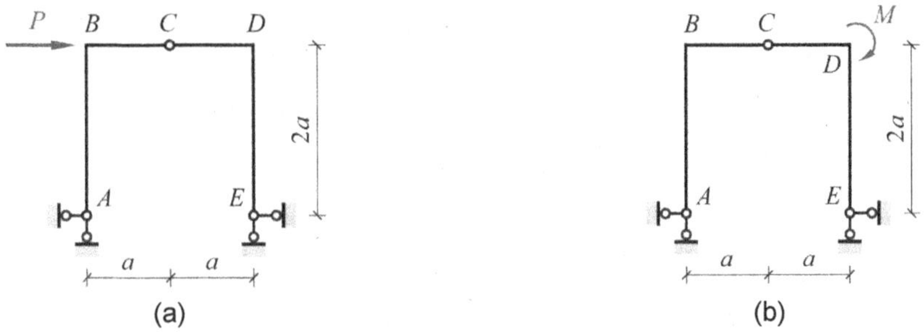

Figure 2.24. Example 2.9. (a) A three-hinged frame subjected to a horizontal concentrated load. (b) A three-hinged frame subjected to an external torque.

Solution: See Figure (2.25) (for problem defined in Figure (2.24a)) and Figure (2.26) (for problem defined in Figure (2.24b)).

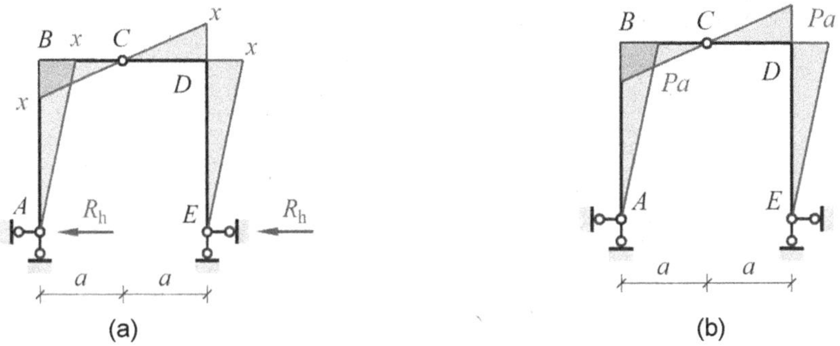

Figure 2.25. Example 2.9. (a) The qualitative moment diagram of the frame shown in Figure 2.24a. (b) The moment diagram of the frame shown in Figure 2.24a.

Figure 2.26. Example 2.9. (a) The qualitative moment diagram of the frame shown in Figure 2.24b. (b) The moment diagram of the frame shown in Figure 2.24b.

2.2 Analysis of Trusses and Composite Structures

2.2.1 Introduction

A truss is a structure that consists of two-force members only, where the members are organized so that the assemblage as a whole behaves as a single object. A "two-force member" is a structural component where force is applied at only two points. Trusses typically comprise five or more triangular units constructed with straight members whose ends are connected at joints referred to as nodes. In this typical context, external forces and reactions to those forces are considered to act only at the nodes and result in forces in the members that are either tensile or compressive. For straight members, moments (torques) are explicitly excluded because, and only because, all the joints in a truss are treated as revolutes, as is necessary for the links to be two-force members. Figure (2.27) illustrates layouts of trusses.

Furthermore, in structural analysis of frames, if each member can sustain both a force and a bending moment, it is referred to as a composite structure, as shown in Figure (2.28).

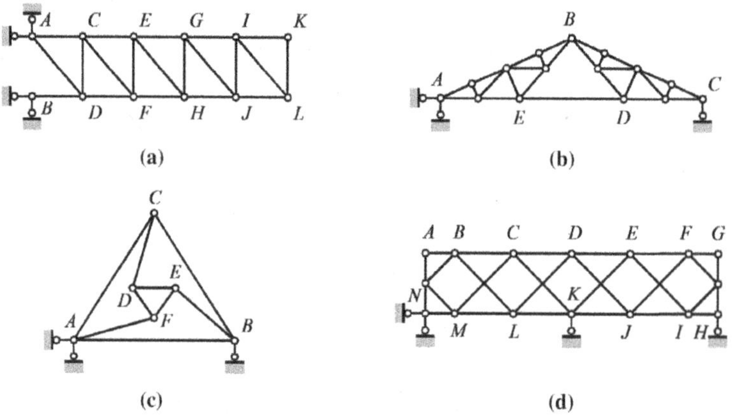

Figure 2.27. (a) A simple truss. (b) A compound truss. (c) A compound truss. (d) A complex truss.

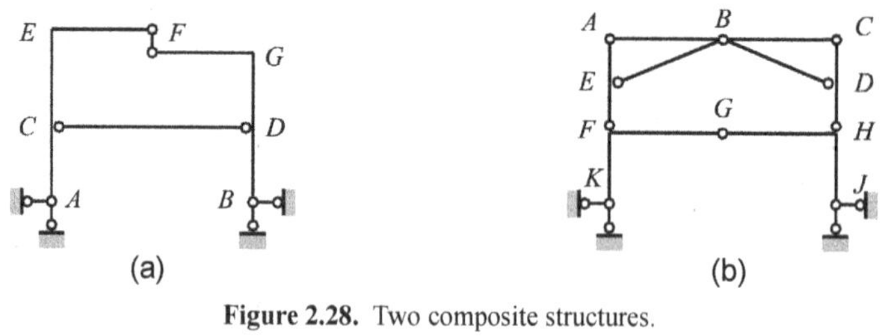

Figure 2.28. Two composite structures.

2.2.2 Analysis of Statically Determinate Trusses

In a statically determinate truss, all the internal member forces can be determined from equilibrium relationships alone. The number of unknown internal forces is equal to the number of equilibrium equations. If a single element is removed the whole truss structure will collapse. The truss structure can be analyzed using either method of joints or method of sections. The method of joints finds the equilibrium of the truss at each joint, while the method of sections establishes the force equilibrium of the truss at an artificial truss "cutting". The following examples in this section illustrate the application of both methods.

EXAMPLE 2.10

Determine the internal forces of the simple truss shown in Figure (2.29a).

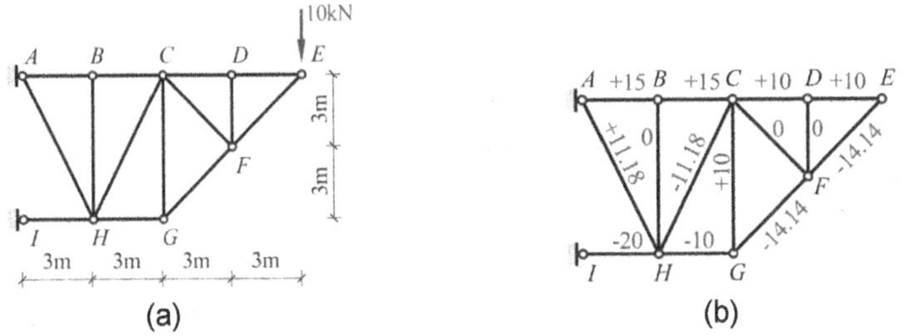

Figure 2.29. Example 2.10. (a) A simple truss subjected to a concentrated load of 10 kN. (b) The internal forces of the truss shown in Figure (2.29a) to the load of 10 kN.

***Solution*:** Use the method of joints to analyze this simple truss.

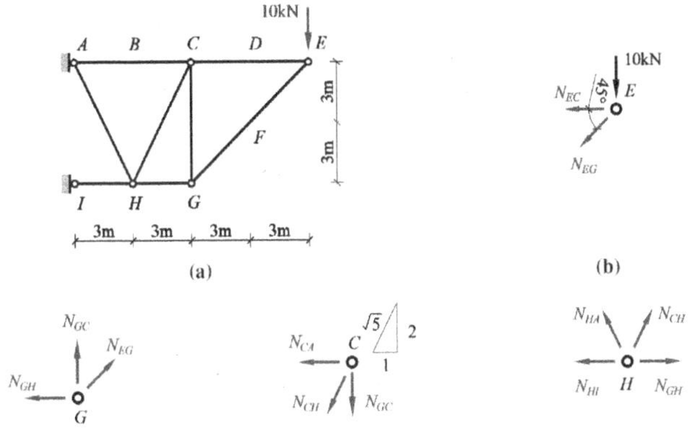

Figure 2.30. Example 2.10. (a) The simplified truss by dropping all the zero-force members. (b)–(e) The free-body diagrams of joints *E*, *G*, *C* and *H*, respectively.

EXAMPLE 2.11

Use the method of sections to determine the internal forces of members *BD*, *DK* and *LK* of the truss shown in Figure (2.31a).

***Solution*:** (1) Find zero-force members and simplify the truss. As shown in Figure (2.31b), the truss can be simplified by removing the zero-forces members *CD*, *DL*, *MD*, *FD*, *DJ* and *DI*.

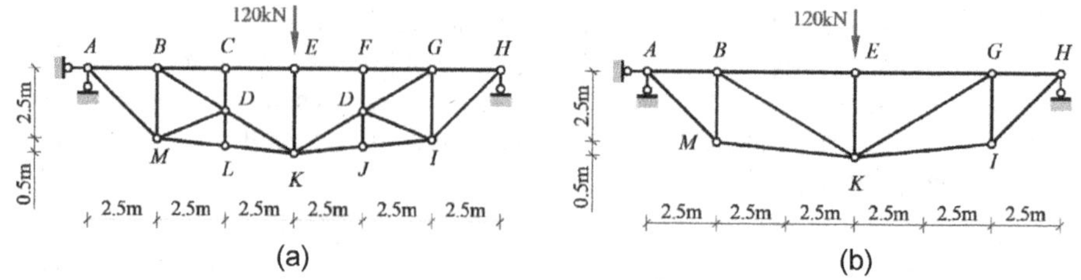

Figure 2.31. Example 2.11. (a) A simply-supported truss. (b) The simplified truss by dropping the zero-force members.

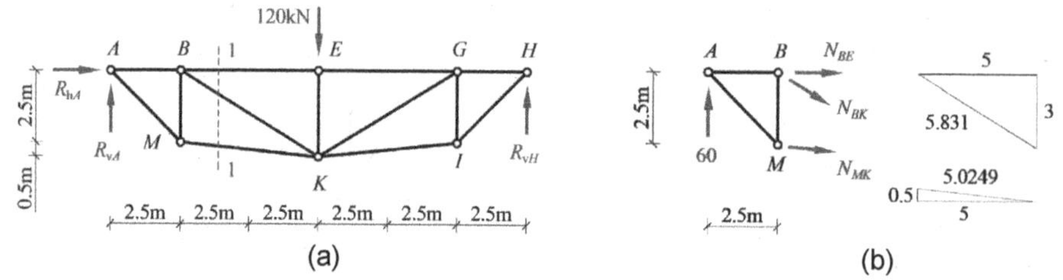

Figure 2.32. Example 2.11. (a) The free-body diagram of the entire structure. (b) The free-body diagram of the part *ABM*.

(2) Determine the support reactions. The force equilibrium in the vertical direction of the entire truss shows that the vertical reaction R_{vA} from support *A* is:

$$R_{vA} = 60\text{kN} \tag{2.16}$$

(3) Draw a free-body diagram. The forces in the members *BE*, *BK* and *MK* can be obtained by finding the equilibrium at an artificial "cutting" 1-1, as shown in Figure (2.32a). The free-body diagram of the left part of this section is shown in Figure (2.32b). $\sum M_B = 0$ provides

$$60 \times 2.5 - \left(N_{MK} \times \frac{5}{5.0249} \right) \times 2.5 = 0 \tag{2.17}$$

which gives

$$N_{MK} = 60.2\text{kN} \tag{2.18}$$

The force equilibrium of the free-body ABM in the vertical direction leads to:

$$N_{BK} \times \frac{3}{5.831} + N_{MK} \times \frac{0.5}{5.0249} - 60 = 0, \tag{2.19}$$

which gives:

$$N_{MK} = 60.2\text{kN} \tag{2.20}$$

Thus, we have:

$$N_{BD} = N_{DK} = 60.2\text{kN}, \quad \text{and} \quad N_{LK} = 104.95\text{kN}. \tag{2.21}$$

2.2.3 *Analysis of Statically Composite Structures*

This section presents an example of analyzing a statically determinate composite structure.

EXAMPLE 2.12

Determine the moments of the flexural members and the axial forces of the links of the composite structure shown in Figure (2.33a).

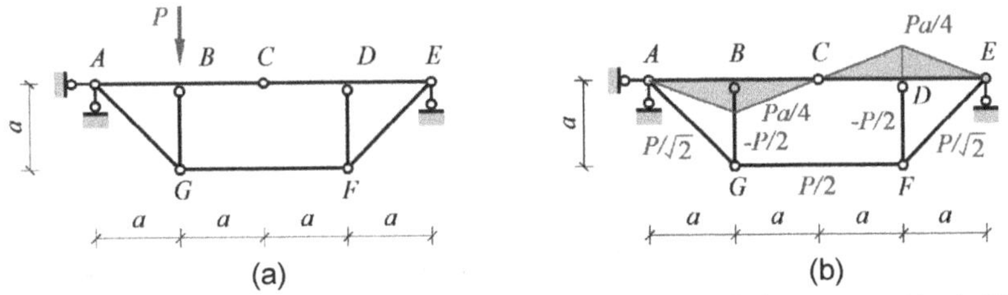

Figure 2.33. Example 2.12. (a) A composite structure subjected to a vertical concentrated load. (b) The internal forces of the composite structure shown in Figure 2.33a.

Solution: The moment diagrams of the flexural members AC and CE may take the forms shown in Figure (2.34a) or Figure (2.34b). The free-body equilibrium of joints G and F in Figure (2.34c) shows that the forces N_{GF} and N_F are the same and can be denoted by N. Because the axial forces N_{GB} and N_{FD} act in the same tension or the same compression, the moment diagram of ACE should take the form shown in Figure (2.34a). Members GB and FD are all in compression, and the equilibrium gives that

$$\frac{1}{4}\times\left(P-|N|\right)\times 2a=\frac{1}{4}\times|N|\times 2a \tag{2.22}$$

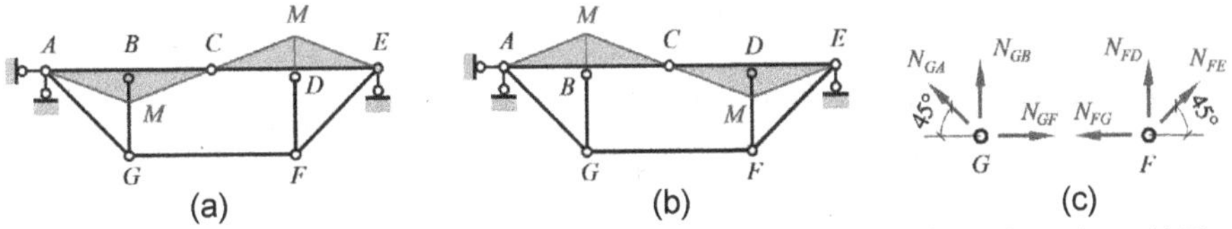

Figure 2.34. Example 2.12. (a)–(b) The possible qualitative moments of the flexural members. (c) The free-body diagrams of joints G and F.

Thus, we have

$$N_{GB}=N_{FD}=N=-\frac{1}{2}P \tag{2.23}$$

The internal forces of the composite structure are indicated in Figure (2.33b).

2.3 Analysis of Statically Determinate Arches

2.3.1 Comparison of Internal Forces in Arches and the Corresponding Beams

A three-hinged arch is shown in Figure (2.35a). Figure (2.35b) shows a simply-supported beam corresponding to the arch, which has the same span and the same loading as the three-hinged arch.

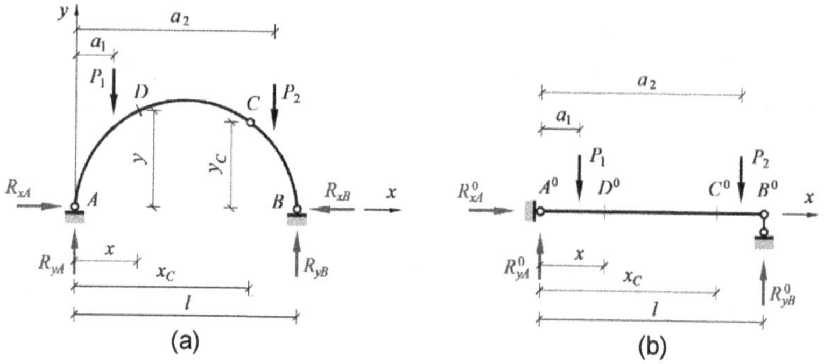

Figure 2.35. Comparison of the internal forces in an arch and the corresponding beam. (a) A three-hinged arch. (b) The beam corresponding to the arch shown in Figure 2.35a.

First, Let's find the support reaction forces in the arch. The moment equilibrium of the free-body of the entire arch shows that the vertical reactions R_{yA} and R_{yB} from supports A and B are:

$$R_{yA} = \frac{1}{l}\left[P_1(l-a_1) + P_2(l-a_2)\right] \tag{2.24}$$

and

$$R_{yB} = \frac{1}{l}\left[P_1 a_1 + P_2 a_2\right] \tag{2.25}$$

The moment equilibrium about pin C of the free-body of the arch segment AC shows that the horizontal reaction from support A is:

$$R_{xA} = \frac{1}{y_C}\left[R_{yA}x_C - P_1(x_C - a_1)\right] \tag{2.26}$$

where x_C and y_C are the x and y coordinates of hinge C, respectively. The horizontal equilibrium of the free-body of the entire arch shows that the horizontal reactions from supports A and B are the same and can be denoted by R_h:

$$R_{xA} = R_{xB} = R_h \tag{2.27}$$

Therefore, unlike a simply-supported beam, there are horizontal reactions on the arch. Designers must be very careful to prevent collapse due to this horizontal thrust at the supports. Usually, arches must have solid foundation abutments.

To determine the internal forces at section D on the arch we draw the free-body diagram of the arch segment AD as shown in Figure (2.36a). Figure (2.36b)

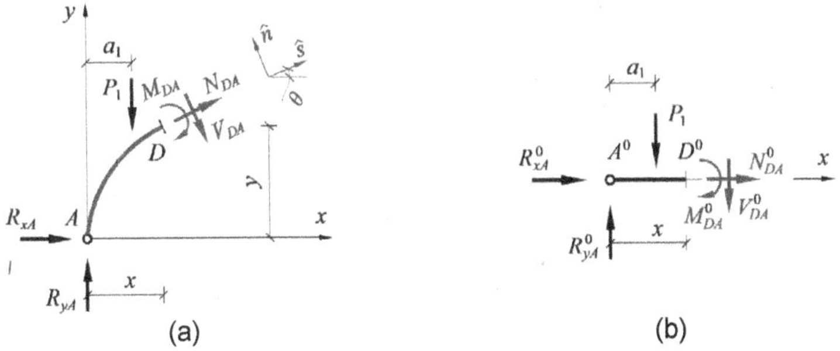

Figure 2.36. (a) The free-body diagram of an arch segment. (b) The free-body diagram of the beam segment corresponding to the arch segment shown in Figure 2.36a.

shows the free-body diagram of the corresponding beam segment A^0D^0. The moment equilibrium about D of the free body AD gives:

$$M_D = R_{yA}x - P_1(x - a_1) - R_h y \tag{2.28}$$

The force equilibrium in the $\hat{s}$ direction gives:

$$N_D = -R_h \cos\theta - R_{xA} \sin\theta + P_1 \sin\theta \tag{2.29}$$

The force equilibrium in the $\hat{n}$ direction gives:

$$V_D = R_{yA} \cos\theta - R_h \sin\theta - P_1 \cos\theta \tag{2.30}$$

Comparison of the internal forces of arches with those of the associate beam shows that the support reactions and moment in arches can be expressed as:

$$R_{yA} = R^0_{yA} \tag{2.31}$$

$$R_{yB} = R^0_{yB} \tag{2.32}$$

$$R_h = \frac{M^0_C}{y_C} \tag{2.33}$$

$$M_D = M^0_D - R_h y \tag{2.34}$$

where R^0_{yA}, R^0_{yB}, M^0_C and M^0_D are the support reactions and moments in the associated beam. The above procedure shows that Equations (2.31)–(2.34) are valid for any vertical load.

All this shows that arches can be used to reduce the bending moments in long-span structures, and may require less material to construct than beams.

Although it requires less material to construct, a three-hinge arch must have large foundation abutments and is not affected by its support settlement or temperature changes. If clearance is not a problem, a tied arch shown in Figure (2.37) can be constructed without the need for large foundation abutments since the tie rod carries the horizontal component of thrust at the supports. The analysis of the tied arch is similar to that of a three-hinged arch and is left as an exercise for the reader because it is similar to that of a three-hinged arch. In engineering practice, the tied arch can be realized as a self-anchored bridge.

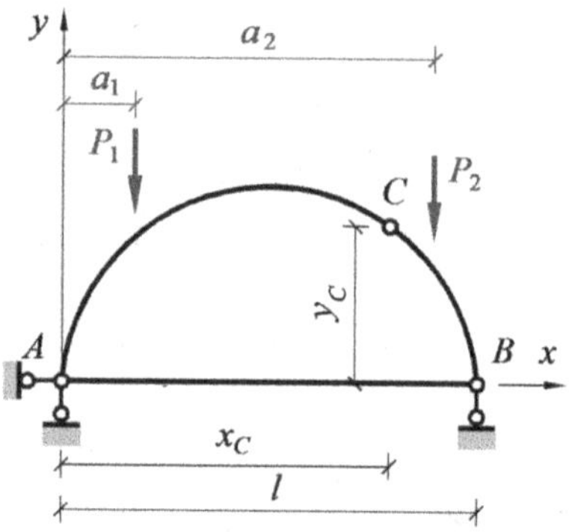

Figure 2.37. A three-hinged tied arch.

2.3.2 *The Ideal Axis of Parabolic Arch*

A parabolic arch is an one whose axis is a parabolic curve, and is commonly used in bridge design, where long spans are needed. This section shows that a parabolic arch is subjected only to axial compression (pure compression) when it supports a vertical uniform load. Such an arch is said to have an ***ideal arch form.***

Consider a parabolic arch shown in Figure (2.38a), whose axis can be expressed as:

$$y(x) = \frac{4y_C}{l^2}x(l-x) \tag{2.35}$$

The moment equilibrium about D of the free-body AD, shown in Figure (2.38b), yields:

$$M_{DA} = \frac{1}{2}q\left(\frac{x}{2}\right)^2 + \frac{ql^2y(x)}{8y_C} - \frac{qlx}{2} \tag{2.36}$$

Substituting Equation (2.35) into Equation (2.36) leads to:

$$M_{DA}(x) = 0 \tag{2.37}$$

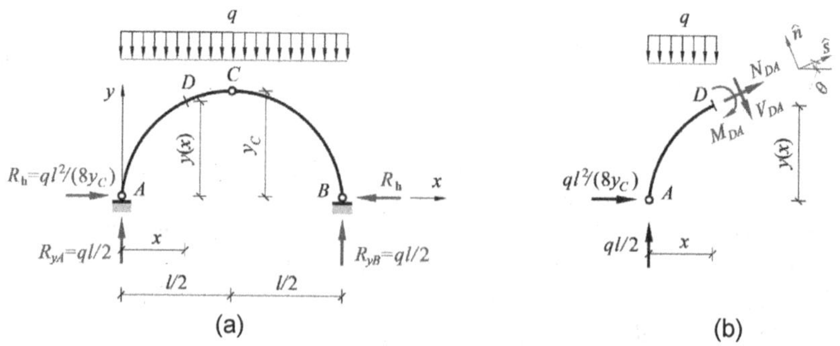

Figure 2.38. (a) A parabolic arch subjected to a vertical uniform load. (b) The free-body diagram of the arch segment AD.

The force equilibrium of the free body *AD* in the $\hat{n}$ direction provides:

$$V_{DA} = \cos\theta\left(\frac{ql}{2} - qx - \frac{ql^2}{8y_C}\tan\theta\right) \tag{2.38}$$

Equation (2.35) shows that the slope of the arch at point *D* can be expressed as:

$$\tan\theta = \frac{\mathrm{d}y}{\mathrm{d}x} = \frac{8xy_C}{l} - \frac{4y_C}{l^2} \tag{2.39}$$

Plugging Equation (2.39) into Equation (2.38) gives:

$$V_{DA}(x) = 0 \tag{2.40}$$

EXERCISES

2.1 Draw the moment diagrams for the flexural members and find the internal forces of the links of the composite structures shown in Figure (2.39).

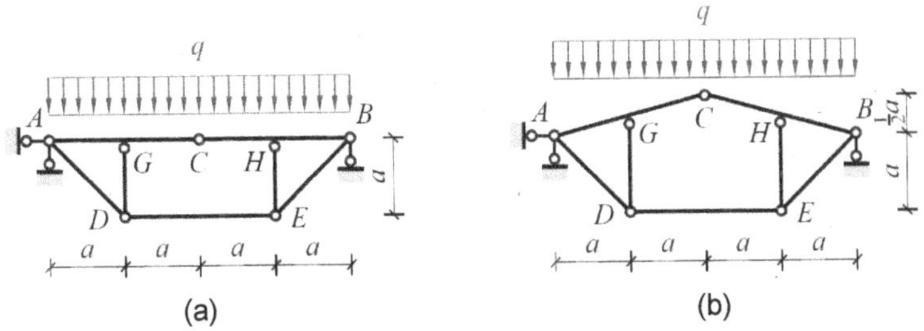

Figure 2.39. Exercise 2.1.

2.2 Draw the moments diagrams for the structures shown in Figure (2.40).

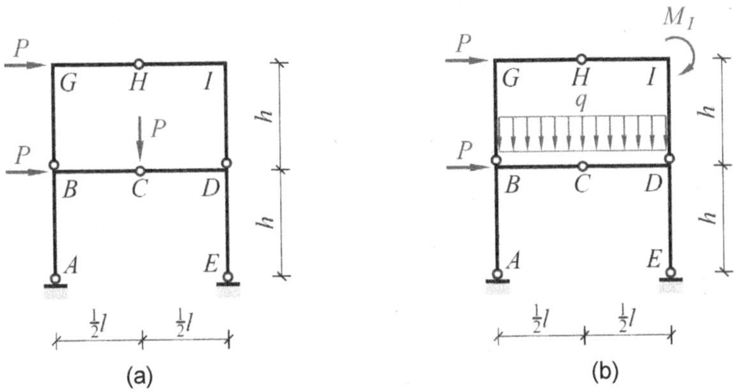

Figure 2.40. Exercise 2.2.

2.3 Draw the moment diagrams for the structures shown in Figure 2.41.

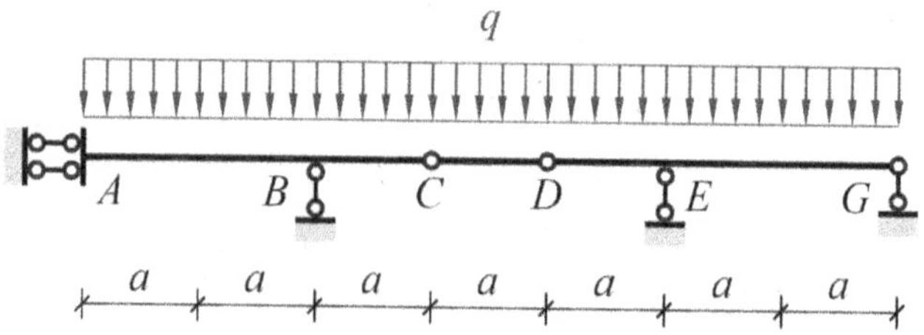

Figure 2.41. Exercise 2.3.

Chapter 3

Deflections of Statically Determinate Structures

3.1 Virtual-Work Principle for Rigid Bodies and Its Applications

3.1.1 Work and Virtual Work

Definition 3.1 (Work) A force $\boldsymbol{F}$ is applied to an object at point p when the point p undergoes a small displacement $\boldsymbol{u}$. The **work**, W, done by the force is defined as:

$$W \equiv \boldsymbol{F} \cdot \boldsymbol{u} \tag{3.1}$$

Definition 3.2 (External Virtual Work) Let $\boldsymbol{F}$ be an external force applied to an object at point p, and $\boldsymbol{u}$ be any small possible displacement at point p. The **external virtual work**, W^*_{ext}, done by the external force $\boldsymbol{F}$ through the displacement $\boldsymbol{u}$ is defined as:

$$W^*_{\text{ext}} \equiv \boldsymbol{F} \cdot \boldsymbol{u} \tag{3.2}$$

The work done, W_{P_C}, W_{M_B}, W_{P_B} and W_{M_C}, by the external loads P_C, M_B, P_B and M_C respectively (Figure 3.1), are:

$$W_{P_C} = -\frac{\sqrt{2}}{2} P_C d, \quad W_{M_B} = 0, \quad W_{P_B} = P_B l\theta, \quad W_{M_C} = M_C\theta \tag{3.3}$$

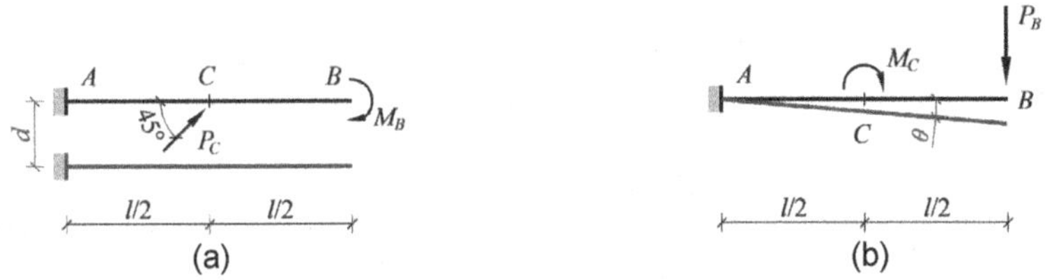

Figure 3.1. (a) Case *I*: A cantilever beam is subjected to a concentrated load, P_C, an external torque, M_B, and a small support settlement, d. (b) Case *II*: The cantilever is subjected to a concentrated load, P_B, an external torque, M_C, and a small support rotation, θ.

The external virtual work done, $W^*_{P_C}$ and $W^*_{M_B}$, by the external load P_C and the torque M_B through the displacements in case *II* are:

$$W^*_{P_C} = -\frac{\sqrt{2}P_C}{2} \times \frac{l\theta}{2}, \text{ and } W^*_{M_B} = M_B\theta \tag{3.4}$$

The statement above is based on the assumption that θ is small.

Similarly, the external virtual work done, $W^*_{P_B}$ and $W^*_{M_C}$, by the external load P_B and the torque M_C through the displacements in Case *I* are:

$$W^*_{P_B} = P_B d, \text{ and } W^*_{M_{BC}} = 0 \tag{3.5}$$

The total external virtual work done by all the external forces applied to the bar *AB* through the displacements in Case *I* is:

$$P_B d + (-P_B d) = 0 \tag{3.6}$$

where the second term on the left-hand side of the equation above is the external virtual work done by the reaction at the fixed end.

3.1.2 Principle of Virtual Displacements for Rigid Bodies

Theorem 3.1 (Virtual-Displacement Principle for Rigid Bodies) *Consider a n-DOF system of rigid bodies subjected to external forces* $\boldsymbol{F}_i$, $n \geq 1$. *The system is in equilibrium if and only if the total external virtual work done by the external loads* $\boldsymbol{F}_i$ *through any kinematically admissible small displacement field* $\boldsymbol{u}^*$ *is zero, namely,*

$$\sum_i F_i \cdot u_i^* = 0 \tag{3.7}$$

here $\boldsymbol{u}_i^*$ *is the virtual displacement at the action point of the external force* $\boldsymbol{F}_i$. *Equation (3.7) is called a* ***virtual-displacement equation*** *of the rigid-body system, while* $\boldsymbol{u}^*$ *is called a* ***virtual displacement field***.

A virtual-displacement equation is equivalent to an equilibrium equation of the system.

To illustrate the use of the principle of virtual displacements, we consider a mechanism shown in Figure (3.2a), and determine the value of the external force P_A such that the mechanism is in equilibrium. P_A can easily be determined by using the moment equilibrium about point *C*. However, here we will utilize Theorem 3.1, the principle of virtual displacements, to solve it.

In the case shown in Figure (3.2b) a virtual displacement is imposed on the mechanism. The virtual-displacement equation can be expressed as:

$$3\theta P_A + (-20 \times 6\theta) = 0 \tag{3.8}$$

where the left-hand side of the total external virtual work done by the forces in the case shown in Figure (3.2a) through the displacements in the case shown in Figure (3.2b). Equation (3.8) yields

$$(3P_A - 20 \times 6)\theta = 0 \tag{3.9}$$

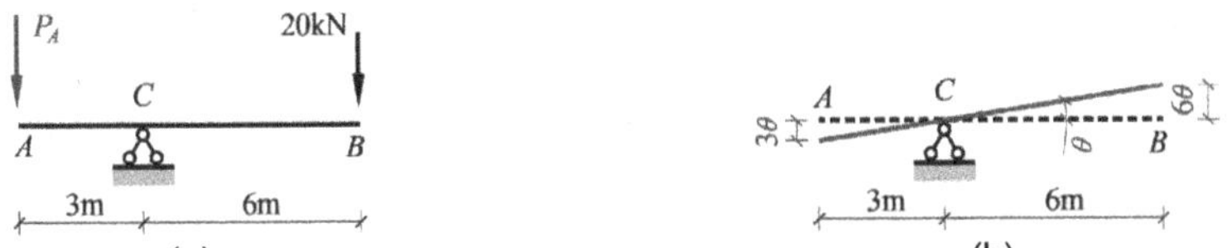

(a) (b)

Figure 3.2. (a) A force case: two concentrated loads are applied to the mechanism. (b) A displacement case: the mechanism undergoes a small rotation—a virtual displacement.

Because θ is arbitrary, Equation (3.9) gives:

$$3P_A - 20 \times 6 = 0 \tag{3.10}$$

which is the moment-equilibrium equation of the mechanism for the force case. Finally, we have:

$$P_A = 40 \text{ (kN)} \tag{3.11}$$

All this shows that the moment-equilibrium equation can be derived from the virtual-displacement Equation (3.8). Since this procedure is reversible, the virtual-displacement equation is equivalent to the equilibrium equation of the system.

EXAMPLE 3.1

Use the principle of virtual displacements to determine the support reaction R_{vC} at C, the internal moment M_B at section B and the shear V_{B+} at the right side of point B of the three-span beam shown in Figure 3.3.

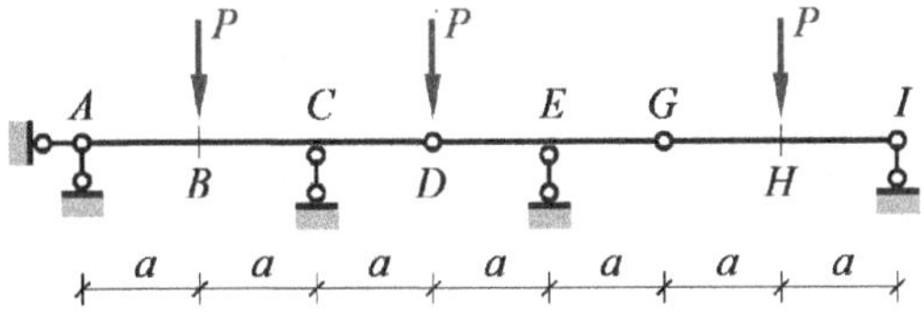

Figure 3.3. Example 3.1. A three-span beam subjected to three concentrated loads.

***Solution*:** (1) Determine the reaction, R_{vC}, from support C. Remove support C and apply its reaction R_{vC} to the beam as shown in Figure (3.4a). A virtual displacement field is imposed on the mechanism as shown in Figure (3.4b). The virtual displacement equation can be written as

$$-a\theta P + 2a\theta R_{vC} - 3a\theta P + 1.5a\theta P = 0 \tag{3.12}$$

which gives

$$(2R_{vC} - 2.5P)\theta = 0 \tag{3.13}$$

Because of the arbitrary nature of θ we conclude from Equation (3.13) that

$$R_{vC} = 1.25P \tag{3.14}$$

(2) Determine the internal moment M_B at section B. A hinge is inserted at point B to release the internal moment M_B. The new mechanism shown in Figure (3.5a) has a single-degree-of-freedom. Impose a virtual displacement field on this mechanism

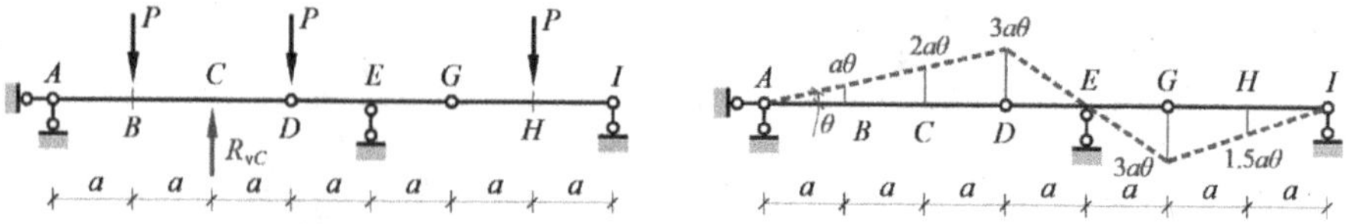

(a) (b)

Figure 3.4. Example 3.1. (a) The force case for calculating the support reaction R_{vC}. (b) A virtual displacement imposed on the mechanism shown in Figure 3.4a.

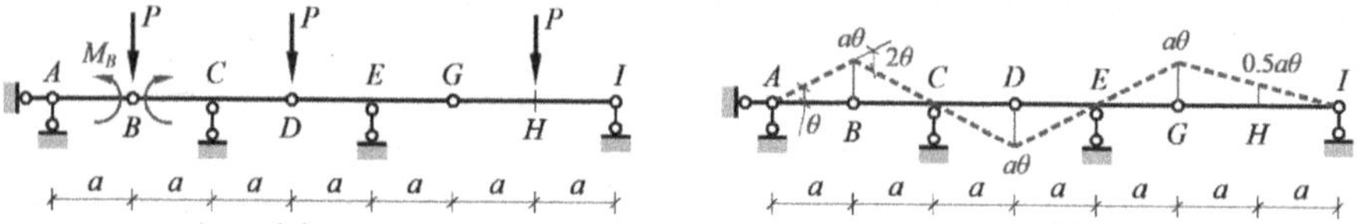

(a) (b)

Figure 3.5. Example 3.1. (a) The force case for calculating the internal moment M_B. (b) A virtual displacement imposed on the mechanism shown in Figure 3.5a.

as shown in Figure (3.5b). In this case, the virtual displacement equation can be written as:

$$M_B 2\theta - a\theta P + a\theta P - 0.5a\theta P = 0 \tag{3.15}$$

which gives

$$(2M_B - 0.5aP)\theta = 0 \tag{3.16}$$

Because θ is arbitrary, the above equation can be satisfied if and only if

$$M_B = 0.25aP \tag{3.17}$$

(3) Determine the shear V_{B+} at the right side of point B. The beam is cut open at a point to the right of point B and a sliding device is inserted at the point.

$$-Pa\theta - 2V_{B+}a\theta - Pa\theta + 0.5Pa\theta = 0 \tag{3.18}$$

Then

$$2aV_{B+} + \frac{3}{2}Pa = 0 \tag{3.19}$$

which gives

$$V_{B+} = -\frac{3}{4}P \tag{3.20}$$

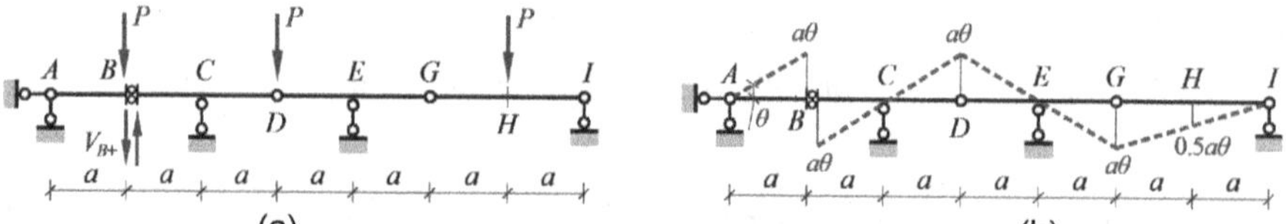

(a) (b)

Figure 3.6. Example 3.1. (a) The force case for calculating the shear. (b) A virtual displacement imposed on the mechanism shown in Figure 3.6a.

3.1.3 Principle of Virtual Forces for Rigid Bodies

Theorem 3.2 (Virtual-Force Principle for Rigid Bodies) *A displacement field* $\boldsymbol{u}$ *imposed on a rigid-body system is kinematically admissible if and only if for arbitrary external forces* $\boldsymbol{P}_i^*$ *applied to the system the total external virtual work done by the external forces* $\boldsymbol{P}_i^*$ *and the support reactions* $\boldsymbol{R}_j^*$ *caused by* $\boldsymbol{P}_i^*$ *is zero:*

$$\sum_i \boldsymbol{P}_i^* \cdot \boldsymbol{u}_i + \sum_i \boldsymbol{R}_j^* \cdot \boldsymbol{u}_j = 0 \tag{3.21}$$

where $\boldsymbol{u}_i$ *are the displacements at the application points of the external forces* $\boldsymbol{P}_i^*$*; and* $\boldsymbol{u}_j$ *are the displacements of the support producing the reactions* $\boldsymbol{R}_j^*$*. Equation (3.21) is called a* ***virtual-force equation*** *for the rigid-body system, while* $\boldsymbol{P}_i^*$ *are called* ***virtual forces****.*

The question of whether a force (or a displacement) should be taken as a virtual quantity depends on the problem. It turns out that forces in the virtual work of a problem involving displacements are said to be virtual. On the other hand, displacements in the virtual work of a problem involving equilibrium are called virtual displacements. Therefore, any force may be taken as a virtual force while any kinematically admissible small displacement may be viewed as a virtual displacement.

A virtual-force equation of a rigid-body system is equivalent to its kinematic constraints which are used to define a system displacement. As an example, we consider a simply-supported beam shown in Figure (3.7a). The settlement of support A causes the beam AB to rotate counterclockwise. Of course, the upward displacement Δ_C at end C can easily be determined by the kinematic constraint using the similarity of two triangles. However, in the following discussion, we will utilize the principle of virtual forces to determine the displacement Δ_C at end C. For this purpose, we apply a virtual force P at end C of the simply-supported beam as shown in Figure (3.7b). The moment equilibrium about point B gives the vertical reaction R_{vA} from support A as follows.

$$R_{vA} = \frac{bP}{a} \tag{3.22}$$

Then, the virtual-force equation can be expressed as:

$$R_{vA}d - P\Delta_C = 0 \tag{3.23}$$

which gives:

$$P\frac{b}{a}d - P\Delta_C = 0 \tag{3.24}$$

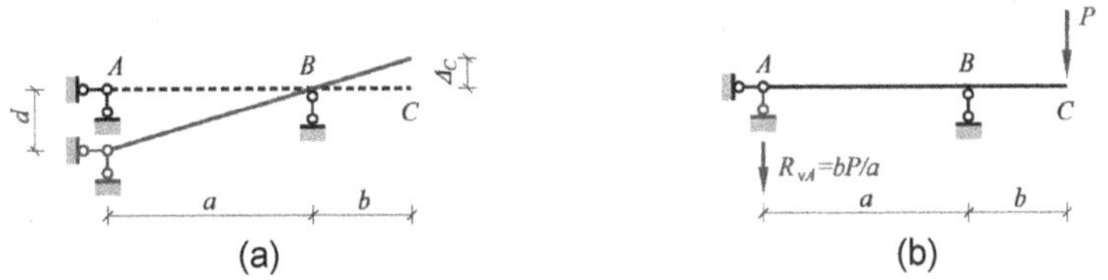

Figure 3.7. (a) A displacement case: a simply-supported beam subjected to a settlement of support A. (b) A force case: a concentrated load P is applied to the simply-supported beam at end C.

Since P is arbitrary, we have:

$$\frac{\Delta_C}{d} = \frac{b}{a} \tag{3.25}$$

namely,

$$\Delta_C = \frac{b}{a} d \tag{3.26}$$

Notice that Equation (3.25) is the kinematic constraint to define the displacement field caused by the support settlement d. It may also be obtained from the similarity of triangles. However, here we derive it from the virtual-force equation.

The principles of virtual displacements and virtual forces are collectively called the ***virtual-work principle***, while the virtual-displacement and virtual-force equations are collectively called the ***virtual-work equation***.

EXAMPLE 3.2

A three-hinged frame, shown in Figure (3.8a), undergoes a settlement d at support A. Utilize the principle of virtual forces to determine the relative rotation between the left and right of pin C and the horizontal displacement at joint D caused by the support settlement d.

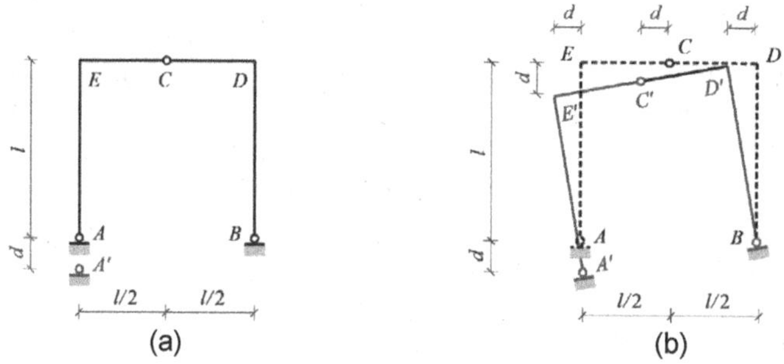

Figure 3.8. Example 3.2. (a) A three-hinged frame undergoing a settlement, d, at support A. (b) The displacement field of the entire frame caused by the support settlement at point A.

Solution: (1) Determine the horizontal displacement, Δ_{hD}, at point D caused by the support settlement d. Apply a horizontal unit force $\bar{F} = 1$ at joint D as shown in Figure (3.9a). In this case, the virtual-force equation can be expressed as:

$$1 \times d + 1 \times \Delta_{hD} = 0 \tag{3.27}$$

Thus,

$$\Delta_{hD} = -d \tag{3.28}$$

(2) Calculate the relative rotation, θ_C, between the left and right of pin C. Unit couple torques $\bar{M} = 1$ are applied at the left and right of pin C. The virtual-force equation can be written as:

$$\frac{1}{l} \times 0 + \frac{1}{l} \times 0 + 1 \times \theta_C = 0 \tag{3.29}$$

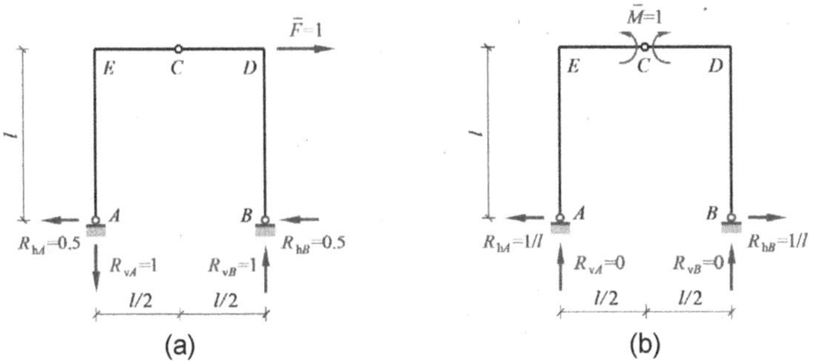

Figure 3.9. Example 3.2. (a) A force case in which a horizontal unit force is applied to the frame at point *D*. (b) A force case in which unit couple moments are imposed at pin *C* of the frame.

Therefore, one obtains:

$$\theta_C = 0 \tag{3.30}$$

Figure (3.8b) shows the displacement field of the entire frame caused by the support settlement.

3.2 Principle of Virtual Forces for Elastic Structures

Definition 3.3 (Internal Virtual Work) *A beam member of length l shown in Figure (3.10a) carries an internal axial force N(x), a moment M(x) and a shear force V(x) in Case I—an internal force case. In Case II, a deformation case, a small segment "dx" of the member at x undergoes a small deformation which can be expressed as a superposition of an axial deformation "du", a bending deformation "dθ", and a shear deformation "dλ" as shown in Figure (3.10b). The **internal virtual work**, W^*_{int}, of the beam member done by the internal forces in Case I through the deformations in Case II is defined as*

$$W^*_{\text{int}} \equiv \int_0^l N(x)\,\mathrm{d}u + M(x)\,\mathrm{d}\theta + V(x)\,\mathrm{d}\lambda \tag{3.31}$$

In most cases, the internal virtual work done by shear is an infinitesimal of higher order. Then, Equation (3.31) can be rewritten as:

$$W^*_{\text{int}} = \int_0^l N(x)\,\mathrm{d}u + M(x)\,\mathrm{d}\theta \tag{3.32}$$

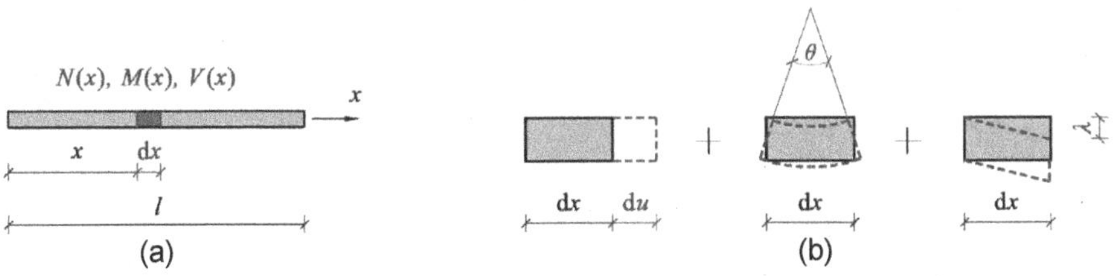

Figure 3.10. Theorem 3.3. (a) Case *I*: A beam member carrying internal forces *N*(*x*), *M*(*x*) and *V*(*x*). Case *II*: A deformation imposed on the beam can be expressed as a superposition of axial, bending and shear deformations.

For a link, the internal virtual work can be expressed as:

$$W_{\text{int}}^{*} = \int_{0}^{l} N(x)\,\mathrm{d}u \tag{3.33}$$

Theorem 3.3 (Virtual-Force Principle for Elastic Structures) *A displacement field* $\boldsymbol{u}$ *imposed on an elastic structure is kinematically admissible if and only if for arbitrary loads* $\boldsymbol{P}_i^*$ *(called* ***virtual loads****) on the structure the total external virtual work done by the virtual loads* $\boldsymbol{P}_i^*$ *and the support reactions* $\boldsymbol{R}_j^*$ *caused by* $\boldsymbol{P}_i^*$ *through the displacement field* $\boldsymbol{u}$ *is equal to the internal virtual work of the structure:*

$$\sum_i \boldsymbol{P}_i^* \cdot \boldsymbol{u}_i + \sum_j \boldsymbol{R}_j^* \cdot \boldsymbol{u}_j = W_{\text{int}}^* \tag{3.34}$$

where W_{int}^* *is the internal virtual work done by the internal forces caused by* $\boldsymbol{P}_i^*$ *through the displacement field* $\mathbf{u}$*. Equation (3.34) is called a* ***virtual-force equation*** *for elastic structures.*

Since the left side of Equation (3.34) is known as the total *external virtual work* of a structure, Equation (3.34) simplifies to:

$$W_{\text{ext}}^* = W_{\text{int}}^* \tag{3.35}$$

As an example, we consider a truss shown in Figure (3.11a). We will utilize Theorem 3.3, the principle of virtual forces, to determine the horizontal displacement Δ_{hB} of point B caused by the vertical load P. A unit horizontal load $\bar{F} = 1$ is imposed on the truss at point B shown in Figure (3.11b), where the internal forces, $\bar{N}_i$, caused by $\bar{F}$ are obtained by statics. The internal forces, N_i^P, of the truss in Case I shown in Figure (3.11a) are indicated in Figure (3.11c). Case I is said to be a real displacement case because we try to find a displacement caused by the vertical load P. From the standpoint of the analysis for the displacements in Case I, the unit force $\bar{F}$ in Case II is a virtual force. Then, the virtual-force equation of this problem can be written as:

$$1 \times \Delta_{hB} = \int_{AB} \bar{N}\mathrm{d}u^P + \int_{BC} \bar{N}\mathrm{d}u^P \tag{3.36}$$

where $\mathrm{d}u^P$ is the deformation of an infinitesimal link element caused by the vertical load P and can be expressed as:

$$\mathrm{d}u^P = \frac{N^P}{EA}\mathrm{d}x \tag{3.37}$$

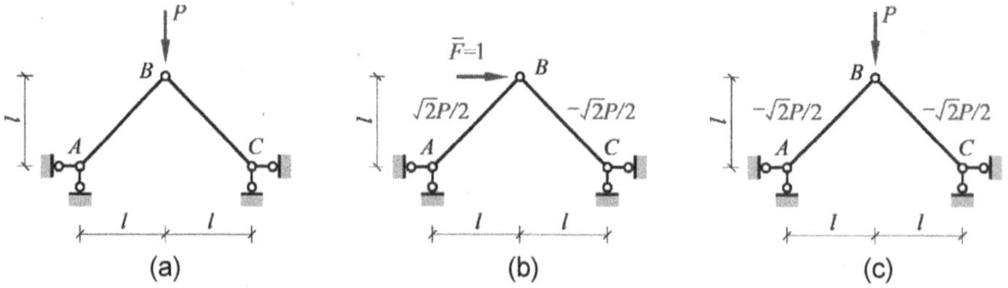

Figure 3.11. (a) Case I: a truss subjected to a vertical load at point B. (b) Case II: a horizontal unit force is applied to point B of the truss. (c) The internal forces of the truss in Case I.

By combining Equation (3.37) and Equation (3.36), one obtains:

$$\begin{aligned}\Delta_{hB} &= \int_{AB} \frac{\bar{N}N^P}{EA}\mathrm{d}x + \int_{BC} \frac{\bar{N}N^P}{EA}\mathrm{d}x \\ &= \bar{N}_{AB}\frac{N^P_{AB}\sqrt{2}l}{EA_{AB}} + \bar{N}_{BC}\frac{N^P_{BC}\sqrt{2}l}{EA_{BC}} \\ &= \frac{\sqrt{2}}{2}\left(-\frac{\sqrt{2}}{2}P\right)\frac{\sqrt{2}l}{EA_{AB}} + \left(-\frac{\sqrt{2}}{2}\right)\left(-\frac{\sqrt{2}}{2}P\right)\frac{\sqrt{2}l}{EA_{BC}} \\ &= \frac{Pl}{2}\left(\frac{1}{EA_{BC}} - \frac{1}{EA_{AB}}\right)\end{aligned} \tag{3.38}$$

where the product EA of the Young's modulus E and the cross-sectional area A of a link is referred to as the ***axial rigidity*** of a link, while EA/l is called the ***axial stiffness*** of a link.

3.3 Deflections Caused by External Loads

3.3.1 Calculation of Structural Displacements to Loads by the Unit-Load Method

The unit-load method developed by John Bernoulli in 1717 provides a means of calculating the displacement of a structure at a given point.

Consider a structure subjected to a lateral concentrated load P and a vertical distributed load q as shown in Figure (3.12a). Let's try to determine the rotation θ_B at joint B of the frame. In this problem, Case (a) shown in Figure (3.12a) is said to be a real displacement case because we attempt to calculate the rotation θ_B in it. In order to establish a virtual-force equation for θ_B, we impose a unit virtual torque, $\bar{M}_B = 1$, at joint B. See Figure (3.12b). The internal virtual work W^*_{int} of the frame done by the internal virtual moment $\bar{M}$ and the internal virtual axial force $\bar{N}$ in Case (b) through the deformations in Case (a) can be expressed as:

$$W^*_{\text{int}} = \sum_{k=1}^{2}\int_{l_k}\left(N^*\mathrm{d}u^P + M^*\mathrm{d}\theta^P\right) = \sum_{k=1}^{2}\int_{l_k}\left(\frac{\bar{N}N^P}{EA}\mathrm{d}x + \frac{\bar{M}M^P}{EI}\mathrm{d}x\right) \tag{3.39}$$

where,

$$\mathrm{d}u^P = \frac{N^P}{EA}\mathrm{d}x, \quad \text{and} \quad \mathrm{d}\theta^P = \frac{M^P}{EI}\mathrm{d}x \tag{3.40}$$

Then the virtual force equation for θ_B can be written as:

$$1 \times \theta_B = \sum_{k=1}^{2}\int_{l_k}\left(\frac{\bar{N}N^P}{EA}\mathrm{d}x + \frac{\bar{M}M^P}{EI}\mathrm{d}x\right) \tag{3.41}$$

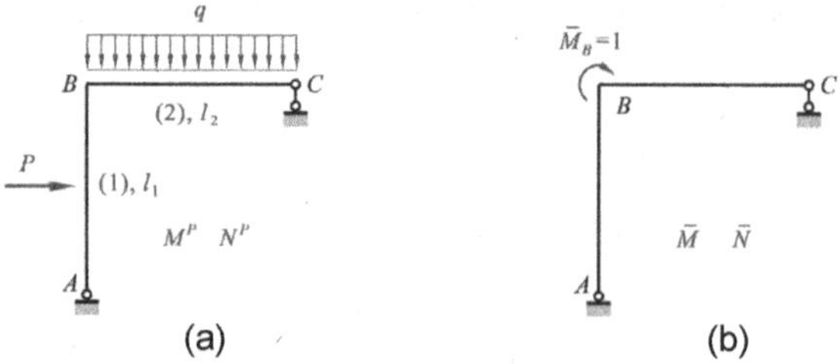

Figure 3.12. (a) A frame having two members (1) and (2) of lengths l_1 and l_2. M^P and N^P denote the internal moment and axial force of the frame to the external loads P and q, respectively. (b) A unit virtual torque $\bar{M}_B = 1$ is applied at point B, and produces the internal moment and axial force $\bar{M}$ and $\bar{N}$ within the frame.

In general, the displacement Δ^P at a specific point caused by external loads can be expressed as:

$$\Delta^P = \sum_k \int_{l_k} \left(\frac{\bar{N}N^P}{EA} \mathrm{d}x + \frac{\bar{M}M^P}{EI} \mathrm{d}x \right) \tag{3.42}$$

where $\bar{N}$ and $\bar{M}$ are the internal virtual forces caused by a unit virtual load applied to the structure. This is the unit-force method for structural deflections to loads.

For a truss (no bending moment), Equation (3.42) can be simplified as:

$$\Delta^P = \sum_k \frac{\bar{N}_k N_k^P}{E_k A_k} l_k \tag{3.43}$$

where $\sum_k$ is the sum over all the links in the truss.

Since the deflections of frames are dominated by the flexural behavior of their members, Equation (3.42) can be approximately expressed as:

$$\Delta^P = \sum_k \int_{l_k} \frac{\bar{M}M^P}{EI} \mathrm{d}x \tag{3.44}$$

For a structure that is influenced by both axial and flexural behavior, we have:

$$\Delta^P = \sum_i \frac{\bar{N}_i N_i^P}{E_i A_i} l_i + \sum_j \int_{l_j} \frac{\bar{M}M^P}{EI} \mathrm{d}x \tag{3.45}$$

where $\sum_i$ is the summation over all the links, and $\sum_j$ is the sum over all the flexural members of the composite structure.

EXAMPLE 3.3

Use the unit-force method to determine the vertical displacement at point C of the truss, shown in Figure (3.13a), to the two vertical loads at points D and E. The axial rigidity of each member of the truss is EA.

***Solution*:** Impose a unit virtual force $\bar{F} = 1$ on the truss at point C. See Figure (3.13b). The internal forces of the truss due to the two real loads and the unit virtual load are illustrated in Figures (3.14a) and (3.14b), respectively. Equation (3.43) gives:

$$\Delta_{vC} = \sum_{i=1}^{7} \frac{\bar{N}_i N_i^P}{E_i A_i} L_i = 6.828 \frac{Pd}{EA}$$

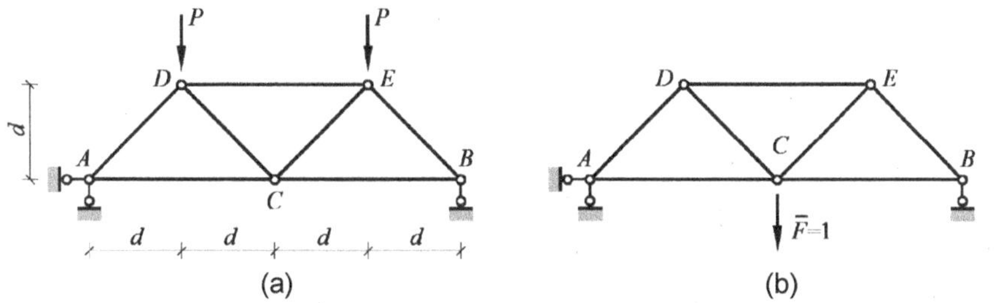

Figure 3.13. Example 3.13. (a) A truss subjected to two vertical loads. (b) A unit virtual force is applied on the truss at point C.

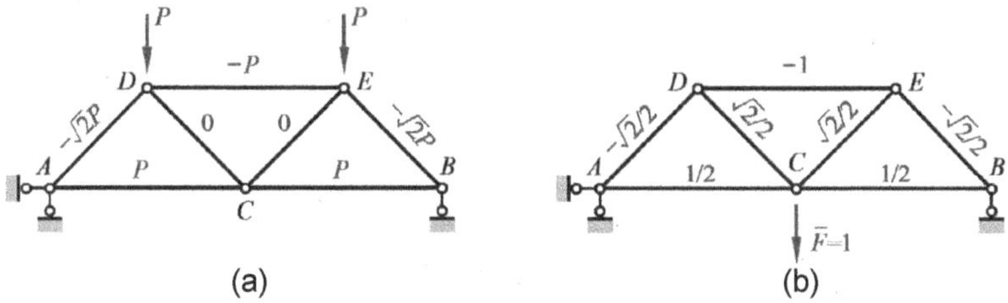

Figure 3.14. (a) The internal forces N_i^P of the truss due to the two vertical loads applied at points D and E. (b) The internal virtual forces $\bar{N}_i$ of the truss to a unit-force $\bar{F} = 1$ at point C.

The positive sign in the equation above means that the displacement Δ_{vC} is downward, i.e., in the same direction as the virtual unit load.

EXAMPLE 3.4

Determine the vertical displacement Δ_{vC} at point C of the frame, shown in Figure (3.15a), to the external torque M_B at joint B. The flexural rigidity of each member of the frame is EI.

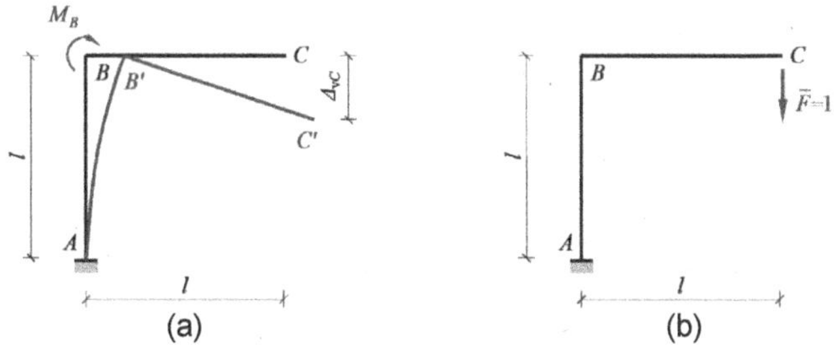

Figure 3.15. Example 3.4. (a) A frame subjected an external torque at point B. (b) A unit-virtual force is imposed on the frame at point C.

Solution: Apply a unit-virtual force $\bar{F} = 1$ to the frame at point C. See Figure (3.15b). The moment diagrams, M^P and $\bar{M}$, of the frame to the real torque M_B and the unit-virtual load $\bar{F}$ are illustrated in Figures (3.16a) and (3.16b), respectively. Equation (3.44) leads to:

$$\Delta_{vC}^{P} = \int_A^B \frac{\bar{M}M^P}{EI}\mathrm{d}x + \int_B^C \frac{\bar{M}M^P}{EI}\mathrm{d}x = \int_A^B \frac{M_B l}{EI}\mathrm{d}x + 0 = \frac{M_B l^2}{EI}$$

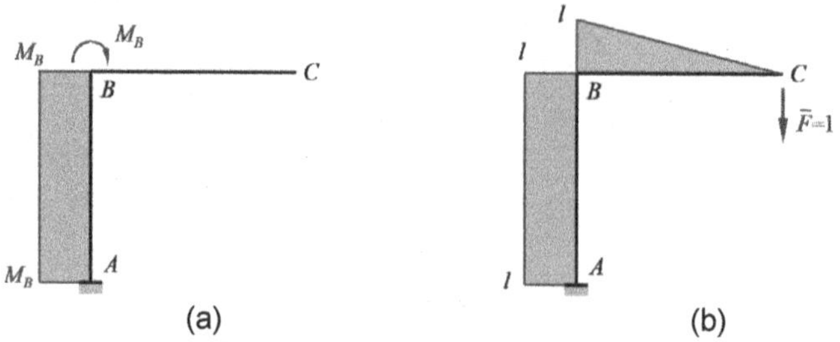

Figure 3.16. Example 3.4. (a) The moment diagram M^P of the frame to M_B. (b) The moment diagram $\overline{M}$ of the frame to the unit-virtual force $\overline{F}$.

3.3.2 Graphic Multiplication and Its Applications

Theorem 3.4 (Graphic Multiplication) *If $f(x)$ and $l(x)$ are positive (or negative) linear functions defined on an interval $[a, b]$, respectively, the following integral can be expressed as:*

$$I \equiv \int_a^b f(x)l(x)\mathrm{d}x \tag{3.46}$$

It can then be evaluated by:

$$I = \pm\omega\left|l(x_c)\right| \tag{3.47}$$

where ω and x_c are the area and centroid of the region under the graph of $y = f(x)$, respectively. See Figure (3.17a). The integral I is positive when $f(x_c)$ and $l(x_c)$ have the same sign. Otherwise, I is negative.

***Proof*:** As shown in Figure (3.17b), we shift the graphs of $y = f(x)$ and $l(x)$ a distance d units to the right such that $L(x)$ can be expressed as:

$$L(x) = x \tan\theta \tag{3.48}$$

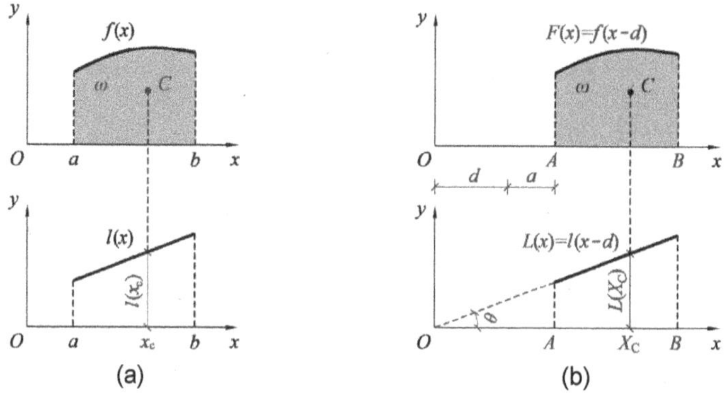

Figure 3.17. Theorem 3.4. (a) $f(x)$ and $l(x)$ are positive (or negative) linear functions, respectively. ω and x_c are the area and centroid of the region under the graph of $y = f(x)$. (b) The graphs of $y = F(x)$ and $y = L(x)$ are the graphs of $y = f(x)$ and $y = l(x)$ shifted d units to the right, respectively.

The integral (3.46) can be rewritten as:

$$I \equiv \int_a^b f(x)l(x)\mathrm{d}x = \int_A^B F(x)L(x)\mathrm{d}x = \tan\theta\int_A^B xF(x)\mathrm{d}x \tag{3.49}$$

With the expression in Equation (3.48), the definition of the centroid provides:

$$\int_A^B (x - X_C)F(x)\mathrm{d}x = 0 \tag{3.50}$$

which gives:

$$\int_A^B xF(x)\mathrm{d}x = X_C\int_A^B F(x)\mathrm{d}x \tag{3.51}$$

By inserting Equation (3.51) into Equation (3.49), one obtains:

$$I = X_C \tan\theta\int_A^B F(x)\mathrm{d}x = \pm\omega\left|L(X_C)\right| = \pm\omega\left|l(x_c)\right| \tag{3.52}$$

Thus, this theorem is true.

The areas and centroids of a right triangle and a parabola are shown in Figure (3.18), which are frequently used in the displacement analysis of structures.

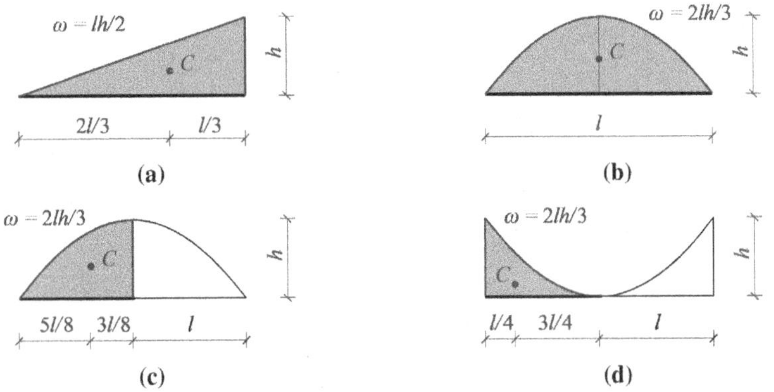

Figure 3.18. Areas and centroids of graphs (a) A right triangle. (b) A parabola. (c) Half a parabola concave downward. (d) Half a parabola concave upward.

EXAMPLE 3.5

Figure (3.19) shows two linear functions $p(x)$ and $q(x)$ defined on $[0, l]$. Use graphic multiplication to evaluate the following integral I_{pq}.

$$I_{pq} \equiv \int_0^l p(x)q(x)\mathrm{d}x \tag{3.53}$$

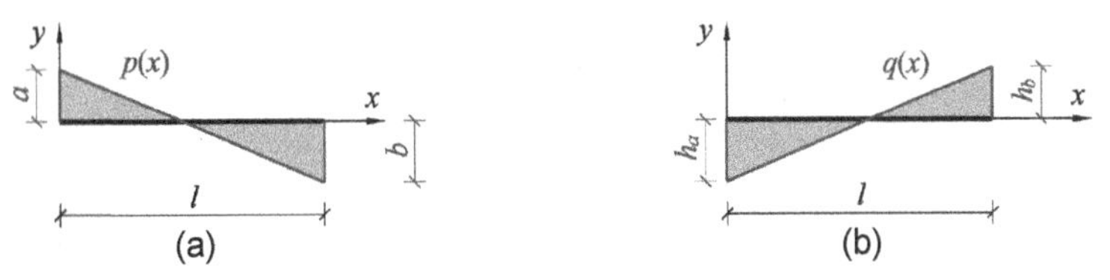

Figure 3.19. Example 3.5. (a) A linear function $p(x)$ defined on $[0, l]$. (b) A linear function $q(x)$ defined on $[0, l]$.

Figure 3.20. Example 3.5. (a) A positive function $f_1(x)$ defined on [0, l]. (b) A negative function $f_2(x)$ defined on [0, l].

***Solution*:** Since $p(x)$ may be taken as a superposition of a positive function $f_1(x)$ and a negative function $f_2(x)$ shown in Figure (3.20), the integral I_{pq} can be expressed as:

$$
\begin{aligned}
I_{pq} &= \int_0^l f_1(x)q(x)\mathrm{d}x + \int_0^l f_2(x)q(x)\mathrm{d}x \\
&= -\frac{al}{2}\times\left(\frac{2h_a}{3}-\frac{h_b}{3}\right)-\frac{bl}{2}\times\left(\frac{2h_b}{3}-\frac{h_a}{3}\right) \\
&= \frac{l}{6}\left[a\left(h_b-2h_a\right)+b\left(h_a-2h_b\right)\right]
\end{aligned}
\tag{3.54}
$$

where we utilized Theorem 3.4 and viewed $q(x)$ as the linear function $l(x)$ in graphic multiplication.

EXAMPLE 3.6

Figure (3.21a) shows a simply-supported beam subjected to a uniform distributed load q. Determine the vertical displacement Δ_{vC} of the beam at its midpoint.

Figure 3.21. Example 3.6. (a) A simply supported beam subjected to a uniform distributed load. (b) A unit virtual load $\bar{F} = 1$ is applied to the simply supported beam at its midpoint.

***Solution*:** To determine the vertical displacement of the beam at its midpoint, a unit virtual force $\bar{F} = 1$ is imposed at the midpoint. Draw the moment diagrams of the simply-supported beam to the uniform distributed load and the unit virtual load as shown in Figure (3.22). The unit-load method gives:

$$
\begin{aligned}
\Delta_{vC} &= \int_0^{2a}\frac{\bar{M}M^P}{EI}\mathrm{d}x = 2\int_0^{a}\frac{\bar{M}M^P}{EI}\mathrm{d}x \\
&= \frac{1}{EI}\left(\frac{2}{3}\times a\times\frac{qa^2}{2}\right)\times\left(\frac{5}{8}\times\frac{a}{2}\right) = \frac{5}{48EI}qa^4
\end{aligned}
$$

where we have taken $\bar{M}$ on half of the beam as the linear function in the graphic multiplication.

Figure 3.22. Example 3.6. (a) The moment diagram, M^P, of the simply-supported beam to the uniform distributed load q. (b) The moment diagram, $\bar{M}$, of the simply-supported beam to the unit virtual load $\bar{F}=1$.

EXAMPLE 3.7

Figure (3.23a) shows a simply-supported beam subjected to a uniform distributed load q and two external torques M_A, M_B at ends A, B. Determine the rotation θ_B of the beam at end B.

***Solution*:** Construct the moment diagrams of the beam to the external loads and the unit virtual torque as shown in Figure (3.24). The unit-load method gives:

$$\begin{aligned}\theta_B &= \int_0^l \frac{M^P \bar{M}}{EI} dx = \frac{1}{EI}\left[M_A l \times \frac{1}{2} + \frac{1}{2} l (M_B - M_A) \times \frac{2}{3} - \frac{2}{3} \times \frac{1}{8} q l^2 \times l \times \frac{1}{2}\right] \\ &= \frac{1}{EI}\left[\frac{1}{6} M_A l + \frac{1}{3} M_B l - \frac{1}{24} q l^3\right]\end{aligned} \tag{3.55}$$

where M^P has been taken as a superposition of a rectangle, a triangle, and a parabola as shown in Figure (3.24a).

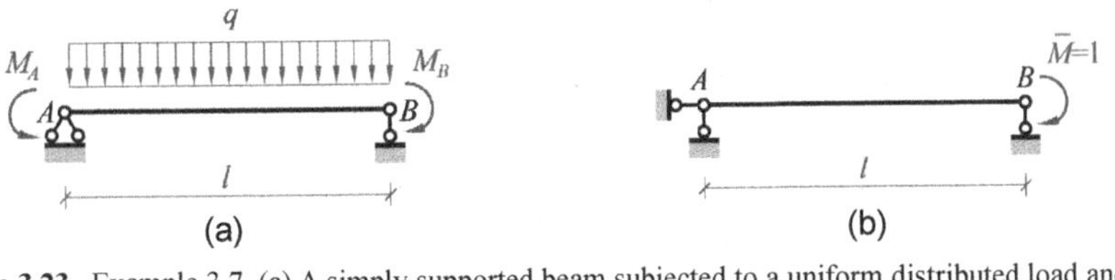

Figure 3.23. Example 3.7. (a) A simply supported beam subjected to a uniform distributed load and two end torques. (b) A unit virtual load $\bar{M}=1$ is imposed on the beam at end B.

Figure 3.24. Example 3.7. (a) The moment diagram, M^P, of the simply-supported beam to the external loads q, M_A and M_B. (b) The moment diagram, $\bar{M}$, of the beam to the unit virtual load $\bar{M}=1$.

EXAMPLE 3.8

Figure (3.25a) shows a composite structure subjected to a uniform distributed load of 10 kN. Determine the relative rotation θ_C between the members at C.

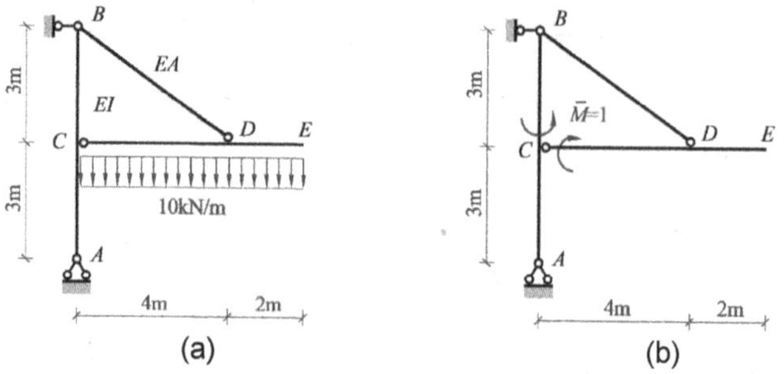

Figure 3.25. Example 3.8. (a) A composite structure subjected to a uniform distributed load of 10 kN. (b) Unit couple torques are applied to the structure at joint C.

***Solution*:** As shown in Figure (3.25b), unit virtual couple torques $\bar{M} = 1$ are imposed at joint C. The unit-load method gives

$$\theta_C = \int_A^B \frac{\bar{M}M^P}{EI}dx + \int_E^C \frac{\bar{M}M^P}{EI}dx + \frac{\bar{N}_{BD}N_{BD}}{EA}l_{BD} = \frac{1}{EI}\left(\frac{1}{2}\times 90\times 3\right)\times\left(\frac{2}{3}\times 1\right)$$

$$+\frac{1}{EI}\left[-\left(\frac{1}{2}\times 20\times 4\right)\times\left(\frac{1}{3}\times 1\right)+\left(\frac{2}{3}\times\frac{1}{8}\times 10\times 4^2\times 4\right)\times\left(\frac{1}{2}\times 1\right)\right]+\frac{75\times\frac{5}{12}}{EA}\times 5 \quad (3.56)$$

$$=\frac{103.33}{EI}+\frac{156.25}{EA}$$

where we used graphic multiplication to calculate the integrals. See Figure 3.26 for the results of the calculated moment diagram.

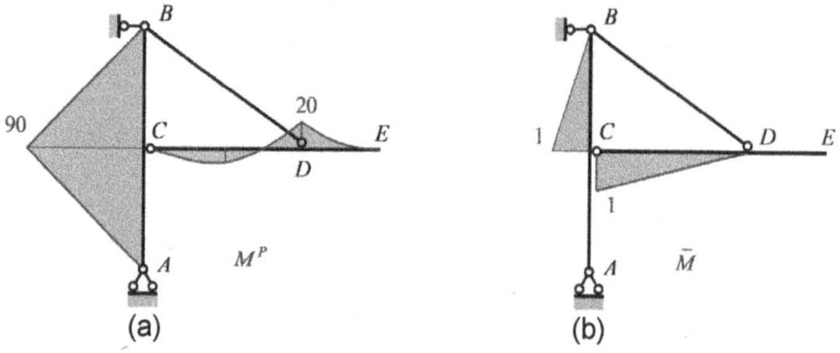

Figure 3.26. Example 3.8. (a) The moment diagram M^P of the composite structure to the uniform load. The moment diagram $\bar{M}$ of the structure to the unit couple torques.

EXERCISES

3.1 Figure 3.27 shows a three-hinged frame subjected to the external loads q, P, and a horizontal displacement d of support A. Determine the vertical displacement at joint C and the relative rotation between the left and right cross sections at joint C.

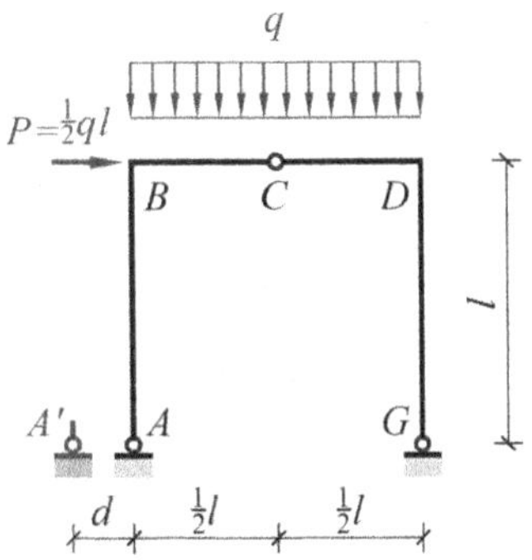

Figure 3.27. Exercise 3.1.

3.2 Figure 3.28 shows a three-hinged frame subjected to the external loads P, $Pl/2$, and a settlement d of support A. Determine the horizontal displacement and the rotation at joint C.

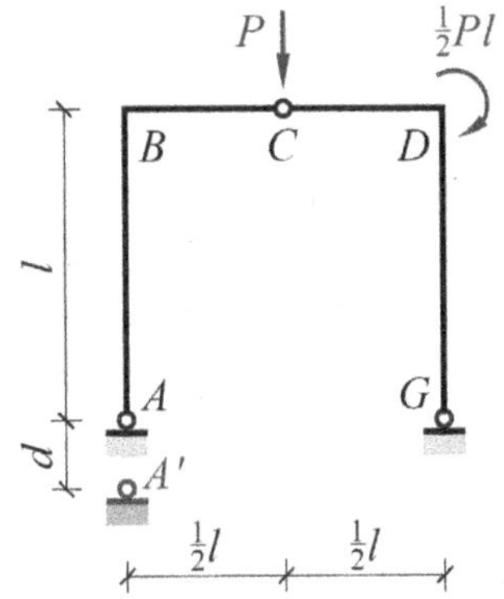

Figure 3.28. Exercise 3.2.

Chapter 4
Force Method

4.1 Statically Indeterminate Structures

A system is said to be **statically indeterminate** if its equilibrium conditions are insufficient for determining its internal forces. For example, consider a system subjected to given loads P_1 and P_2 as shown in Figure (4.1a). The actual structure is transformed into a frame, shown in Figure (4.1b), subjected to the external loads P_1, P_2 and X_C, where X_C is unknown. Since the frame shown in Figure 4.1b is statically determinate, its internal forces would be uniquely determined by its equilibrium conditions when X_C was known. Unfortunately, X_C is unknown. This implies that equilibrium conditions are insufficient for determining the internal forces of the actual frame shown in Figure (4.1a). The reaction X_C is referred to as a **redundant force** or a **redundant**. In general, we have the following proposition.

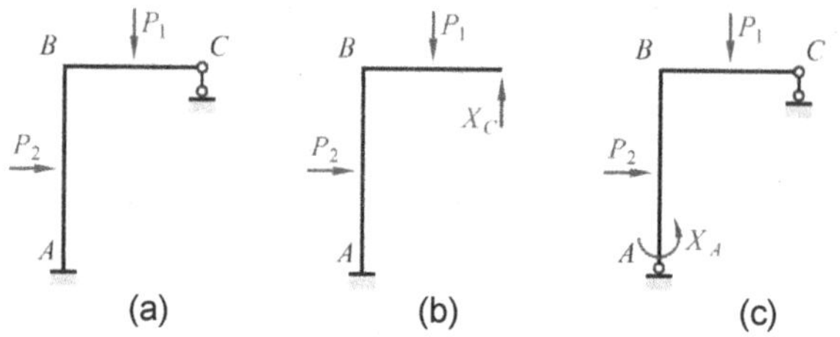

Figure 4.1. Analysis of a statically indeterminate structure.

Theorem 4.1 *A geometrically stable system having n redundant constraints is statically indeterminate and has unknown redundant forces which cannot be determined by equilibrium conditions.*

It is noticed that the choice of the extent of redundancy is arbitrary. For example, the statically indeterminate frame shown in Figure (4.1a) may be transformed into a statically determinate frame shown in Figure (4.1c) by inserting a hinge at A and releasing moment X_A at A. In this case, moment X_A is the unknown redundant force. However, the extent of redundancy of a statically indeterminate structure is a constant and is referred to as the **degree of indeterminacy** of the structure.

EXAMPLE 4.1

Determine the degree of indeterminacy of the structures shown in Figure (4.2).

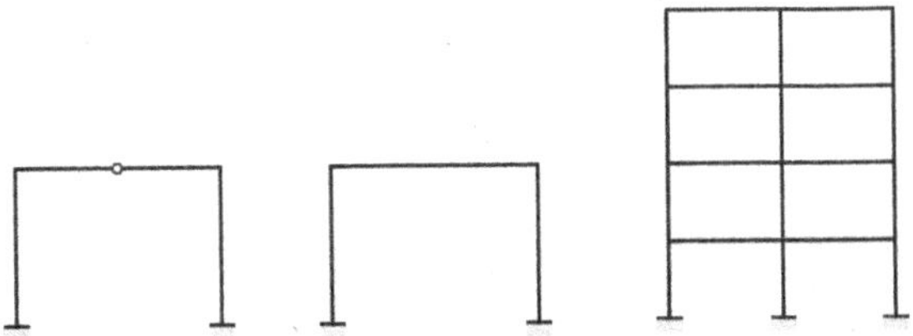

Figure 4.2. Example 4.1.

***Solution*:** See the figure below for the process to determine the degree of indeterminacy.

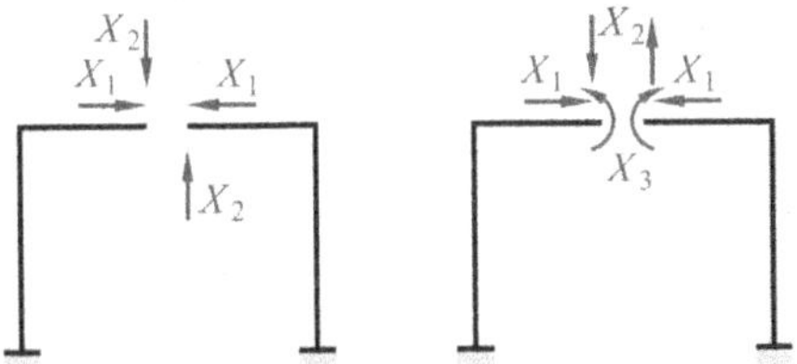

Figure 4.3. Solutions of Example 4.1.

EXAMPLE 4.2

Determine the degree of indeterminacy of the structures shown in Figure 4.4.

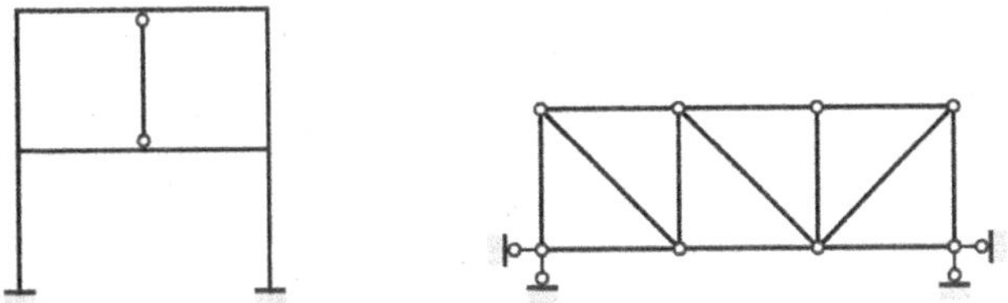

Figure 4.4. Example 4.3.

***Solution*:** See the figure below for the process to determine the degree of indeterminacy.

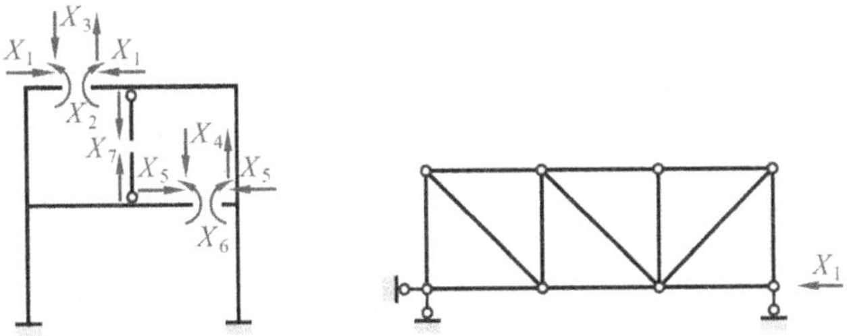

Figure 4.5. Solutions of Example 4.2.

4.2 General Procedure of the Force Method

The left side of Figure (4.6) shows a statically indeterminate beam with a single degree of indeterminacy. In order to analyze this beam, we remove support *B* to release its reaction X_1, and obtain a statically determinate beam as shown in the right side of the figure. The statically determinate beam subjected to q and X_1 is referred to as a ***primary system*** associated with the actual system shown in the left side of Figure (4.6). The structure in the primary system is called a ***primary structure***, which is statically determinate. The primary system is equivalent to the actual system if the displacement at point B of the primary system in the direction of X_1 is zero:

$$\Delta_1 = \Delta_{11} + \Delta_{1q} = 0 \tag{4.1}$$

in which Δ_1 is the displacement at point B of the primary system; Δ_{11} and Δ_{1q} are the displacements at point B of the primary structure under X_1 and q, respectively. Equation (4.1) is called a ***compatibility condition*** for the equivalence of the actual and

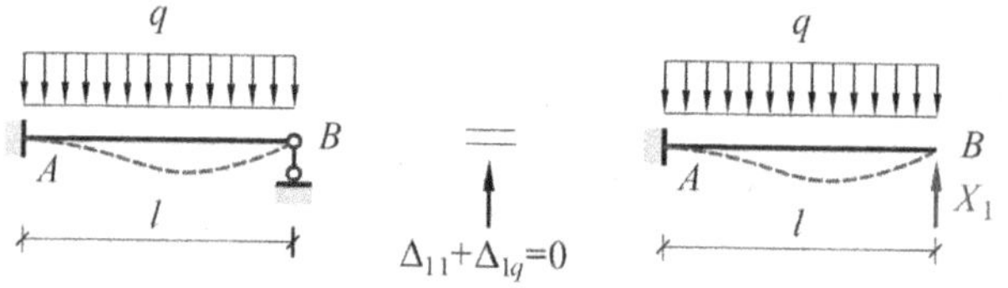

Figure 4.6. The equivalence of the actual and primary systems.

primary systems. If δ_{11} denotes the displacement at point B of the primary structure under a unit load in the direction X_1, then the ***compatibility equation*** can be rewritten as:

$$\delta_{11} X_1 + \Delta_{1q} = 0 \tag{4.2}$$

where δ_{11} is a flexibility coefficient of the primary structure. Equation (4.2) is the governing equation of the force method for solving this problem. In order to calculate δ_{11} and Δ_{1q}, draw the moment diagram as shown in Figure (4.7).

$$\delta_{11} = \int_0^l \frac{\overline{M}_1 \overline{M}_1}{EI} \mathrm{d}x = \frac{1}{EI}\left[\left(\frac{1}{2} \times l \times l\right) \times \left(\frac{2}{3} \times l\right)\right] = \frac{l^3}{3EI} \tag{4.3}$$

and

$$\Delta_{1q} = \int_A^B \frac{\overline{M}_q \overline{M}_1}{EI} \mathrm{d}x = -\frac{1}{EI}\left[\left(\frac{1}{3} \times l \times \frac{ql^2}{2} \times \left(\frac{3}{4} \times l\right)\right)\right] = -\frac{ql^4}{8EI} \tag{4.4}$$

By substituting Equations (4.3) and (4.4) into Equation (4.2), we get,

$$\frac{l^3}{3EI} X_1 - \frac{ql^4}{8EI} = 0 \tag{4.5}$$

which yields the redundant force:

$$X_1 = \frac{3ql}{8} \tag{4.6}$$

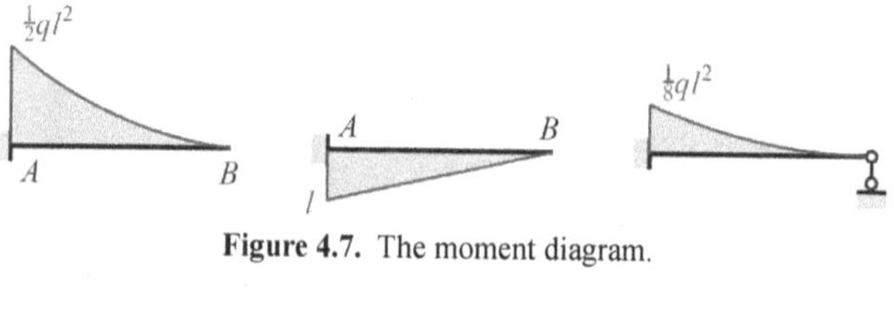

Figure 4.7. The moment diagram.

Since X_1 satisfies the compatibility equation, the deflections and internal forces of the actual system are equal to those of the primary system. This shows that the moment M of the actual system can be expressed as:

$$M = M_q + \bar{M}_1 X_1 \tag{4.7}$$

where $\bar{M}_1$ and M_q are the moments of the primary structure caused by $X_1 = 1$ and q, respectively.

Next, consider a frame with two degrees of indeterminacy as shown in Figure (4.8a). We obtain a primary system, shown in Figure (4.8b), for solving this problem by removing the sliding support C and releasing the reactions X_1 and X_2. The compatibility conditions for the equivalence of the actual system and the primary system are:

$$\begin{cases} \Delta_1 = 0 \\ \Delta_2 = 0 \end{cases} \tag{4.8}$$

where

$$\begin{cases} \delta_{11} X_1 + \delta_{12} X_2 + \Delta_{1q} = 0 \\ \delta_{21} X_1 + \delta_{22} X_2 + \Delta_{2q} = 0 \end{cases} \tag{4.9}$$

Equation (4.9) is the governing equation of the force method for this problem. Notice that the primary system for a problem is not unique. For example, we may take the system shown in Figure (4.8c) as the primary system for solving this problem.

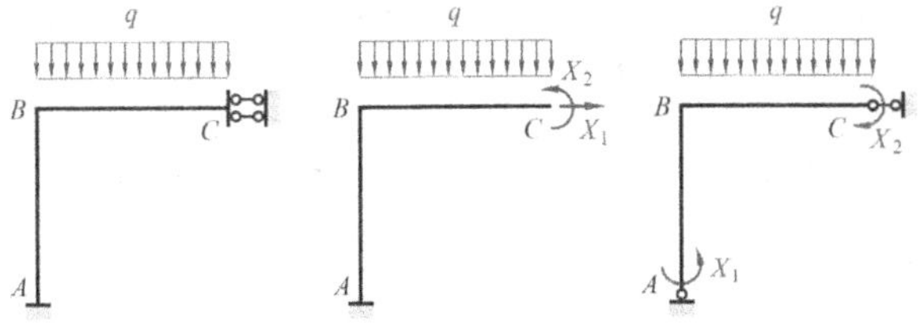

Figure 4.8. A frame with two degrees of indeterminacy.

In general, the governing equation of the force method for a structure with n-degrees of indeterminacy can be expressed as

$$\begin{cases} \delta_{11} X_1 + \delta_{12} X_2 + \cdots + \delta_{1n} X_n + \Delta_{1P} = 0 \\ \delta_{21} X_1 + \delta_{22} X_2 + \cdots + \delta_{2n} X_n + \Delta_{2P} = 0 \\ \vdots \\ \delta_{n1} X_1 + \delta_{n2} X_2 + \cdots + \delta_{nn} X_n + \Delta_{nP} = 0 \end{cases} \tag{4.10}$$

where, Equation (4.10) can be rewritten in matrix form as

$$[\delta]\{X\} + \{\Delta_P\} = \{0\} \tag{4.11}$$

where $[\delta]$ is the flexibility matrix of the structure. The reciprocal-displacement theorem shows that $\delta_{ij} = \delta_{ji}$, namely, the flexibility matrix $[\delta]$ is symmetric.

After computation of each redundant X_i from Equation (4.10), the internal force F of the actual structure can be calculated by:

$$F = F_P + \sum_{i=1}^{n} X_i \bar{F}_i \tag{4.12}$$

where $\bar{F}_i$ is the internal force of the primary structure due to $X_i = 1$; F_P is the internal force of the primary structure under the external loads. The term internal force is used in a general sense to include the moment, axial force, and shear of the structure.

4.3 Analysis of Statically Indeterminate Structures under Loads

Consider the frame shown in Figure (4.9a). We shall draw the moment diagram of this statically indeterminate frame by the force method.

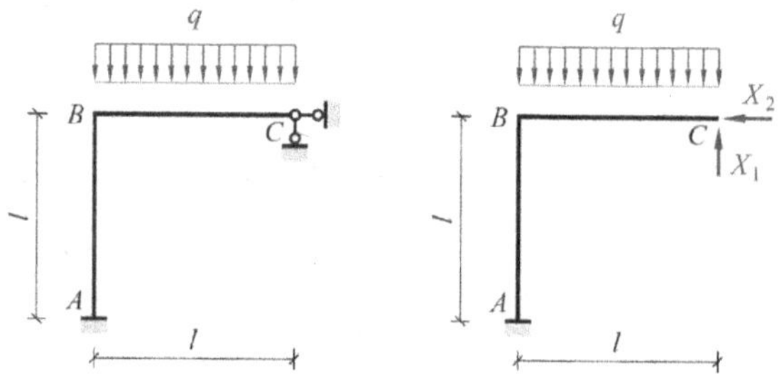

Figure 4.9. A statically indeterminate frame.

First, we remove pins to obtain the primary system shown in Figure (4.9b). The compatibility conditions at point C lead to the governing equation of the force method for this problem as follows.

$$\begin{cases} \delta_{11}X_1 + \delta_{12}X_2 + \Delta_{1q} = 0 \\ \delta_{21}X_1 + \delta_{22}X_2 + \Delta_{2q} = 0 \end{cases} \tag{4.13}$$

where δ_{ij} and the Δ are the flexibility coefficients. In order to determine flexibility coefficients δ_{ij} and the Δ, we draw as shown in Figure 4.10. The definition of δ_{ij} and the graphic multiplication lead to:

$$\begin{cases} \delta_{11} = \dfrac{1}{1.5EI}(l\times l\times l) + \dfrac{1}{2EI}\left(\dfrac{1}{2}\times l\times l\times\dfrac{2l}{3}\right) = \dfrac{5l^3}{6EI} \\ \delta_{12} = \dfrac{1}{1.5EI}\left(\dfrac{1}{2}\times l\times l\times l\right) = \dfrac{l^3}{3EI} \\ \delta_{21} = \delta_{12} \\ \delta_{22} = \dfrac{1}{1.5EI}\left(\dfrac{1}{2}\times l\times l\times\dfrac{2l}{3}\right) = \dfrac{2l^3}{9EI} \\ \Delta_{1q} = -\dfrac{1}{1.5EI}\left(\dfrac{ql^2}{2}\times l\times l\right) - \dfrac{1}{2EI}\left(\dfrac{1}{3}\times\dfrac{ql^2}{2}\times l\times\dfrac{3l}{4}\right) = -\dfrac{19ql^4}{48EI} \\ \Delta_{2q} = -\dfrac{1}{1.5EI}\left(\dfrac{ql^2}{2}\times l\times\dfrac{l}{2}\right) = -\dfrac{ql^4}{6EI} \end{cases} \tag{4.14}$$

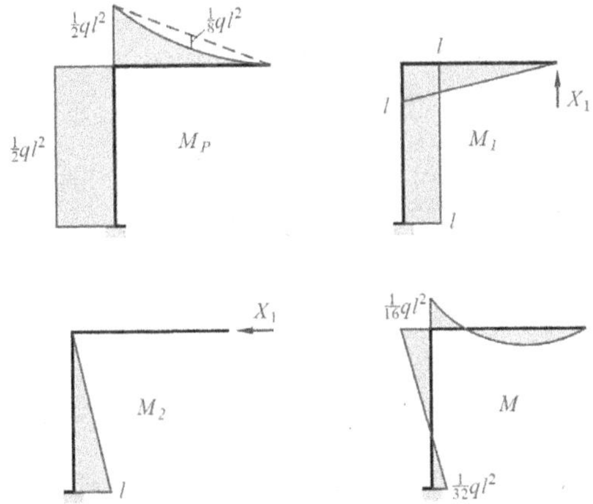

Figure 4.10. Solutions of Figure 4.9.

Substituting Equation (4.14) into Equation (4.13) gives:

$$\begin{cases} \frac{5}{6}X_1 + \frac{1}{3}X_2 = \frac{19}{48}ql \\ \frac{1}{3}X_1 + \frac{2}{9}X_2 = \frac{1}{6}ql \end{cases} \tag{4.15}$$

Solving for X_1 and X_2 the equations above yield:

$$\begin{cases} X_1 = \frac{7}{16}ql \\ X_2 = \frac{3}{32}ql \end{cases} \tag{4.16}$$

Since the moment M of the actual frame can be expressed as:

$$M = M_P + X_1\bar{M}_1 + X_2\bar{M}_2 \tag{4.17}$$

we get the moment diagram for the actual frame as shown in Figure (4.10d).

As a second example, we analyze the internal forces of the truss as shown in Figure (4.11).

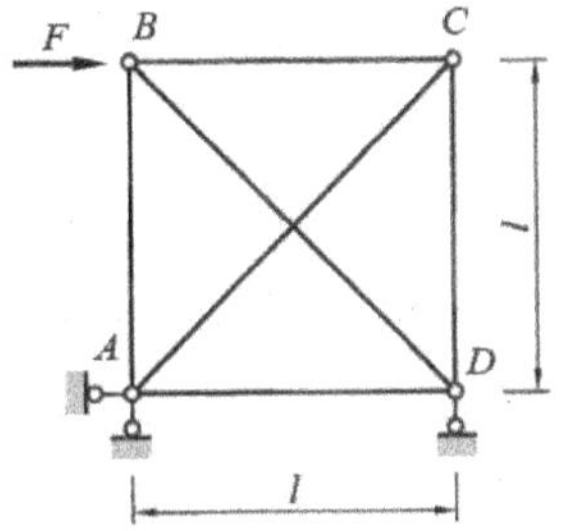

Figure 4.11. A truss structure subjected to a horizontal force F.

First, we obtain the primary system for this problem by cutting link BC and releasing the internal force X_1, see Figure (4.12). Then, the governing equation of the force method for this problem can be written as:

$$\delta_{11}X_1 + \Delta_{1P} = 0 \tag{4.18}$$

Determine δ_{11} and Δ_{1P} as follows:

$$\begin{aligned}\delta_{11} &= \sum_{i=1}^{6}\frac{\overline{N}_i^2}{EA}l_i \\ &= \frac{2}{EA}\left[1^2\times l + 1^2\times l + \left(-\sqrt{2}\right)^2\times\sqrt{2}l\right] = \frac{4\left(1+\sqrt{2}\right)l}{EA}\end{aligned} \tag{4.19}$$

$$\begin{aligned}\Delta_{1P} &= \sum_{i=1}^{6}\frac{\overline{N}_i N_i^P}{EA}l_i = \frac{1}{EA}[1\times F\times l + 1\times F\times l + (-\sqrt{2})\times(-\sqrt{2}F)\times\sqrt{2}l] \\ &= \frac{2(1+\sqrt{2})Fl}{EA}\end{aligned} \tag{4.20}$$

$$X_1 = -\frac{F}{2} \tag{4.21}$$

Since,

$$N_i = N_i^P + X_i\overline{N}_i,\ i = 1, 2, ..., 6 \tag{4.22}$$

we can obtain the internal forces of the actual truss as shown in Figure (4.12).

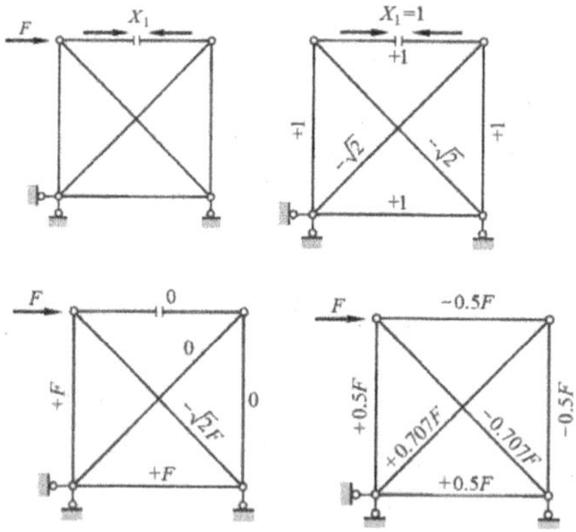

Figure 4.12. Solutions of Figure 4.11.

4.4 Symmetric Structures and Their Half-Structures

4.4.1 Symmetry of Structures, Loads and Responses

If, for a structure, there exists an axis such that the left is a reflected counterpart (a mirror image) of the right about the axis, then the structure is symmetric about it. The symmetry of a structure requires the material composition, geometry, and

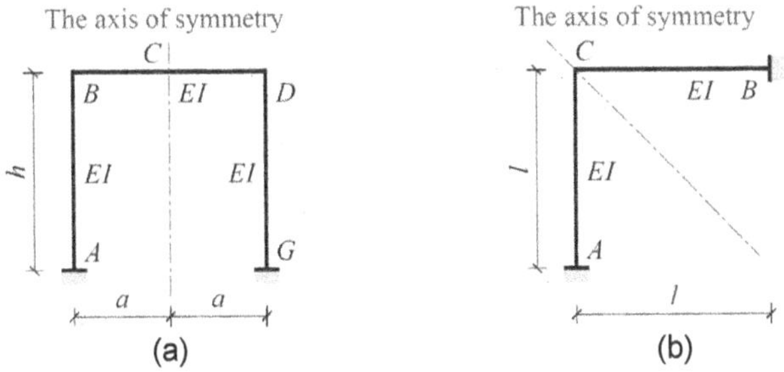

Figure 4.13. Two symmetric structures. (a) A symmetric frame with a vertical axis of symmetry. (b) A symmetric frame having an inclined axis of symmetry.

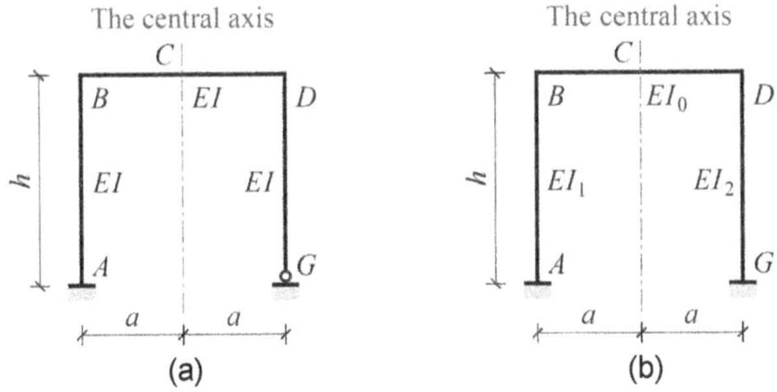

Figure 4.14. Two unsymmetrical structures. (a) The supports of the frame are unsymmetrical. (b) The material composition of the frame is unsymmetrical.

the support to be the same on each side of the structure. The material properties include axial rigidity *EA*, and the flexural rigidity *EI*. Another way to think about a symmetric structure is that if it were to be folded in half over the axis, the two halves would be identical. For example, the frames shown in Figure (4.13) are symmetric, while the frames shown in Figure (4.14) are unsymmetrical.

The loading applied to a symmetric structure may be symmetric, antisymmetric, or unsymmetrical. A loading is said to be symmetric if the reflected counterpart of its right side is identical to its left side. Similar to a symmetric structure, if a set of symmetric loads is folded in half over the axis, the two halves are each other's mirror images. For example, the loading shown in Figure (4.15a) is symmetric. A loading is said to be antisymmetric if the reflected counterpart of its right side has the same magnitude as the left but is in the opposite direction of the left. If a set of antisymmetric loads are folded in half over the axis, the two halves have the same magnitude, but are in the opposite directions. As shown in Figure (4.15b), the torque M_c applied at point *C* on the axis of symmetry is antisymmetric. It is equivalent to the load case shown in Figure (4.15c). Figure (4.15d) shows that the mirror image of the torque $M_c/2$ to the right of *C* has the same magnitude as the torque applied at the left of *C* and is in the opposite direction. Thus, it is antisymmetric. A load is said to be unsymmetrical if it is neither symmetric nor antisymmetric.

A response of a symmetric structure is said to be symmetric if the response at any point of the structure is equal to its reflection at the point of reflection. On the other hand, a response of a symmetric structure is said to be antisymmetric if the response

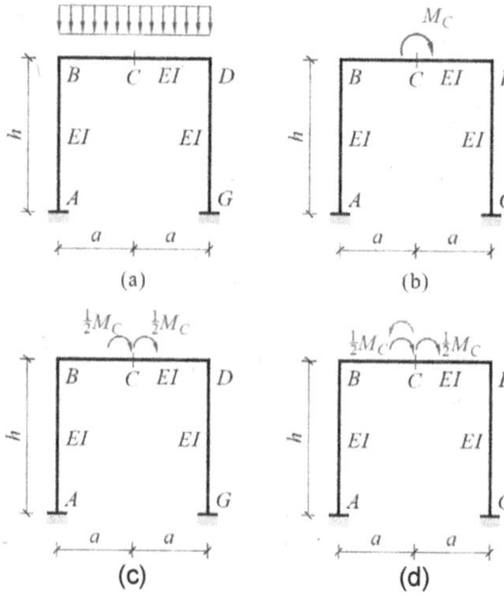

Figure 4.15. (a) A symmetric loading on a symmetric frame. (b) An antisymmetric torque. (c) An equivalence of case (b). (d) The mirror image of the torque at the right of C has the same magnitude and the opposite direction as the torque at the left of C (a, b, c, and d are upper left, lower left, upper right, and lower right figures, respectively).

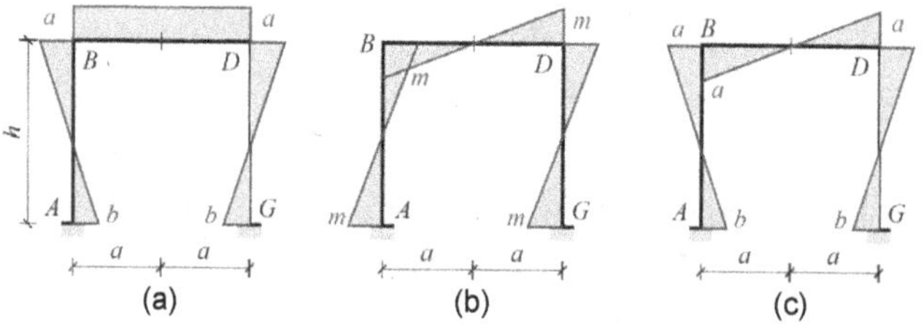

Figure 4.16. (a) left: symmetric response (b) mid: antisymmetric response; (c) right: symmetric response between AB and GD, but antisymmetric response for BD.

at any point of the structure has the same magnitude but the opposite direction to its reflection at the point of reflection. Figure (4.16) shows different types of responses.

If a loading on a symmetric structure is unsymmetrical, then it can be decomposed into symmetric and antisymmetric components. Figure (4.17) shows the decomposition of an unsymmetrical loading on a symmetric frame.

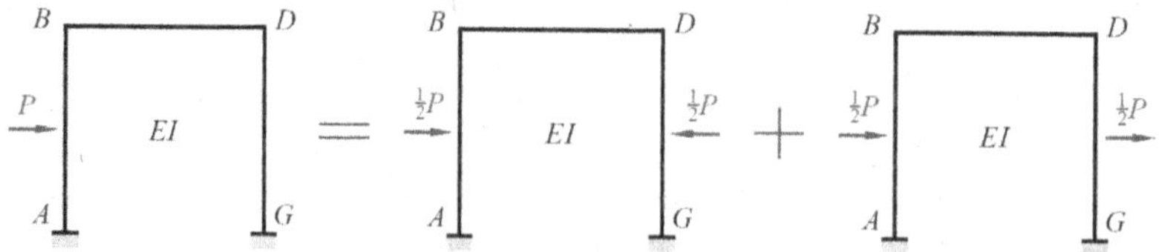

Figure 4.17. An unsymmetrical loading P on a symmetric frame is decomposed into the symmetric and antisymmetric components.

4.4.2 Half-Structures of Symmetric Structures

Two important properties of symmetric structures are frequently used in the analysis of structures and may be stated as the following theorems.

Theorem 4.2 *Any response of a symmetric structure to symmetric loads is symmetric.*

The term response is used in a general sense including any response quantity, such as displacement, internal force, or reaction of the structure. Theorem 4.2 shows that the antisymmetric component of a response vanishes for symmetric loads.

Theorem 4.3 *Any response of a symmetric structure to antisymmetric loads is antisymmetric.*

Theorem 4.3 shows that the symmetric component of a response vanishes for antisymmetric loads. Theorems (4.2) and (4.3) indicate that provided a structure is symmetric and its loading is either symmetric or antisymmetric, then the static analysis will only have to be performed based on half the structure.

It should be noted that for a dynamic analysis, due to the reason that different orders of vibration modes can be either symmetric or antisymmetric, Theorems (4.2) and (4.3) may not be valid.

Consider a symmetric frame over a single span subjected to a set of symmetric loads as shown in Figure (4.18a). Because the shear V_A, the rotation θ_A and the horizontal displacement Δ_{Ah} at point A are antisymmetric, Theorem 4.2 leads to:

$$V_A = 0, \quad \theta_A = 0, \quad \Delta_{Ah} = 0 \tag{4.23}$$

Equation (4.23) shows that the actual problem shown in Figure (4.18a) is equivalent to the problem shown in Figure (4.18b), in which the structure is called the ***half-structure*** for solving this problem.

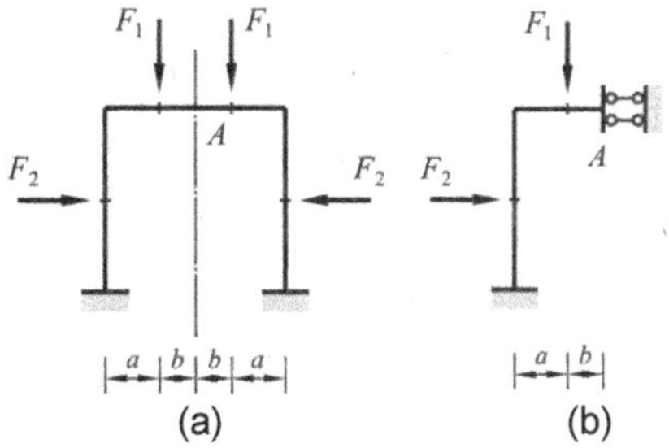

Figure 4.18. (a) A symmetric frame over a single span subjected to a set of symmetric loads. (b) The half-structure equivalent to the actual structure.

Consider a symmetric frame over two spans subjected to a set of symmetric loads as shown in Figure (4.19a). Because the rotation θ_A and horizontal displacement Δ_{Ah} at point A are symmetric, Theorem 4.2 gives:

$$\theta_A = 0, \quad \Delta_{Ah} = 0 \tag{4.24}$$

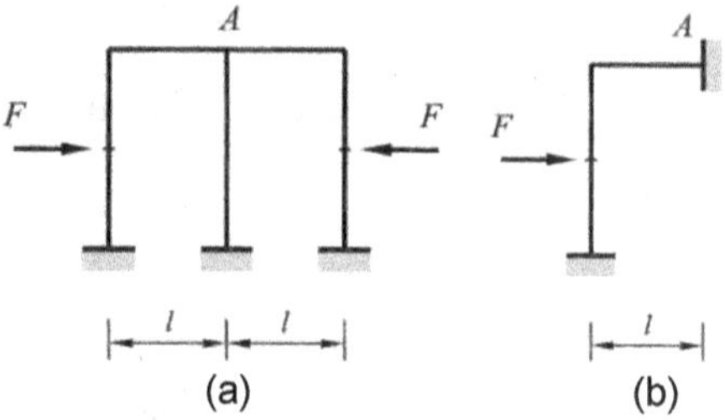

Figure 4.19. (a) A symmetric frame over two spans subjected to a set of symmetric loads. (b) The half-structure equivalent to the actual structure.

Since the axial deformation of a flexural member can be neglected, the vertical displacement at point A is zero, namely,

$$\Delta_{Av} = 0 \tag{4.25}$$

Equations (4.24) and (4.25) show that, for solving this problem, one can analyze the half-structure, shown in Figure (4.19b), instead of the actual frame shown in Figure (4.19a).

Consider a symmetric frame over a single span subjected to antisymmetric loads as shown in Figure (4.20a). Since the moment M_A, the axial force N_A and the vertical displacement at point A are symmetric, Theorem 4.3 leads to:

$$M_A = 0, \quad N_A = 0, \quad \Delta_{Av} = 0 \tag{4.26}$$

Thus, the half-structure for solving this problem is shown in Figure (4.20b).

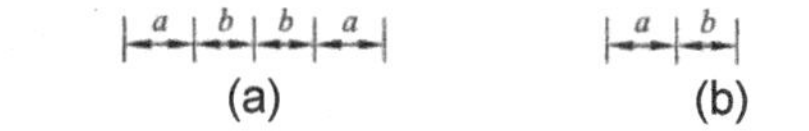

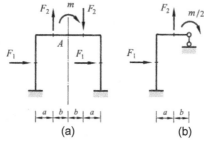

Figure 4.20. (a) A symmetric frame over a single span subjected to antisymmetric loads. (b) The half-structure equivalent to the actual structure.

Consider a symmetric frame over two spans subjected to antisymmetric loads as shown in Figure (4.21a). The actual problem is equivalent to the problem shown in Figure (4.21b), in which the structure is a three-span frame with an infinitesimal mid-span. Similar to the analysis of the problem shown in Figure (4.19a), we can analyze the problem shown in Figure (4.21c) instead of the problem shown in Figure (4.21b). Since the axial deformation of a flexural member can be neglected, the vertical displacement of point A is zero. This implies that the roller support at A is an inefficient constraint and can be removed. Finally, the actual problem shown in Figure (4.21a) is equivalent to the problem shown in Figure (4.21d).

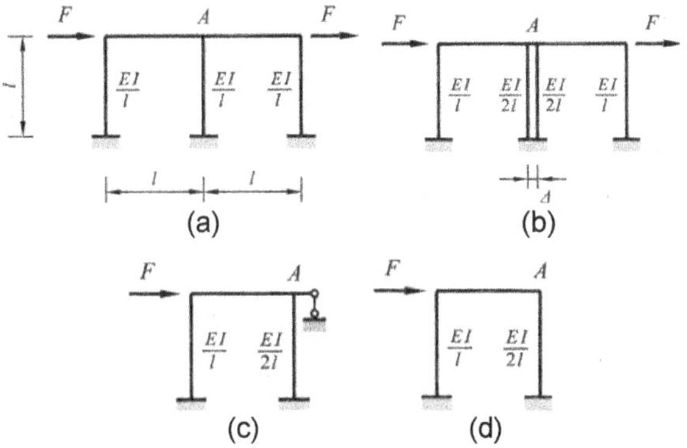

Figure 4.21. (a) a symmetric frame over two spans subjected to antisymmetric loads. (b) The half-structure equivalent to the actual structure. (c) Equivalent structure of (b). (d) Equivalent structure of (c).

EXAMPLE 4.1

Draw the moment diagram of the frame shown in Figure (4.22a).

Solution: (1) Transformation of the unsymmetrical loading.

The unsymmetrical loading applied to the actual frame can be transformed into its symmetric and antisymmetric components as shown in Figures (4.22b and 4.22c). As the axial deformation of a flexural member can be neglected, the moment in the symmetric case shown in Figure (4.22c) is zero. We shall only analyze the antisymmetric case shown in Figure (4.22b).

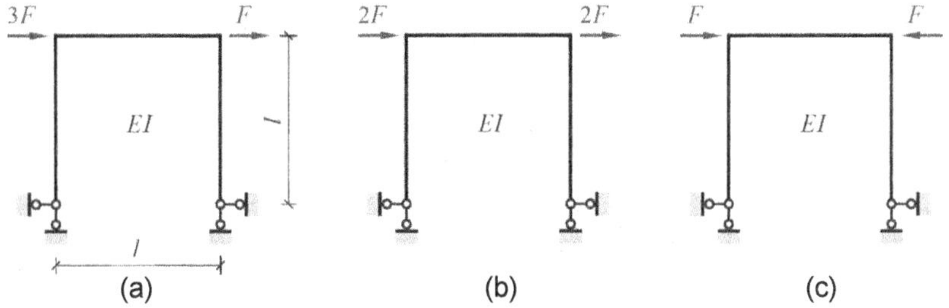

Figure 4.22. Example 4.1.

(2) The half-structure of the antisymmetric case shown in Figure (4.22b) is shown in Figure (4.23a).

(3) The half-structure is a statically determinate frame, whose moment diagram is shown in Figure (4.23b).

(4) The moment diagram of the actual system according to Theorem 4.2 shows that the moment in the case shown in Figure (4.23) is antisymmetric. Therefore, one obtains the moment diagram as shown in Figure (4.23(c)).

EXAMPLE 4.2

Draw the moment diagram of the frame shown in Figure (4.24a) and find the axial force of the column.

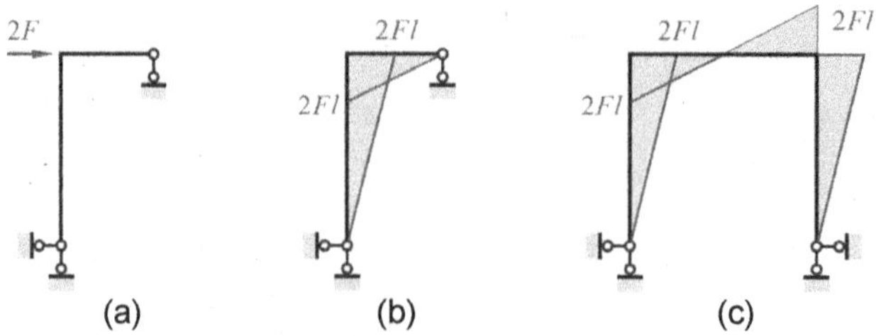

Figure 4.23. Solution of Example 4.1.

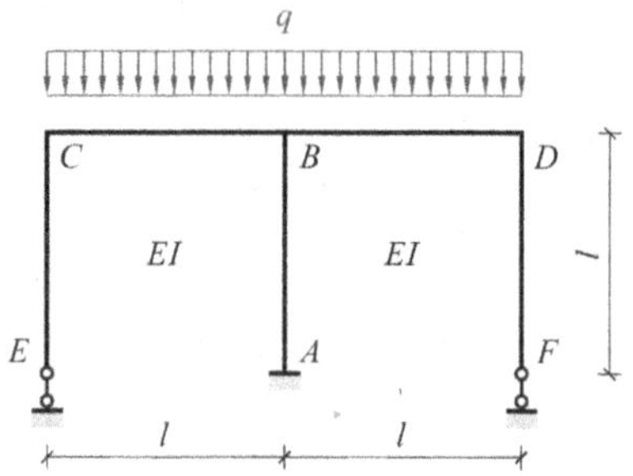

Figure 4.24. Example 4.2.

Solution: (1) Construct the half-structure.

The half-structure for the actual system is shown in Figure (4.25a).

(2) Select a primary system and formulate the corresponding governing equation of the force method. We can obtain a primary system shown in Figure (4.25b) by removing the roller at E and releasing the reaction, which gives:

$$\delta_{11}X_1 + \Delta_{1P} = 0 \tag{4.27}$$

(3) Determine the flexibility coefficients and the displacements under the applied loading and find the redundant.

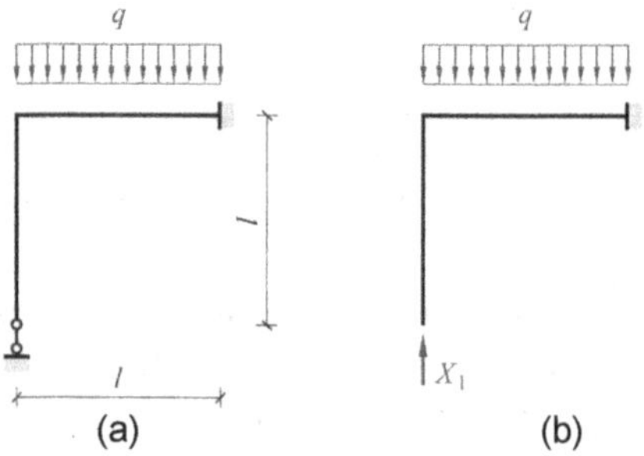

Figure 4.25. Construct the half-structure.

Draw the moment diagram $\bar{M}_1$ and M_P as shown in Figure (4.26). The graphic multiplication of $\bar{M}_1$ gives:

$$\delta_{11} = \frac{1}{EI}\left(\frac{1}{2}\times l\times l\times\frac{2l}{3}\right) = \frac{l^3}{3EI} \tag{4.28}$$

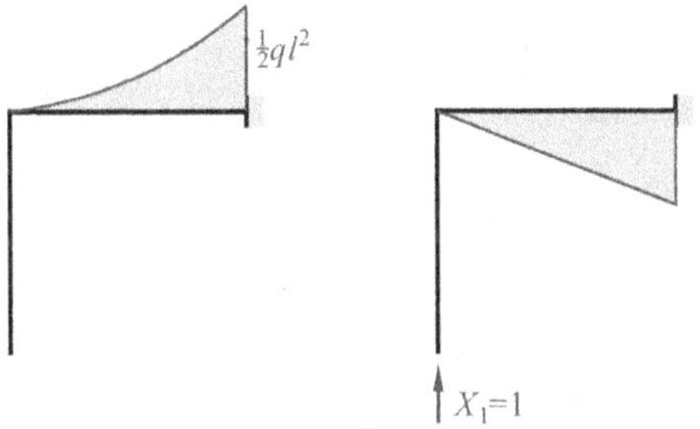

Figure 4.26. Moment diagram.

Similarly, the graphic multiplication of $\bar{M}_1$ and M_P provides:

$$\Delta_{1P} = -\frac{1}{EI}\left(\frac{1}{3}\times\frac{ql^2}{2}\times l\times\frac{3l}{4}\right) = -\frac{ql^4}{8EI} \tag{4.29}$$

Substituting Equations (4.28) and (4.29) into Equation (4.27) gives:

$$\frac{l^3}{3EI}X_1 = \frac{ql^4}{8EI} \tag{4.30}$$

which yields:

$$X_1 = \frac{3}{8}ql \tag{4.31}$$

(4) Draw the moment diagram of the actual structure.

$$M = \bar{M}_1 X_1 + M_P \tag{4.32}$$

The moment diagram of the half-structure is shown in Figure (4.27a), while the moment diagram of the actual structure is shown in Figure (4.27b).

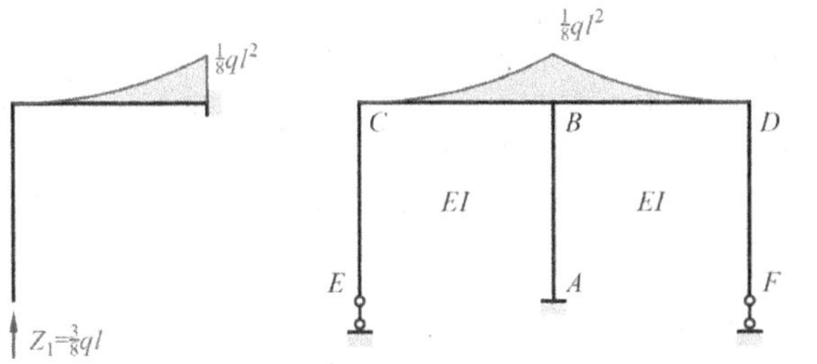

(a) (b)

Figure 4.27. (a) The moment diagram of the half-structure. (b) The moment diagram of the actual structure.

(5) Determine the internal force of column *AB*.

Draw the free-body diagram of the beam segment *CB* as shown in Figure (4.28a). $\sum_C = 0$ gives,

$$V_{BC} = -\frac{1}{l}\left(\frac{ql^2}{2}+\frac{ql^2}{8}\right) = -\frac{5ql}{8} \tag{4.33}$$

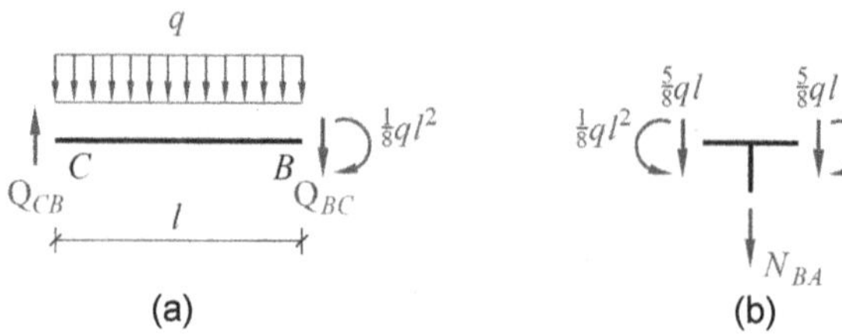

Figure 4.28. (a) Free-body diagram of the beam segment *CB*. (b) Internal forces.

Draw the free-body diagram, shown in Figure (4.28b), of joint *B*. $\sum F_{By} = 0$ provides

$$N_{BA} = -\frac{5ql}{4} \tag{4.34}$$

EXAMPLE 4.3

Draw the moment diagram of the frame shown in Figure (4.29), and find the axial force in the adjacent members *BC* and *CD*.

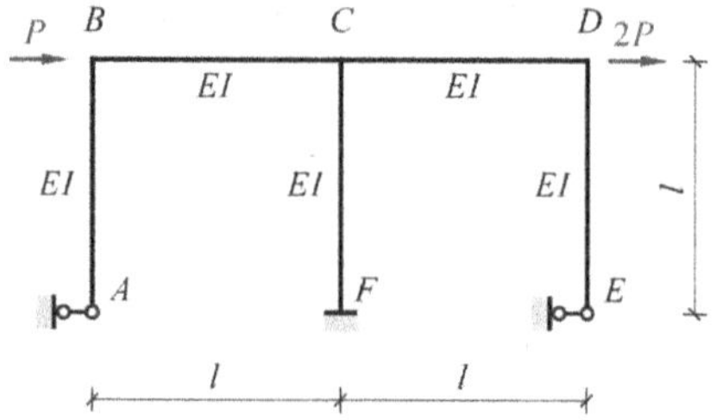

Figure 4.29. Example 4.3.

Solution: (1) Transform the unsymmetrical loading to symmetric and antisymmetric components. The unsymmetrical loading applied to the actual frame can be transformed into symmetric and antisymmetric loads shown in Figure (4.30). No moment exists in the symmetric case shown in Figure (4.30b). We shall only analyze the antisymmetric case shown in Figure (4.30a).

(2) The half-structure of the antisymmetric case is given in Figure (4.31).

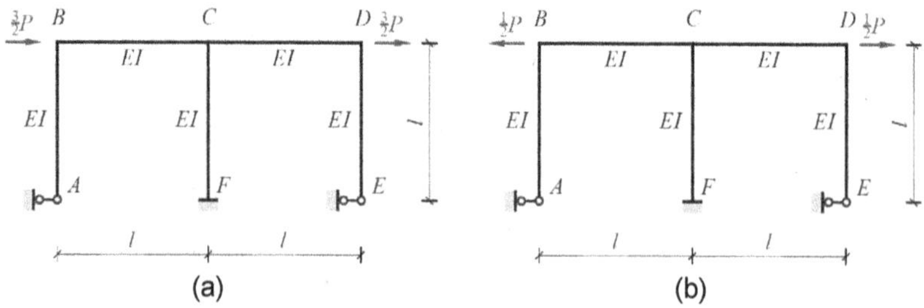

Figure 4.30. The unsymmetrical loading applied to the actual frame can be transformed into symmetric (a) and antisymmetric loads (b).

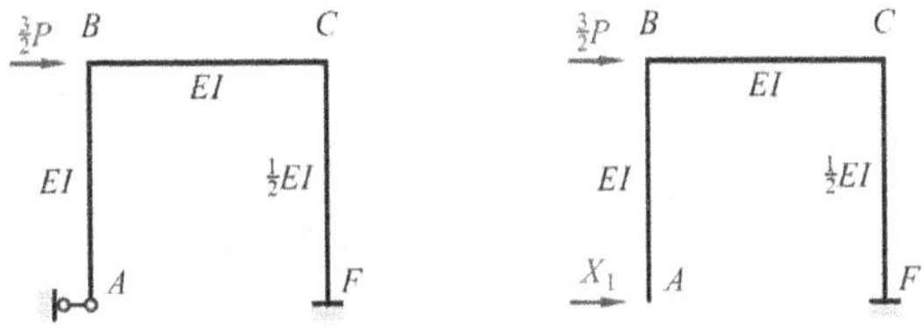

Figure 4.31. The half structure.

(3) Construct a primary structure and formulate the governing equation of the force method.

$$\delta_{11}X_1 + \Delta_{1P} = 0 \tag{4.35}$$

(4) Determine the flexibility coefficients, the displacements under loading, and find the redundant. Draw the moment diagrams $\bar{M}_1$ and M_P as shown in Figure (4.32).

$$\delta_{11} = \frac{1}{EI}\left(l \times l \times l + \frac{1}{2} \times l \times l \times \frac{2l}{3}\right) + \frac{1}{0.5EI}\left(\frac{1}{2} \times l \times l \times \frac{2l}{3}\right) = \frac{2l^3}{EI} \tag{4.36}$$

$$\Delta_{1P} = -\frac{1}{0.5EI}\left(\frac{1}{2} \times 1.5Pl \times l \times \frac{l}{3}\right) = -\frac{Pl^3}{2EI} \tag{4.37}$$

$$\frac{2l^3}{EI}X_1 = \frac{Pl^3}{2EI} \tag{4.38}$$

$$X_1 = \frac{P}{4} \tag{4.39}$$

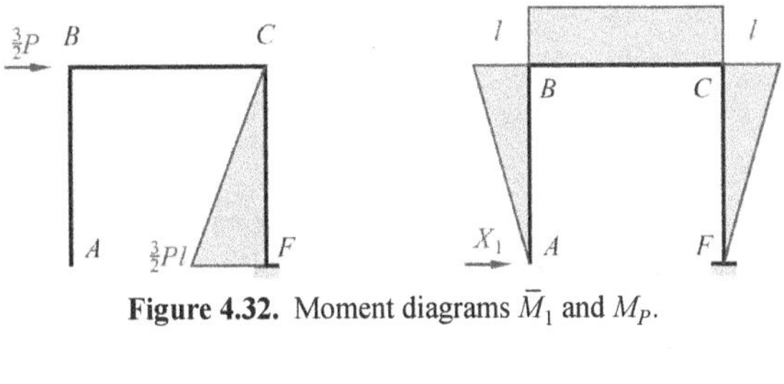

Figure 4.32. Moment diagrams $\bar{M}_1$ and M_P.

(5) Draw the moment diagram of the actual structure.

$$M = M_P + \frac{P}{4} \times \bar{M}_1 \tag{4.40}$$

(6) Determine the axial forces, N_{BC} and N_{CD}, in the adjacent members BC and CD.

To determine the internal force in the antisymmetric case, draw the free-body diagrams shown in Figure (4.34).

$\sum F_v = 0$ gives

$$V_{BA}^{\text{Ant}} = -\frac{P}{4} \tag{4.41}$$

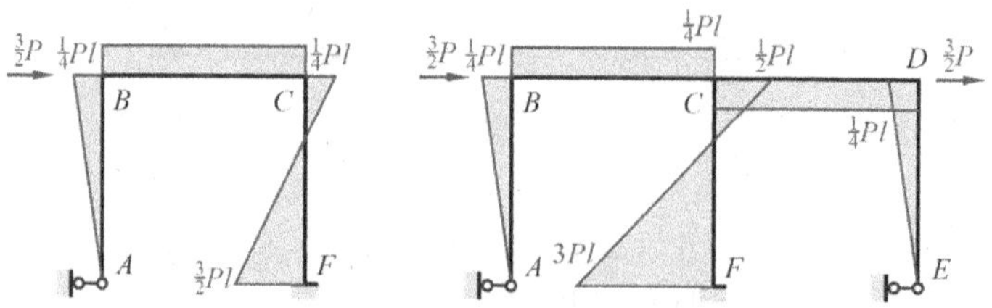

Figure 4.33. Moment diagram of the actual structure.

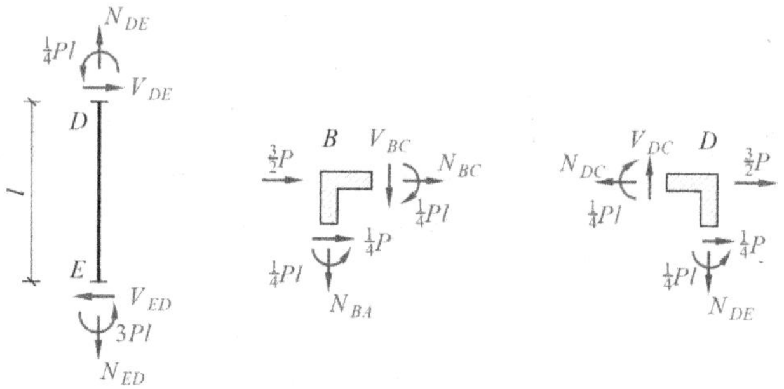

Figure 4.34. Determine the internal forces.

$$N_{BC}^{\text{Ant}} = -\frac{7P}{4} \quad \text{and} \quad N_{CD}^{\text{Ant}} = \frac{7P}{4} \tag{4.42}$$

Thus, axial forces of the actual structure are:

$$N_{BC} = N_{BC}^{\text{Ant}} + N_{BC}^{\text{Sym}} = -\frac{7P}{4} + \frac{P}{2} = -\frac{5P}{4} \tag{4.43}$$

$$N_{DC} = N_{DC}^{\text{Ant}} + N_{DC}^{\text{Sym}} = \frac{7P}{4} + \frac{P}{2} = \frac{9P}{4} \tag{4.44}$$

4.5 Analysis of Statically Indeterminate Structures Having Thermal Changes, Fabrication Errors and Support Settlements

Consider the frame shown in Figure (4.35(a)).

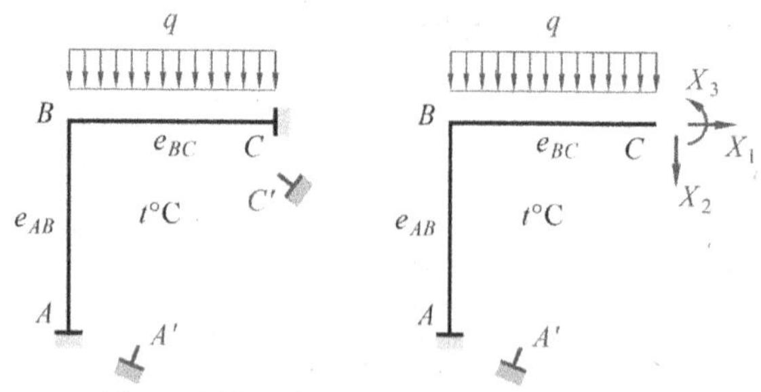

Figure 4.35. A frame subject to lateral line load.

Select a primary system for this problem as shown in Figure (4.35b). The compatibility condition for the equivalence of the actual system and the primary system is that the displacements at point C of the primary structure due to the redundant force X_i, the external load q, the temperature change t°C and the displacements of support A are equal to the displacements of support C of the actual structure. Thus, the governing equations of the force method for this problem can be expressed as:

$$\begin{cases} \delta_{11}X_1 + \delta_{12}X_2 + \delta_{13}X_3 + \Delta_{1q} + \Delta_{1t} + \Delta_{1s} = \Delta_1 \\ \delta_{21}X_1 + \delta_{22}X_2 + \delta_{23}X_3 + \Delta_{2q} + \Delta_{2t} + \Delta_{2s} = \Delta_2 \\ \delta_{31}X_1 + \delta_{32}X_2 + \delta_{33}X_3 + \Delta_{3q} + \Delta_{3t} + \Delta_{3s} = \Delta_3 \end{cases} \tag{4.45}$$

where Δ_{iq}, Δ_{it} and Δ_{is} are the displacements of the primary structure caused by the load, temperature change and the support displacements, respectively; Δ_i is the displacement of support C of the actual structure.

The moment M of the actual structure can be expressed as:

$$M = M_q + X_1\bar{M}_1 + X_2\bar{M}_2 + X_3\bar{M}_3 \tag{4.46}$$

EXAMPLE 4.4

Consider the frame shown in Figure (4.36(a)). $h = l/10$. Find the axial forces.

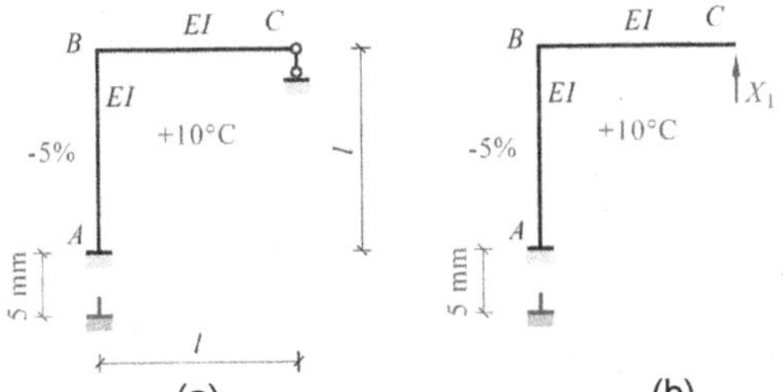

Figure 4.36. Frame structure (a) subjected to temperature and its equivalent structure (b).

Solution: (1) Construct the primary system shown in Figure (4.37b). The governing equation of the force method is:

$$\delta_{11}X_1 + \Delta_{1t} + \Delta_{1s} + \Delta_{1e} = 0 \tag{a}$$

(2) Determine δ_{11}, Δ_{1t}, Δ_{1s}, and find the redundant X_1. Draw the moment and axial force diagrams under $X_1 = 1$ as shown in Figure (4.37). The graphic multiplication of $\bar{M}_1$ and $\bar{M}_1$ gives

$$\delta_{11} = \sum\int\frac{\bar{M}_1^2}{EI}\mathrm{d}x = \frac{1}{EI}\left(l\times l\times l + \frac{1}{2}\times l\times l\times\frac{2l}{3}\right) = \frac{4l^3}{3EI} \tag{b}$$

$$\Delta_{1t} = \sum\alpha t_0\omega_{\bar{N}} + \sum\frac{\alpha\Delta t}{h}\omega_{\bar{M}} = 5\alpha l + \frac{10\alpha}{h}\left(l^2 + \frac{l^2}{2}\right) = 155\alpha l \tag{c}$$

where we used $h = l/10$.

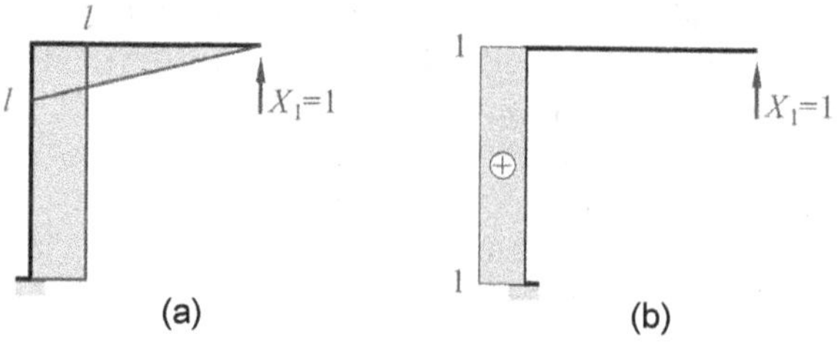

Figure 4.37. The moment (a) and axial force (b) diagrams.

$$\Delta_{1s} = -0.005 \tag{d}$$

$$\Delta_{1e} = -0.05l \tag{e}$$

Substituting Equations (b)–(e) into Equation (a) and solving it provides,

$$X_1 = \frac{3EI}{4l^3}(0.005 + 0.05l - 155\alpha l)$$

For each member, the axial force N can be expressed as:

$$N = N_P + X_1\bar{N}_1$$

where $N_P = 0$. Thus, we have

$$\begin{cases} N_{AB} = \dfrac{3EI}{4l^3}(0.005 + 0.05l - 155\alpha l) \\ N_{BC} = 0 \end{cases}$$

4.6 Deflections of Statically Indeterminate Structures

Consider the frame shown in Figure 4.38a, whose moment diagram under the distributed load q is shown in Figure (4.38b). Let us find the rotation θ_B of joint B. The crude approach for solving this problem is as follows.

1. Apply a virtual-force case by exerting a unit torque at joint B as shown in Figure (4.39(a)).
2. Draw the moment diagram of the actual frame under the unit torque as shown in Figure (4.39(b)).

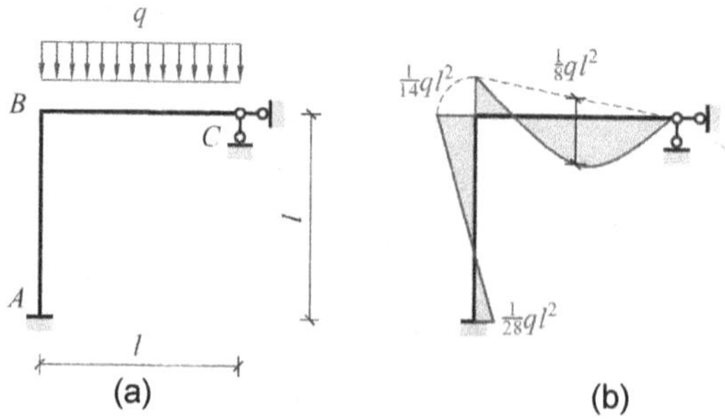

Figure 4.38. (a) Frame subjected to the distributed load q (b) Moment diagram.

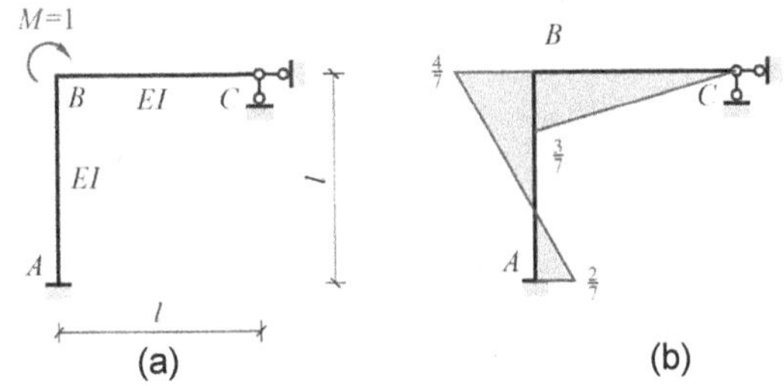

Figure 4.39. (a) Apply a virtual-force case by exerting a unit torque at joint *B*. (b) The moment diagram of the actual frame under the unit torque.

3. Use the unit-force method to calculate the rotation of joint *B* by the following equation:

$$\theta_B = \sum_{i=1}^{2} \int_{l_i} \frac{\overline{M}^{\mathrm{Ac}} M_P^{\mathrm{Ac}}}{EI} \mathrm{d}x \tag{4.47}$$

The major disadvantage of the above approach is that one must analyze the actual frame with two degrees of indeterminacy in order to draw the moment diagram $\overline{M}$. To avoid analyzing the statically indeterminate frame twice, an alternative and better approach is proposed as follows.

1. Select the primary system for the actual system as shown in Figure (4.40a).
2. The equivalence of the primary system and the actual system shows that the rotation θ_B at joint *B* the actual system is equal to the rotation at joint *B* of the primary structure caused by the distributed loading *q* and the redundancies X_1, X_2. Now, the problem is transformed into the problem to find the rotation at joint *B* of the primary structure caused by the distributed loading q and the redundancies X_1, X_2. Notice that the primary structure is a statically determinate frame.
3. Apply a virtual-force case by applying a unit torque at joint *B* of the primary structure as shown in Figure (4.40b). The moment diagram of the primary structure due to the unit torque is shown in Figure (4.40c).
4. Use the unit-force method to determine θ_B, namely,

$$\theta_B = \sum_{i=1}^{2} \int_{l_i} \frac{\overline{M}^{\mathrm{Pr}} M_P^{\mathrm{Ac}}}{EI} \mathrm{d}x \tag{4.48}$$

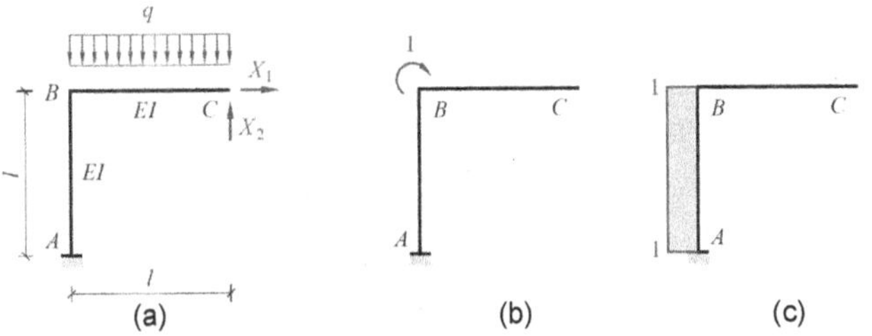

Figure 4.40. (a) The primary system for the actual system. (b) Apply a virtual-force case by exerting a unit torque at joint *B*. (c) The moment diagram of the primary structure due to the unit torque at joint *B*.

where $\bar{M}$ is the moment shown in Figure (4.40c); and M_P is the moment shown in Figure (4.38b) because the primary system shown in Figure (4.40a) is equivalent to the actual system shown in Figure (4.38a). Thus, we have:

$$\theta_B = \frac{1}{EI}\left(\frac{1}{2}\times l\times\frac{ql^2}{14} - \frac{1}{2}\times l\times\frac{ql^2}{28}\times 1\right) = \frac{ql^3}{56EI} \tag{4.49}$$

Note that the primary system for calculating the displacement is arbitrary.

EXAMPLE 4.5

Consider a composite structure shown in Figure (4.41a). If no load acts on the structure, what would be the rotation at the free end C if link GH were e (m) shorter?

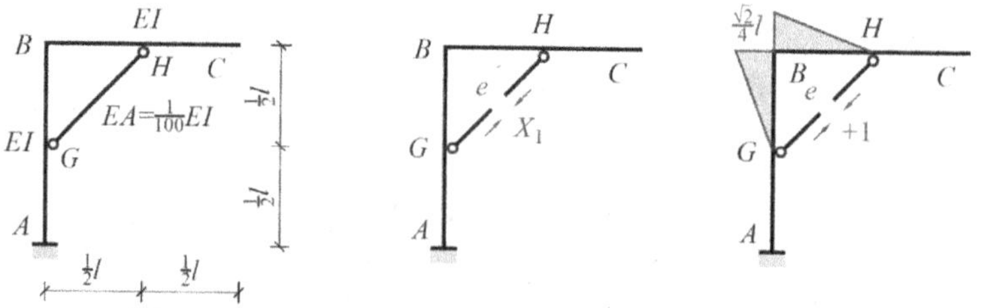

Figure 4.41. (a) The structure system. (b) The link GH was e (m) shorter. (c) $\bar{M}_1$ and $\bar{N}_1$.

Solution: (1) Select a primary system, shown in Figure (4.41b), to determine the internal forces of the actual structure due to the fabrication error.

$$\delta_{11}X_1 = e \tag{a}$$

(2) Determine δ_{11} and X_1

$$\delta_{11} = \sum_{i=1}^{2}\int_{l_i}\frac{\bar{M}_1^2}{EI}dx + \sum_{i=1}^{1}\frac{\bar{N}_1^2}{EA}l_i$$
$$= \frac{2}{EI}\left(\frac{1}{2}\times l\times\frac{\sqrt{2}l}{4}\right)\times\left(\frac{2}{3}\times\frac{\sqrt{2}l}{4}\right) + \frac{1}{EA}\times 1\times 1\times\frac{\sqrt{2}l}{2} \tag{b}$$
$$= \frac{l^3}{12EI} + \frac{\sqrt{2}l}{2EA}$$

$$X_1 = \left(\frac{l^3}{12EI} + \frac{\sqrt{2}l}{2EA}\right)^{-1} e \tag{c}$$

(3) Draw the internal-force diagram of the actual system

$$M = X_1\bar{M}_1 \quad \text{and} \quad N = X_1\bar{N}_1$$

(4) Determine the rotation θ_C

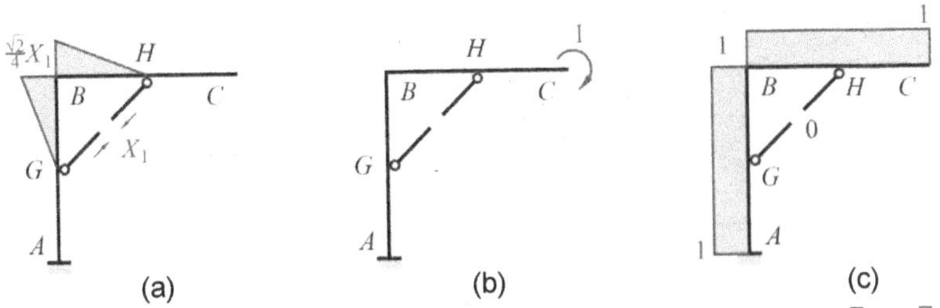

Figure 4.42. (a) M_P and N_P (b) The virtual-force case. (c) Internal-force diagram $\bar{M}$ and $\bar{N}$.

Invert a virtual-force case shown in Figure (4.42b) and draw its internal-force diagram shown in Figure (4.42c).

$$\theta_C = \sum_{i=1}^{2} \int_{l_i} \frac{\bar{M}M_P}{EI} \mathrm{d}x + \sum_{i=1}^{1} \frac{\bar{N}N_P}{EA} l_i = \frac{2}{EI}\left(\frac{1}{2} \times \frac{\sqrt{2}lX_1}{4} \times \frac{l}{2} \times 1\right) + 0 = \frac{\sqrt{2}el^2}{\frac{2l^2}{3} + \frac{\sqrt{2}}{2}\frac{EI}{EA}}$$

EXAMPLE 4.6

Test whether the moment diagram shown in Figure (4.43) is correct or not.

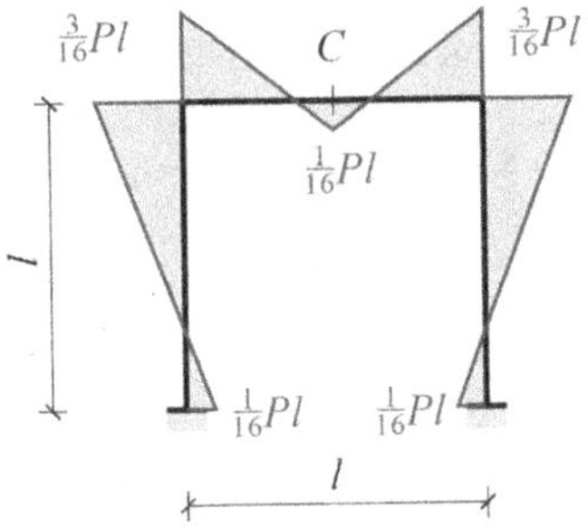

Figure 4.43. A moment diagram.

***Solution*:** Select a primary structure as shown in Figure (4.44(a)).

$$\theta_C = \sum_{i=1}^{3} \int_{l_i} \frac{\bar{M}M^{\mathrm{Ac}}}{EI} \mathrm{d}x = \frac{1}{EI} \sum_{i=1}^{3} \int_{l_i} M^{\mathrm{Ac}} \mathrm{d}x \neq 0$$

Thus, the moment diagram is incorrect.

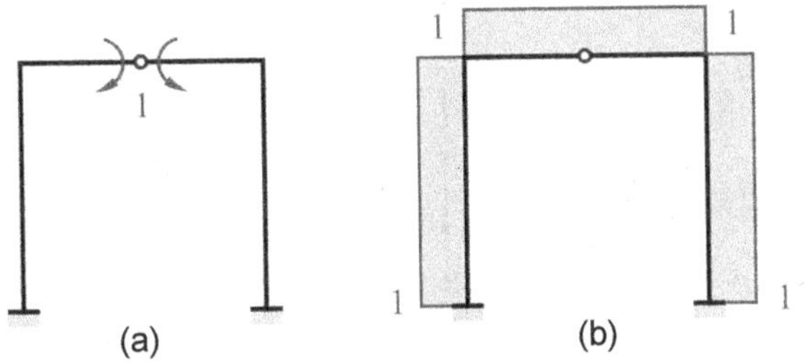

Figure 4.44. A primary structure (a) and its moment diagram (b).

PROBLEMS/EXERCISE

4.1a Why are the moment diagrams shown in Figure (4.45) wrong?

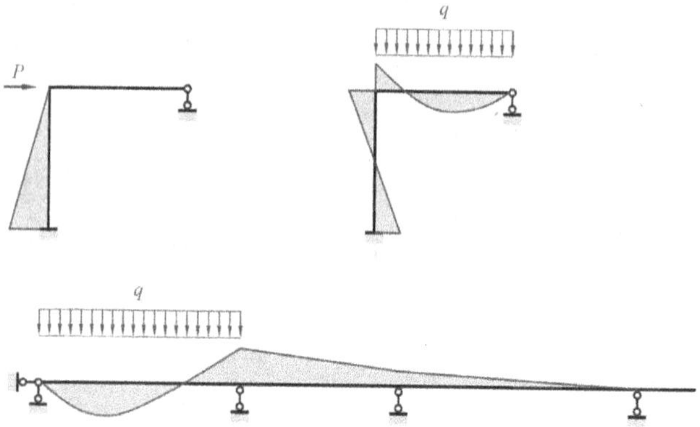

Figure 4.45. A few moment diagrams for justification.

4.1b Consider a frame shown in Figure 4.46, in which the rigidity of each member is *EI*. The structure is subjected to two different thermal environments. The increase in temperature inside the frame is $\Delta t = 10°C$, while the temperature at the outer surface remains unchanged. The co-efficient of thermal expansion for each member is $\alpha = 10^{-5}/°C$. Each of members has a rectangular cross section of height $h = 0.5$ m. Use force method to determine the base moment M_A at point.

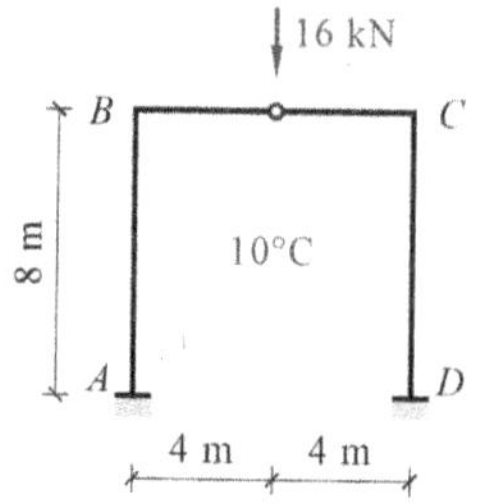

Figure 4.46. Exercise 4.1.

4.2 Consider a frame shown in Figure (4.47), use force method to determine the moment diagram.

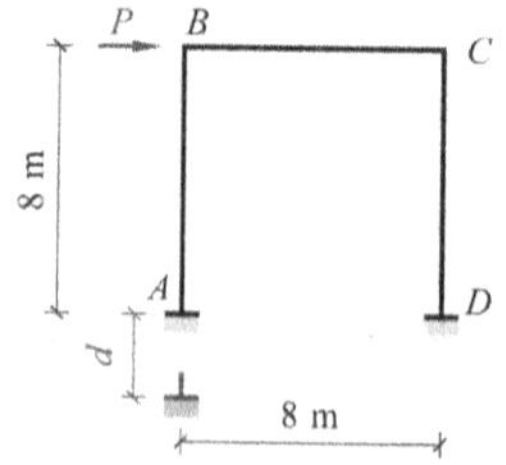

Figure 4.47. Exercise 4.2.

Chapter 5
Displacement Method

5.1 Beams with Support Displacements and Slope-Deflection Equations

Analysis of beams with support displacements is an essential prerequisite for developing the displacement method.

Figure (5.1a) shows a statically indeterminate beam with a support rotation θ_A. The configuration of the beam can be uniquely described by θ_A according to the Euler-Bernoulli beam theorem. θ_A is said to be an *elastic degree of freedom* to define the configuration of the beam. Insert a pin at A and release the moment M_{ab}^{Δ} due to the support rotation as shown in Figure (5.1b). One can then use the force method to draw the moment diagram of the beam with the support rotation, as in Figure (5.1c). The end moment M_{ab}^{Δ} of the beam can be expressed as:

$$M_{ab}^{\Delta} = 3i\theta_A \tag{5.1}$$

where i is the stiffness of the beam, given by:

$$i \equiv \frac{EI}{l} \tag{5.2}$$

Equation (5.1) is referred to as the ***slope-deflection equation*** of the clamped-roller beam with the support rotation.

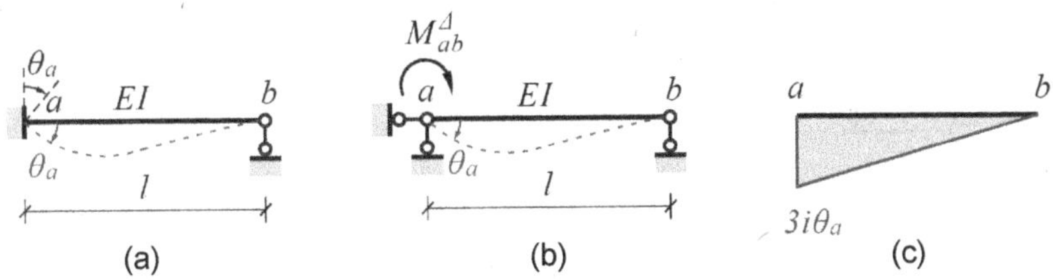

Figure 5.1. (a) A clamped-roller beam with a support rotation. (b) Insert a pin at A and release the moment M_{ab}^{Δ} due to the support rotation. (c) The moment diagram of the beam.

Figure (5.2) shows the clamped-roller with an angular rotation θ_A and a linear displacement Δ of its supports. The configuration of the beam can be described by the two elastic DOFs, θ_A and Δ. The moment diagram shown in Figure (5.2(c)) yields:

$$M_{ab}^{\Delta} = 3i\theta_A - 3i\frac{\Delta}{l} \tag{5.3}$$

where M_{ab}^{Δ} (given in Figure 5.2b) is the end moment of the beam due to the support displacements θ_A and Δ. Equation (5.3) is called the slope-deflection equation of the clamped-roller beam having two elastic DOFs.

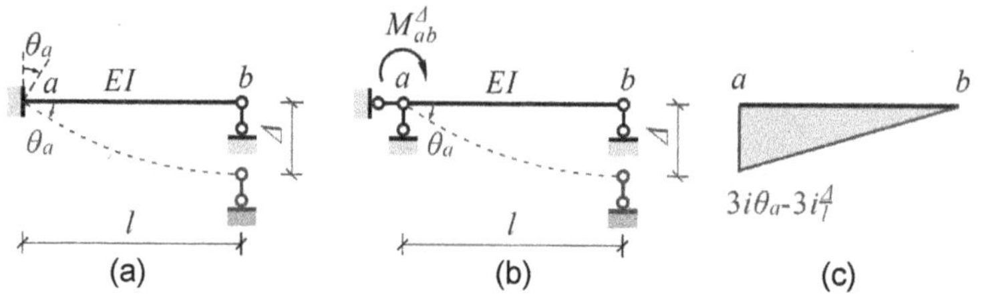

Figure 5.2. (a) A clamped-roller beam with two elastic DOFs. (b) Insert a pin at A and release the moment M_{ab}^{Δ} due to the support rotation. (c) The moment diagram of the beam.

Figure (5.3a) shows a clamped-sliding beam with a fixed-support rotation. The moment diagram shown in Figures (5.3b) and (5.3c) provides:

$$\begin{cases} M_{ab}^{\Delta} = i\theta_a \\ M_{ba}^{\Delta} = -i\theta_a \end{cases} \tag{5.4}$$

Figure (5.4a) shows a fixed-end beam with five independent support displacements, v_a, u_a, θ_a, v_b and θ_b. The horizontal end displacements are identical, namely, $u_b = u_a$. The moment diagram shown in Figure (5.4) yields:

$$\begin{cases} M_{ab}^{\Delta} = 4i\theta_a + 2i\theta_b - 6i\frac{\Delta}{l} \\ M_{ba}^{\Delta} = 2i\theta_a + 4i\theta_b - 6i\frac{\Delta}{l} \end{cases} \tag{5.5}$$

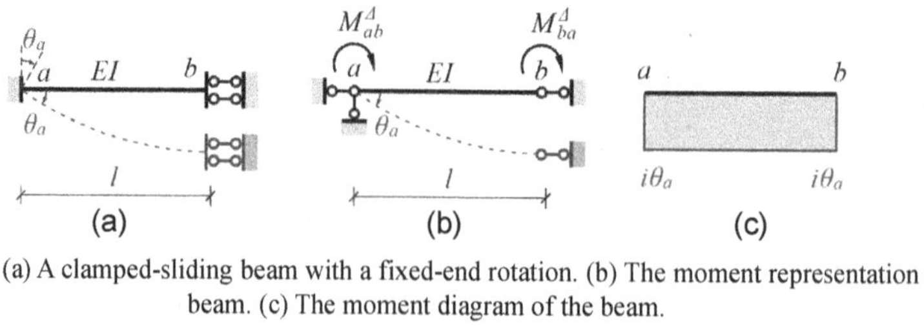

Figure 5.3. (a) A clamped-sliding beam with a fixed-end rotation. (b) The moment representation of the beam. (c) The moment diagram of the beam.

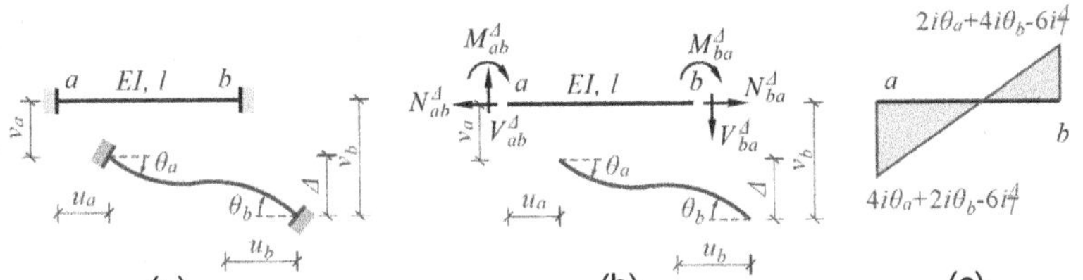

Figure 5.4. (a) A fixed-end beam with five independent support displacements. (b) The moment and force representations of the beam. (c) The moment diagram of the beam.

5.2 Displacement Method for Analyzing Frames under Nodal Loads

5.2.1 The Procedure Using Slope-Deflection Equations

As shown in Figure (5.5a) a frame with 3-degrees of indeterminacy is subjected to a torque of 16 kN·m at node *B*. Its unknown configuration shown in Figure (5.5b) can be *completely* described by the rotation Z_1 at joint *B*, which is said to be an unknown degree of freedom to describe the configuration of the system. To analyze the frame, draw the free-body diagrams of column *AB*, beam *BC* and joint *B* as shown in Figure (5.6). The slope-deflection Equation (5.5) yields:

$$\begin{cases} M^{\Delta}_{AB} = 2iZ_1 \\ M^{\Delta}_{BA} = 4iZ_1 \end{cases} \quad (5.6)$$

and

$$\begin{cases} M^{\Delta}_{BC} = 4iZ_1 \\ M^{\Delta}_{CB} = 2iZ_1 \end{cases} \quad (5.7)$$

where $i \equiv \frac{EI}{l}$ is the stiffness of the flexural members.

Figure (5.6c) shows that the equilibrium condition for the DOF Z_1 can be expressed as:

$$16 - M^{\Delta}_{BA} - M^{\Delta}_{BC} = 0 \quad (5.8)$$

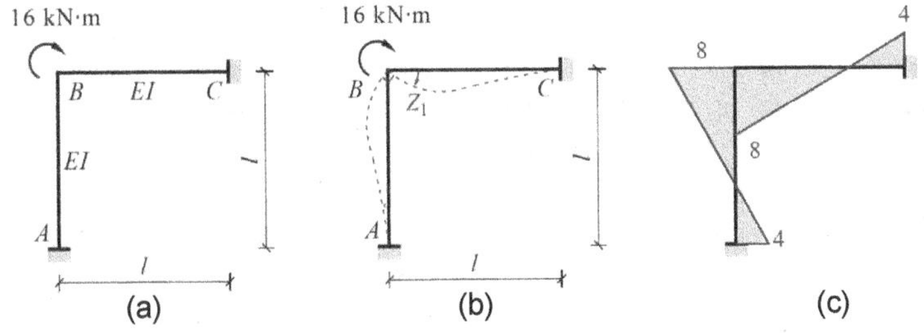

Figure 5.5. (a) A frame with 3-degrees of indeterminacy is subjected to a torque of 16 kN·m at node B. (b) The unknown configuration of (a) described by the rotation Z_1 at joint B. (c) The moment diagram.

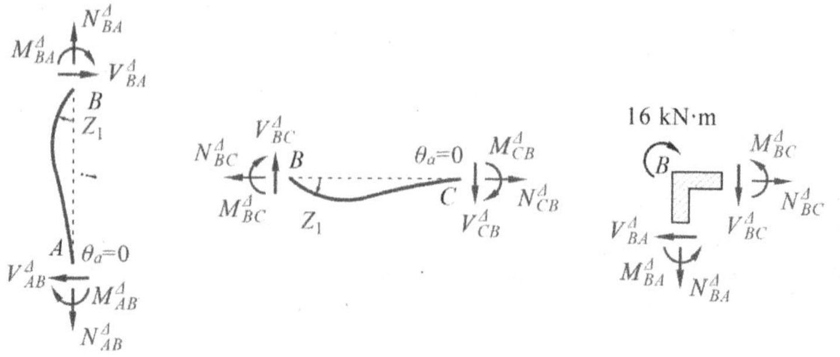

Figure 5.6. The free-body diagrams of AB (left), BC (middle), and forces at B (right).

Substituting Equations (5.6)–(5.7) into Equation (5.8) provides

$$8iZ_1 = 16 \tag{5.9}$$

which gives

$$Z_i = \frac{2}{i} \tag{5.10}$$

Equation (5.10) is the *governing equation of the displacement method* for this problem.

By inserting Equation (5.10) into Equations (5.6)–(5.7), one obtains:

$$\begin{cases} M_{AB}^{\Delta} = 4 \\ M_{BA}^{\Delta} = 8 \\ M_{BC}^{\Delta} = 8 \\ M_{CB}^{\Delta} = 4 \end{cases} \tag{5.11}$$

which leads to the moment diagram of the frame, see Figure (5.5c).

5.2.2 The Procedure Directly Using Primary Systems

An alternative procedure of the displacement method is to formulate the governing equation by constructing a primary system.

Consider again the system shown in Figure (5.5a), which is called the actual system. The system shown in the figure to the right of Figure 5.7, called a ***primary system*** of the displacement method, is constructed by adding a ***virtual constraint*** (or ***virtual support***) at the unknown DOF of the left actual system. Let Z_1 be the angular displacement of the virtual constraint, and R_1 be the reaction from the virtual constraint to the frame caused by the rotational displacement Z_1. The actual system on the left and the primary system on the right are equivalent to each other, if we find a virtual-support displacement Z_1 such that the reaction of the virtual constraint is equal to the given torque of 16 kN·m, i.e.,

$$R_1 = 16 \text{ kN} \cdot \text{m} \tag{5.12}$$

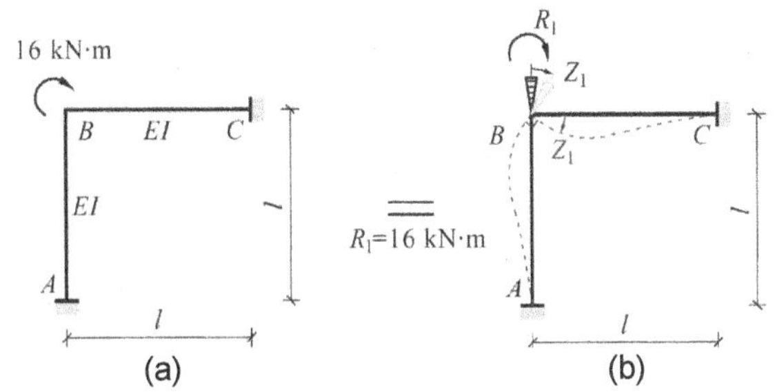

Figure 5.7. (a) The actual frame subjected to a given torque of 16 kN·m at node B. (b) A primary system corresponding to the actual system, which is obtained by adding a virtual constraint at the unique unknown DOF.

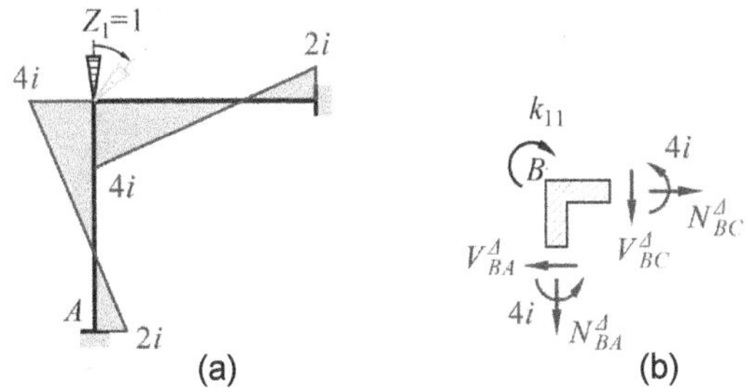

Figure 5.8. (a) The moment diagram $\bar{M}_1$ of the primary system due to the unit rotation of the virtual constraint. (b) The free-body diagram of node B in the primary system for $Z_1 = 1$.

To formulate the reaction R_1, we draw the moment diagram $\bar{M}_1$ of the primary system due to a unit rotation of the virtual constraint as shown in Figure (5.8a), in which we used the slope-deflection Equation (5.5). Figure (5.8b) shows the free-body diagram of node B in the primary system for $Z_1 = 1$, in which k_{11} is the reaction on the structure due to the unit virtual-support displacement, which gives

$$k_{11} = 8i \tag{5.13}$$

Therefore, Equation (5.12) may be rewritten as:

$$k_{11}Z_1 = P_1 \tag{5.14}$$

where P_1 is the load in the DOF Z_1, which also leads to the governing Equation (5.9).

Due to the equivalence of the actual and primary systems, the moment M of the actual frame can be expressed as:

$$M = Z_1 \bar{M}_1 \tag{5.15}$$

which gives the moment diagram shown in Figure (5.5).

5.2.3 Unknown Degrees of Freedom of the Displacement Method

Consider the frame shown in Figure (5.9a). Its unknown forces can be completely described by the horizontal displacement Z_1 and rotation Z_2 at node C by using the solutions of the clamped beam and clamped-roller beam. Z_1 and Z_2 are said to be the

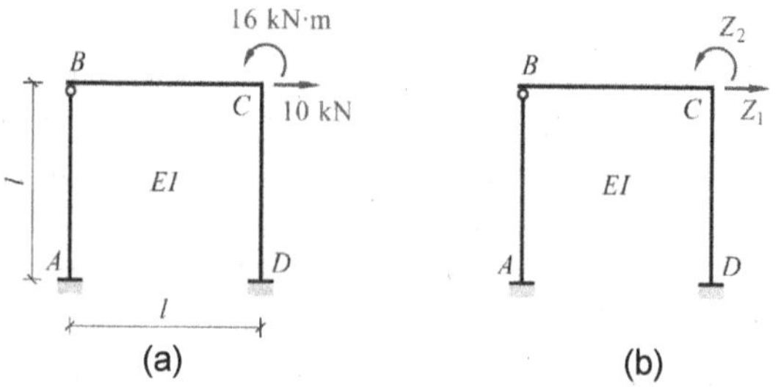

Figure 5.9. (a) A frame structure subject to forces at C. (b) The first and second unknown DOFs of the system Z_1 and Z_2 at node C.

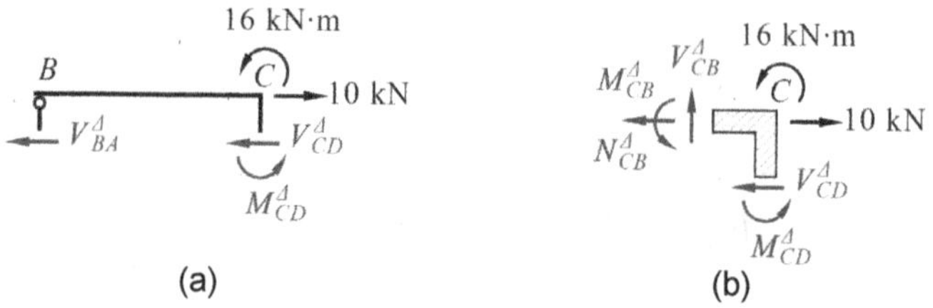

Figure 5.10. The free-body diagrams of BC (left) and node C (right).

first and second unknown DOFs of the system, respectively. Specifically, the end moments of each member can be expressed as:

$$\begin{cases} M_{AB}^{\Delta} = -3i\dfrac{Z_1}{l} \\ M_{BA}^{\Delta} = 0 \\ M_{BC}^{\Delta} = 0 \\ M_{CB}^{\Delta} = 3iZ_2 \\ M_{CD}^{\Delta} = 4iZ_2 - 6i\dfrac{Z_1}{l} \\ M_{DC}^{\Delta} = 2iZ_2 - 6i\dfrac{Z_1}{l} \end{cases} \tag{5.16}$$

To formulate the governing equations for this system, we draw the free-body diagrams of beam BC and node C as shown in Figure (5.10). The equilibrium conditions in the unknown DOFs, Z_1 and Z_2, can be rewritten as

$$\begin{cases} V_{BA}^{\Delta} + V_{CD}^{\Delta} = 10 \\ M_{CB}^{\Delta} + M_{CD}^{\Delta} = -16 \end{cases} \tag{5.17}$$

Equations given in (5.16) yield:

$$\begin{cases} V_{BA}^{\Delta} = \dfrac{3iZ_1}{l^2} \\ V_{CD}^{\Delta} = -\dfrac{1}{l}\left(6iZ_2 - 12i\dfrac{Z_1}{l}\right) = 12i\dfrac{Z_1}{l^2} - \dfrac{6iZ_2}{l} \end{cases} \tag{5.18}$$

Inserting Equations (5.16) and (5.18) into Equation (5.17) gives:

$$\begin{cases} \frac{15i}{l^2}Z_1 - \frac{6i}{l}Z_2 = 10 \\ -\frac{6i}{l}Z_1 + 7iZ_2 = -16 \end{cases} \tag{5.19}$$

Equation (5.19) is the governing equation of the displacement method for this problem. Its solution is:

$$Z_1 = \frac{70l^2 - 96l}{69i} \quad \text{and} \quad Z_2 = \frac{20l - 80}{23i} \tag{5.20}$$

The end moment M_{ij} of the actual system can be expressed as:

$$M_{ij} = M_{ij}^{\Delta} \tag{5.21}$$

which gives:

$$\begin{cases} M_{AB}^{\Delta} = -3i\frac{Z_1}{l} = \frac{70l^2 - 96l}{23} \\ M_{BA}^{\Delta} = 0 \\ M_{BC}^{\Delta} = 0 \\ M_{CB}^{\Delta} = 3iZ_2 = \frac{60l - 240}{23} \\ M_{CD}^{\Delta} = 4iZ_2 - 6i\frac{Z_1}{l} = -\frac{60l + 128}{23} \\ M_{DC}^{\Delta} = 2iZ_2 - 6i\frac{Z_1}{l} = \frac{32 - 100l}{23} \end{cases} \tag{5.22}$$

Alternatively, one can formulate the governing equation by constructing a primary system as follows. As shown in Figure (5.11a), we construct the primary structure by adding two virtual constraints which control the horizontal displacement and rotation at node C, respectively. Let R_1 be the reaction on the structure from the first virtual constraint due to Z_1, and R_2 be the reaction on the structure from the second constraint due to Z_2. The equivalence of the actual and the primary systems requires that:

$$\begin{cases} R_1 = 10 \\ R_2 = -16 \end{cases} \tag{5.23}$$

If k_{ij} denotes the reaction on the structure from the ith virtual constraint due to $Z_j = 1$, then R_i can be expressed as:

$$\begin{cases} R_1 = k_{11}Z_1 + k_{12}Z_2 \\ R_2 = k_{21}Z_2 + k_{22}Z_2 \end{cases} \tag{5.24}$$

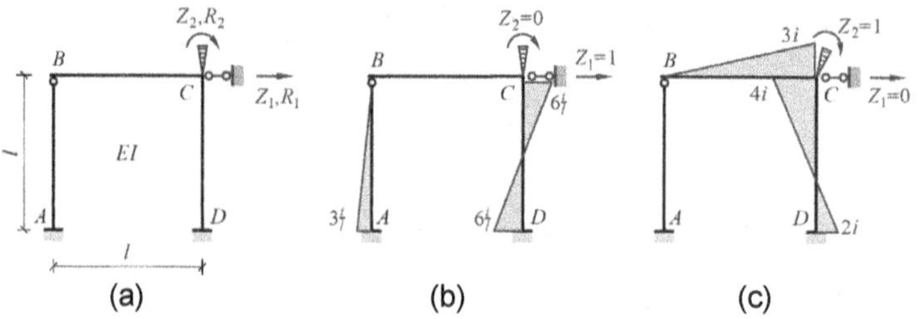

Figure 5.11. (a) Constructing a primary system by adding two virtual constraints which control the horizontal displacement and rotation at node C. (b) The moment diagram due to Z_1. (c) The moment diagram due to Z_2.

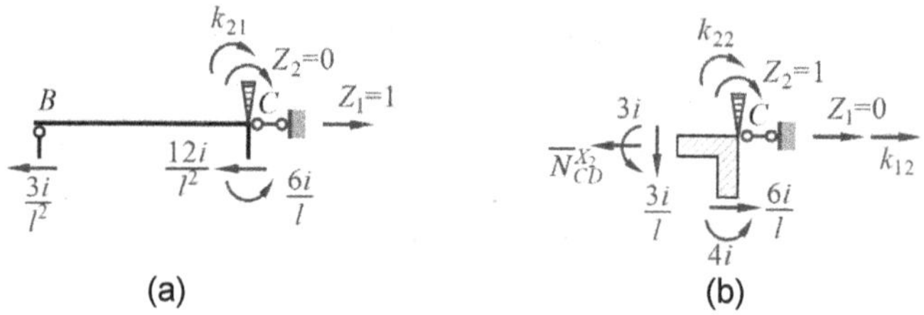

Figure 5.12. (a) The free-body diagrams of beam BC. (b) the free-body diagrams at node C.

To get the expression of k_{ij}, we construct the moment diagrams, $\bar{M}_1$ and $\bar{M}_2$, which are caused by $Z_1 = 1$ and $Z_2 = 1$, respectively. See Figures (5.11b and c). To formulate R_1 and R_2, Figure (5.12) shows the free-body diagrams of beam BC and node C. Thus, one has:

$$\begin{cases} k_{11} = \dfrac{3i}{l^2} + \dfrac{12i}{l^2} = \dfrac{15i}{l^2} \\ k_{12} = k_{21} = -\dfrac{6i}{l} \\ k_{22} = 3i + 4i = 7i \end{cases} \tag{5.25}$$

where we used the reciprocal theorem $k_{12} = k_{21}$.

Substitution of Equations (5.24) and (5.25) gives the governing equation. After obtaining Z_1 and Z_2, the moment M of the actual system can be expressed as:

$$M = Z_1 \bar{M}_1 + Z_2 \bar{M}_2 \tag{5.26}$$

EXAMPLE 5.2

Select the proper unknown DOFs for the analysis of the structures shown in Figure (5.13).

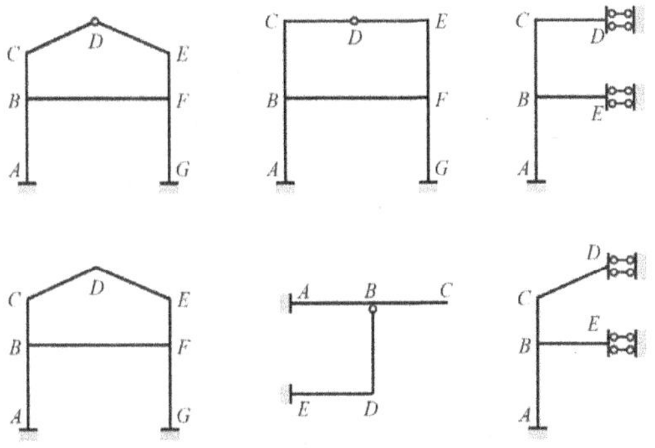

Figure 5.13. Structures for analysis.

Solution: See Figure 5.14 for solution.

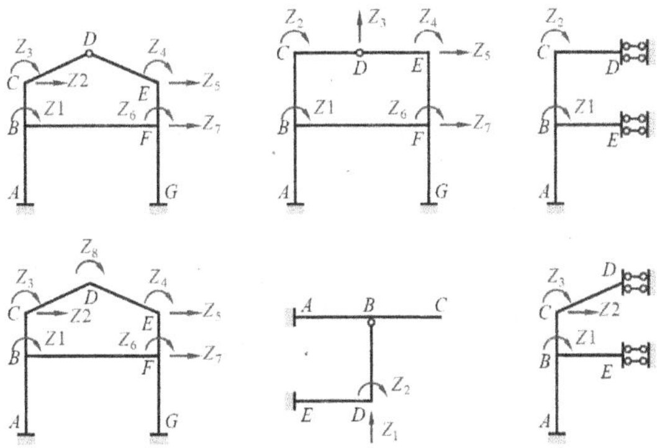

Figure 5.14. Solutions.

5.3 The Analysis of Frames Under In-Span Loading

In the above case, the frame is assumed to be subjected only to a joint load. In general, however, a structure may be subjected to not only moments at its joints, but also loadings acting on the span of the member.

5.3.1 Fixed-End Forces

Figure (5.15a) shows that a fixed-end beam subjected to a uniformly distributed line load q. Figure (5.15b) indicates the *positive directions* of the end moments and shears of the beam. Its bending moment distribution is given in Figure (5.15c) and expressed as:

$$M_{ab}^{F} = -\frac{ql^2}{12} \quad \text{and} \quad M_{ba}^{F} = \frac{ql^2}{12} \tag{5.27}$$

Figure 5.16 to Figure 5.20 show the moment diagrams of the beam with different support conditions and load types.

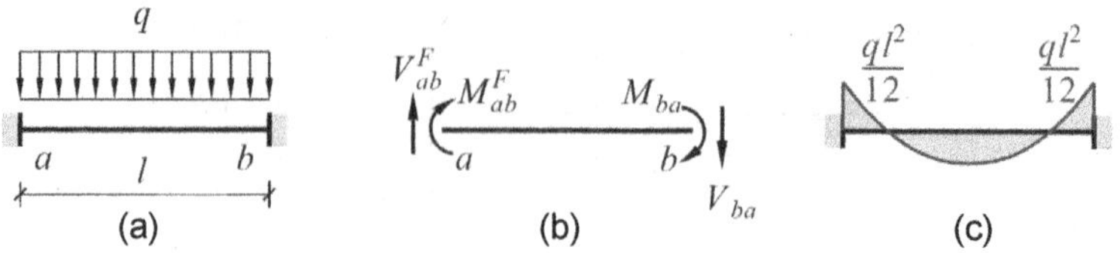

Figure 5.15. (a) A fixed-end beam subjected to a uniformly distributed line load q. (b) The positive directions of the end shears and moments. (c) The moment diagram of the beam.

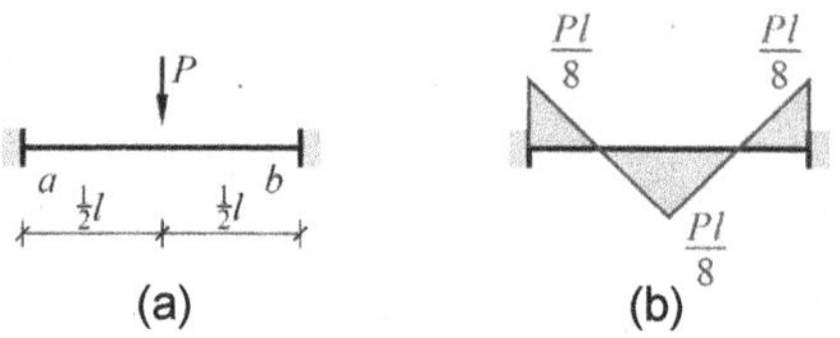

Figure 5.16. (a) A fixed-end beam subjected to concentrated load P. (b) The moment diagram of the beam.

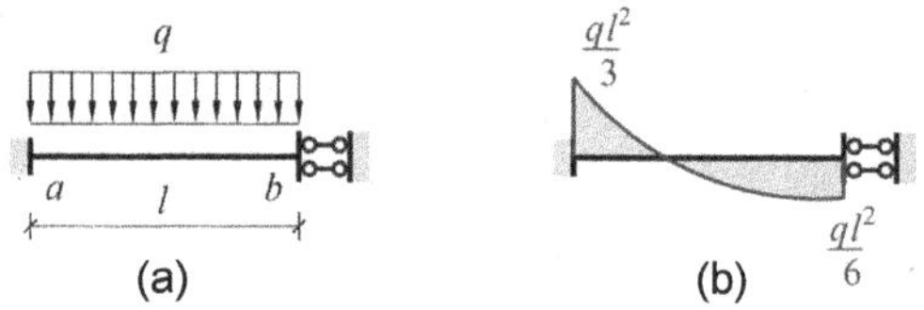

Figure 5.17. (a) A beam subjected to a uniformly distributed line load q. (b) The moment diagram of the beam.

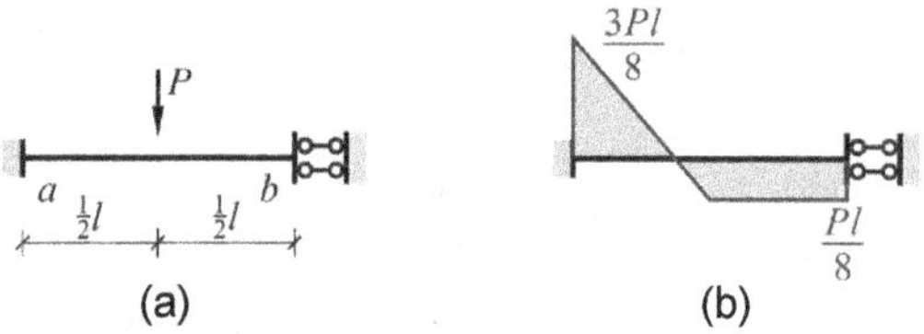

Figure 5.18. (a) A beam subjected to concentrated load P. (b) The moment diagram of the beam.

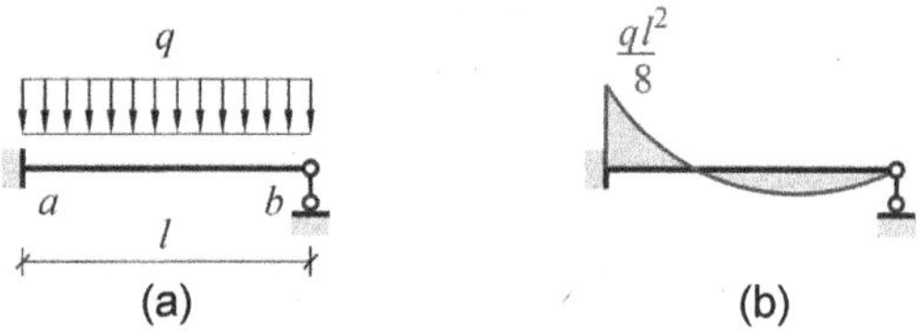

Figure 5.19. (a) A beam subjected to a uniformly distributed line load q. (b) The moment diagram of the beam.

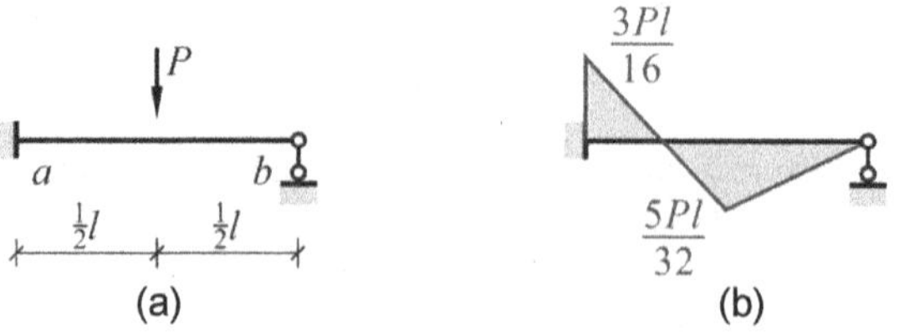

Figure 5.20. (a) A beam subjected to concentrated load P. (b) The moment diagram of the beam.

5.3.2 Processing of In-Span Loads and Nodal Equivalent Loads

The frame shown in Figure (5.21(a)) is subjected to a uniform load of 5 kN/m on member BC, a concentrated load of 10 kN on member AB, and a torque of 20 kN·m at node B. The actual case shown in Figure (5.21(a)) can be considered to be a superposition of the two cases shown in Figures 5.21(b) and (c), in which M_B^F is referred to as a ***fixed-DOF torque*** (or a ***fixed-DOF load***) because it is applied to prevent the rotation of node B under the in-span loads; M_B^E is defined as:

$$M_B^E \equiv -M_B^F \tag{5.28}$$

M_B^E is termed a ***nodal equivalent torque*** (or a ***nodal equivalent load***). The frame shown in Figure (5.21b) is subjected only to nodal loads, and can be analyzed by the previous approach, as long as the nodal equivalent load M_B^E is determined. The responses of the actual system shown in Figure (5.21a) is the superposition of the systems shown in Figures 5.21b and c. Specifically, the moment M of the actual system can be expressed as:

$$M = M^{\Delta} + M^F \tag{5.29}$$

where M^{Δ} and M^F are the moments of systems as in Figures 5.21b and c, respectively.

One can construct the moment diagram M^F of system (Figure 5.21c) by using the fixed-end moments, see Figure (5.22), given by:

$$\begin{cases} M_{AB}^F = \dfrac{10l}{8} \\ M_{BA}^F = -\dfrac{10l}{8} \end{cases} \text{and} \begin{cases} M_{BC}^F = -\dfrac{5l^2}{12} \\ M_{CB}^F = \dfrac{5l^2}{12} \end{cases} \tag{5.30}$$

The free-body diagram of joint B is shown in Figure (5.22b). The equilibrium in the DOF Z_1 gives:

$$M_B^F = M_{BA}^F + M_{BC}^F = -\frac{10l}{8} - \frac{5l^2}{12} \tag{5.31}$$

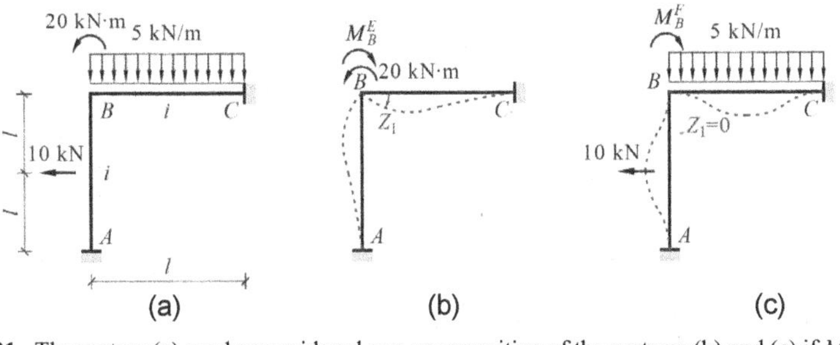

Figure 5.21. The system (a) can be considered as a superposition of the systems (b) and (c) if $M_B^E = -M_B^F$. (a) A frame subjected to nodal and in-span loads. (b) The frame is subjected only to the actual nodal torque and the nodal equivalent torque. (c) The frame is subjected to the in-span loads and a fixed-DOF load such that the unknown DOF Z_1 is fixed.

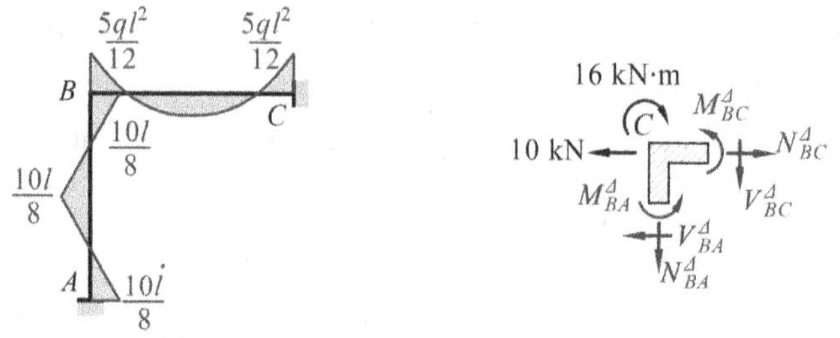

Figure 5.22. Establish the moment diagram M^F by using the (a) fixed-end moments and (b) the forces at node B.

Then, the nodal efficient torque M_B^E is:

$$M_B^E = -M_B^F = \frac{10l}{8} + \frac{5l^2}{12} \tag{5.32}$$

The slope-deflection equation of the fixed-end beam gives:

$$\begin{cases} M_{BA}^{\Delta} = 4iZ_1 \\ M_{BC}^{\Delta} = 4iZ_1 \end{cases} \tag{5.33}$$

Thus, the equation governing Z_1 of the actual system can be expressed as:

$$8iZ_1 = M_B + M_B^E \tag{5.34}$$

or

$$8iZ_1 = \frac{5l^2}{12} + \frac{10l}{8} - 20 \tag{5.35}$$

This is the stiffness equation of the frame subjected to in-span loads.

EXAMPLE 5.4

Determine the base moment M_A of column AB of the frame shown in Figure 5.23(a).

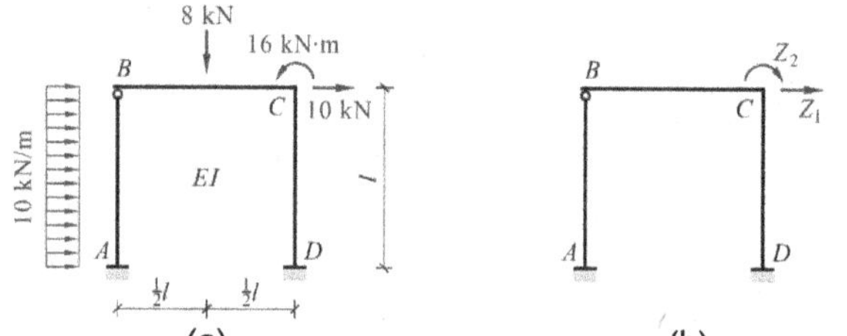

Figure 5.23. (a) The frame structure. (b) The horizontal displacement Z_1 and the rotation Z_2 at node C are taken as the first and second unknown DOFs, respectively.

***Solution*:**

(1) Select the unknown DOFs.

The horizontal displacement Z_1 and the rotation Z_2 at node C are taken as the first and second unknown DOFs, respectively, as shown in Figure (5.23b).

(2) Determine the nodal equivalent loads.

The actual system shown in Figure (5.23a) is considered as a superposition of the systems (a) and (b) shown in Figure (5.24). To determine the fixed-DOF loads P_1^F and P_2^F, draw the free-body diagrams of members AB and BC, and joint C of the system shown in Figure 5.24(a), see Figure (5.25). Thus, one has:

$$V_{BA}^F = -\frac{15l}{4} \tag{5.36}$$

$$P_1^F = -\frac{15l}{4} \tag{5.37}$$

$$P_2^F = \frac{3l}{2} \tag{5.38}$$

$$\begin{cases} P_1^E = -P_1^F = \dfrac{15l}{4} \\ P_2^E = -P_2^F = -\dfrac{3l}{2} \end{cases} \tag{5.39}$$

(3) Formulate the equations governing Z_1 and Z_2, and determine the unknown DOFs Z_1 and Z_2.

$$\begin{cases} \dfrac{15i}{l^2} Z_1 - \dfrac{6i}{l} Z_2 = 10 + \dfrac{15l}{4} \\ -\dfrac{6i}{l} Z_1 + 7iZ_2 = -16 - \dfrac{3l}{2} \end{cases} \tag{5.40}$$

which yields:

$$Z_1 = \frac{105l^3 + 244l^2 - 384l}{276i} \quad \text{and} \quad Z_2 = \frac{15l^2 + 25l - 160}{46i}, \tag{5.41}$$

(4) Determine the moment M of the actual system:

$$M_A = M_A^F + M_A^\Delta = -\frac{5l^2}{4} - 3i \times \frac{105l^3 + 244l^3 - 384l}{276i}$$
$$= \frac{384l - 350l^2 - 105l^3}{92}$$

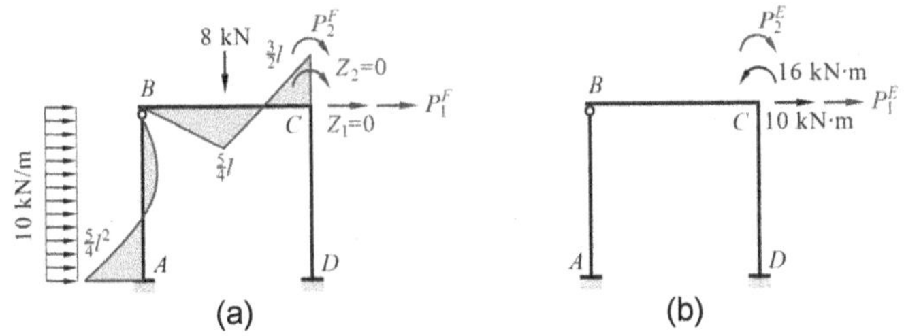

Figure 5.24. (a) A system whose unknown DOFs are "locked" by fixed-DOF loads. (b) A system subjected to the actual and equivalent nodal loads.

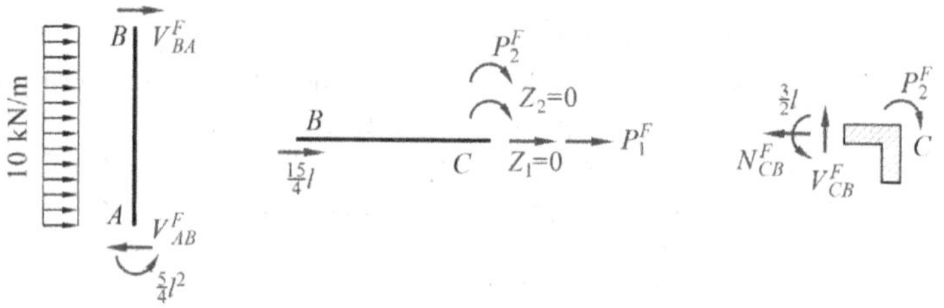

Figure 5.25. The free-body diagrams of members *AB* (left) and *BC* (middle), and joint *C* (right) of the system.

5.4 Examples of Frames with In-Span Loads

EXAMPLE 5.5

Determine the free-body diagrams of members for the frame shown in Figure (5.26).

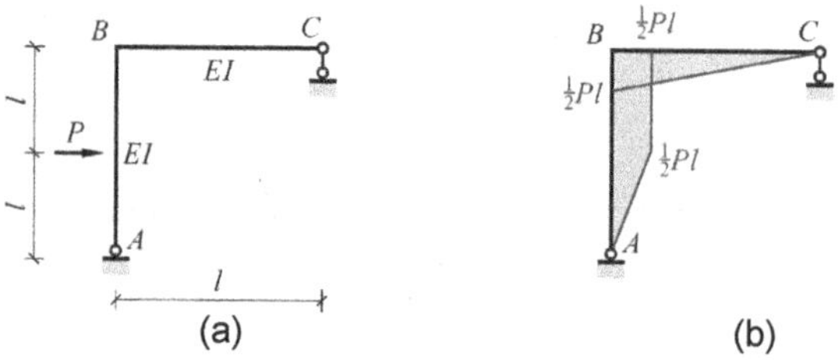

Figure 5.26. (a) A frame subject to a concentrated load *P*, (b) Bending moment diagram.

Solution I: (1) Select the unknown Degrees Of Freedom. One may take the rotation, Z_1, and horizontal displacement, Z_2, at node *B* as the first and second unknown Degrees Of Freedom, respectively, as shown in Figure (5.27a).

(2) Determine the nodal equivalent loads. The actual case is a superposition of cases (b) and (c) shown in Figure (5.27), in which the unknown DOFs in case (b) are locked to be zero. The fixed-DOF loads in case (b) are determined by using the fixed-end moments and shears.

(3) Analyze the frame with nodal loads as shown in Figure (5.27c). Construct the primary system shown in Figure (5.28a), which leads to the following equations governing the unknown DOFs, Z_1 and Z_2.

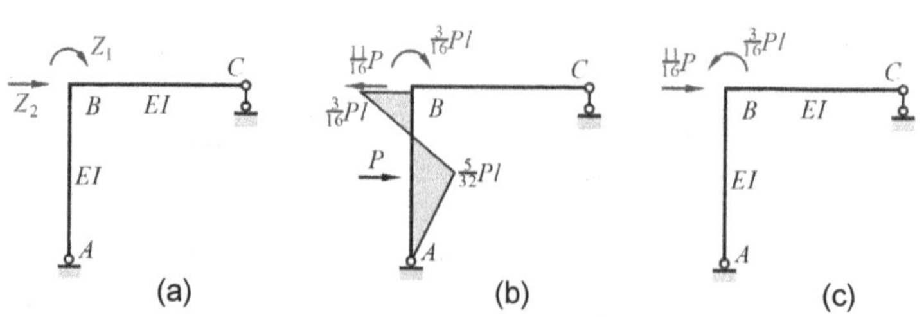

Figure 5.27. Solutions.

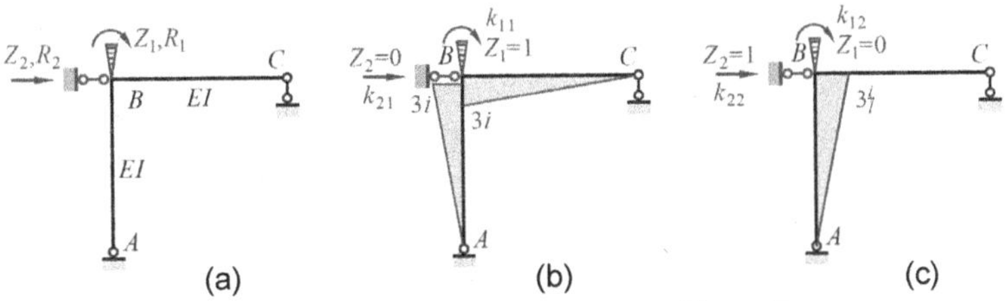

Figure 5.28. (a) The primary system. (b) $\bar{M}_1$. (c) $\bar{M}_2$.

$$\begin{cases} R_1 = k_{11}Z_1 + k_{12}Z_2 = -\dfrac{3Pl}{16} \\ R_2 = k_{21}Z_1 + k_{22}Z_2 = \dfrac{11P}{16} \end{cases} \tag{5.42}$$

To determine k_{ij}, draw $\bar{M}_1$ and $\bar{M}_2$ shown in Figures 5.28b and c by using the slope-deflection equations. Figure (5.28b) gives:

$$k_{11} = 3i + 3i = 6i \tag{5.43}$$

Figure (5.28c) gives:

$$\begin{cases} k_{22} = \dfrac{3i}{l^2} \\ k_{12} = -\dfrac{3i}{l} \end{cases} \tag{5.44}$$

The reciprocal theorem shows that:

$$k_{21} = k_{12} \tag{5.45}$$

Thus, the governing equation can be written as:

$$\begin{cases} 6iZ_1 - \dfrac{3i}{l}Z_2 = -\dfrac{3Pl}{16} \\ -\dfrac{3i}{l}Z_1 + \dfrac{3i}{l^2}Z_2 = \dfrac{11P}{16} \end{cases} \tag{5.46}$$

which leads to:

$$\begin{cases} Z_1 = \dfrac{Pl}{6i} \\ Z_2 = \dfrac{19Pl^2}{48i} \end{cases} \tag{5.47}$$

(4) Determine the end moments of the actual structure.

The moment M of the actual frame can be expressed as:

$$M = M^F + Z_1\bar{M}_1 + Z_2\bar{M}_2$$

Thus, one has the end moments, M_{BA} and M_{BC}, of the actual frame, which can be expressed as:

$$M_{BA} = \frac{3Pl}{16} + 3i \times \frac{Pl}{6i} + \left(-\frac{3i}{l}\right) \times \frac{19Pl^2}{48i} = -\frac{Pl}{2}$$

which leads to the moment diagram shown in Figure 5.26(b).

Solution I can be written in the procedure directly using slope-deflection equation. This is left as an exercise for the reader.

Solution II: (1) Select the unknown DOFs. We may take the rotation Z_1 of joint B as the single DOF for the analysis as shown in Figure 5.29(a).

(2) Determine the nodal equivalent loading. The actual case shown in Figure 5.26(a) is considered to be a superposition of cases (b) and (c) shown in Figure (5.29), in which the rotation of joint B in case (b) is zero. The nodal equivalent loading is the clockwise torque of $\frac{Pl}{2}$ at node B.

(3) Analyze the case shown in Figure 5.29(c).

The slope-deflection equation of a clamped-roller beam shows that:

$$M_{BC}^{\Delta} = 3iZ_1 \tag{5.48}$$

To formulate M_{BA}^{Δ}, draw the free-body diagram of member AB, see Figure (5.30a). The horizontal equilibrium shows that the shear of member AB is zero. This implies the moment of member AB is a constant. Since, $M_{BA}^{\Delta} = 0$, one has:

$$M_{BA}^{\Delta} = 0 \tag{5.49}$$

Figure (5.30b) shows the possible deformation for this case. To formulate the governing equation, draw the free-body diagram of joint B, see Figure (5.30c). The moment equilibrium yields:

$$M_{BA}^{\Delta} + M_{BC}^{\Delta} = \frac{Pl}{2} \tag{5.50}$$

or

$$3iZ_1 = \frac{Pl}{2} \tag{5.51}$$

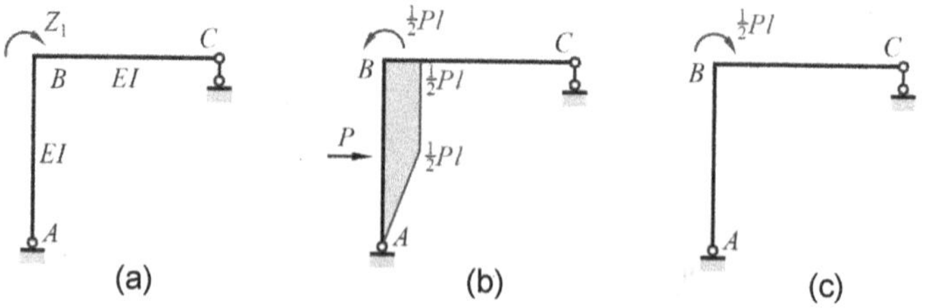

Figure 5.29. (a) Z_1 is taken as the unique unknown DOF for the analysis. (b) M^F. (c) The frame subjected to the nodal equivalent torque.

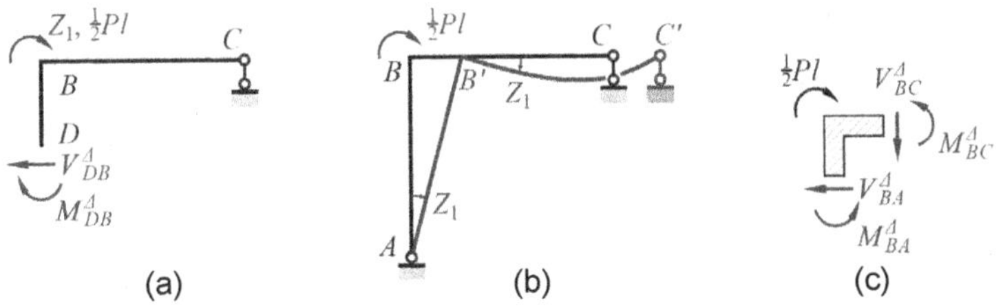

Figure 5.30. Solutions.

where

$$Z_1 = \frac{Pl}{6i} \tag{5.52}$$

(4) Determine the end moments of the actual system.

The end moments of the actual system can be expressed as:

$$\begin{cases} M_{BA} = M_{BA}^{F} + M_{BA}^{\Delta} = -\dfrac{Pl}{2} + 0 = -\dfrac{Pl}{2} \\ M_{BC} = M_{BC}^{F} + M_{BC}^{\Delta} = 0 + 3iZ_1 = \dfrac{Pl}{2} \end{cases} \tag{5.53}$$

which lead to the moment diagram shown in Figure (5.26b).

Solution II can be written in the procedure using a primary system. This is left as an exercise for the reader.

EXAMPLE 5.6

Determine the end moment M_{CB} of beam BC, the base moment M_{DC} and the base shear V_{DC} of column DC of the system shown in Figure (5.31a).

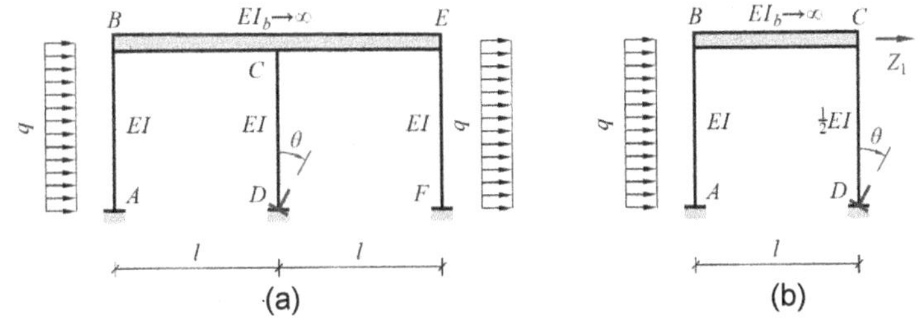

Figure 5.31. (a) A symmetric frame subjected to wind loads q and a support settlement d. (b) The half-structure and the unknown DOF, Z_1, for analyzing the half-structure by the displacement method.

Solution: (1) Construct the half-structure for analyzing the actual system.

(2) Select the unknown DOFs for analyzing the half-structure by the displacement method.

As the beam is a rigid body, the unknown configuration of the half-structure can be completely described by it lateral displacement, Z_1, shown in Figure 5.31(b), which is taken as the unique unknown DOF for the analysis.

(3) Determine the nodal equivalent loading.

The half-structure shown in Figure (5.32b) may be considered to be a superposition of systems (a), (b) and (c) shown in Figure (5.32).

(4) Analyze the half-structure subjected to the nodal equivalent load to find the unknown DOF Z_1.

Figure (5.33a) shows the primary structure for analyzing the system shown in Figure (5.32c), which leads to the following governing equation.

$$k_{11}Z_1 = \frac{ql}{2} + \frac{3i\theta}{l}$$

To determine k_{11}, draw the moment diagram $\bar{M}_1$ shown in Figure (5.33b), which shows that

$$k_{11} = \frac{18i}{l^2}$$

by combining the two equations above, one obtains:

$$Z_1 = \frac{ql^3}{36i} + \frac{\theta l}{6}$$

(5) Determine the end moments M_{CB} and M_{DE}, and shear V_{CD} of the actual structure.

$$M_{CB}^h = M_{CB,q}^F + M_{CB,\theta}^F + Z_1\bar{M}_{CB} = 0 - i\theta + \left(\frac{ql^3}{36i} + \frac{\theta l}{6}\right) \times \frac{3i}{l}$$

$$= \frac{ql^2}{12} - \frac{i\theta}{2}$$

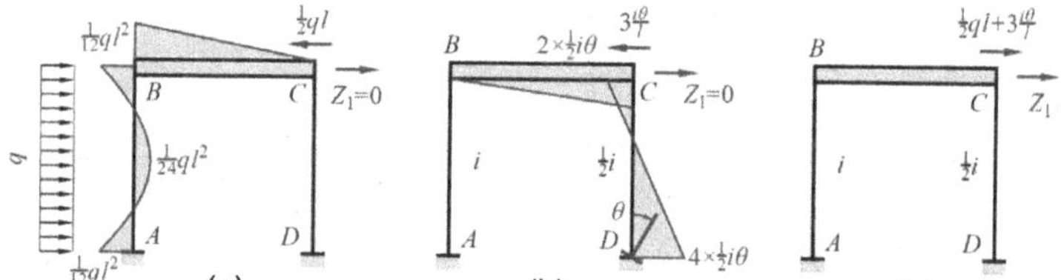

(a) (b) (c)

Figure 5.32. (a) The fixed-end moment M_q^F caused by the in-span load of q. (b) The fixed-end moment M_θ^F caused by the support rotation θ. (c) The nodal equivalent load.

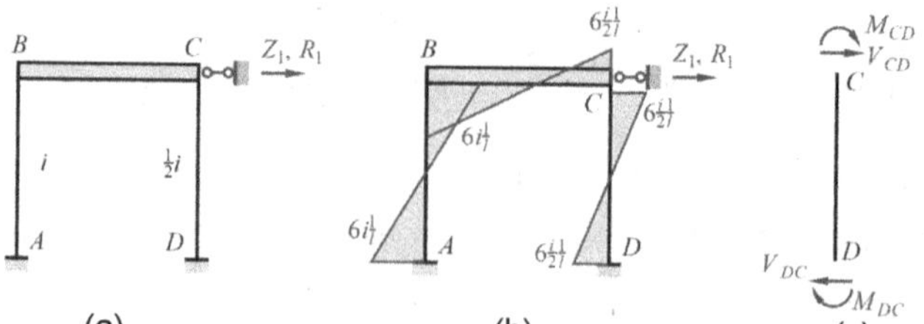

(a) (b) (c)

Figure 5.33. (a) Primary structure for analyzing the system shown in Figure 5.32(c). (b) Moment diagram $\bar{M}_1$. (c) Free-body diagram of column *DC*.

and

$$M^h_{CD} = M^F_{CD,q} + M^F_{CD,\theta} + Z_1\bar{M}_{CD} = 0 + i\theta + \left(\frac{ql^3}{36i} + \frac{\theta l}{6}\right) \times \left(-\frac{3i}{l}\right)$$

$$= \frac{i\theta}{2} - \frac{ql^2}{12}$$

and

$$M^h_{DC} = M^F_{DC,q} + M^F_{DC,\theta} + Z_1\bar{M}_{DC} = 0 + 2i\theta + \left(\frac{ql^3}{36i} + \frac{\theta l}{6}\right) \times \left(-\frac{3i}{l}\right)$$

$$= \frac{3i\theta}{2} - \frac{ql^2}{12}$$

Thus, the end moments, M_{CB}, M_{CD} and M_{DC}, of the actual frame can be expressed as

$$M_{CB} = M^h_{CB} = \frac{ql^2}{12} - \frac{i\theta}{2}$$

$$M_{CD} = 2M^h_{CD} = i\theta - \frac{ql^2}{6}$$

$$M_{DC} = 2M^h_{DC} = 3i\theta - \frac{ql^2}{6}$$

To determine the base shear V_{DC}, draw the free-body diagram of column DC, see Figure (5.33c).

$$V_{DC} = -\frac{M_{CD} + M_{DC}}{l} = \frac{ql}{3} - \frac{4i\theta}{l}$$

5.5 Moment Distribution Approach

5.5.1 Moment Distribution Approach for SDOF Structures

Determine the end moments of the beam shown in Figure (5.34a). If θ_B denotes the rotation of the beam at joint B, then its moment diagram is shown in Figure (5.34b), which gives:

$$M^{\Delta}_{BA} = \frac{4i\theta_B}{4i\theta_B + 6i\theta_B} \times 20 = 0.4 \times 20 = 8 \quad (5.54)$$

$$M^{\Delta}_{BC} = \frac{6i\theta_B}{4i\theta_B + 6i\theta_B} \times 20 = 0.6 \times 20 = 12 \quad (5.55)$$

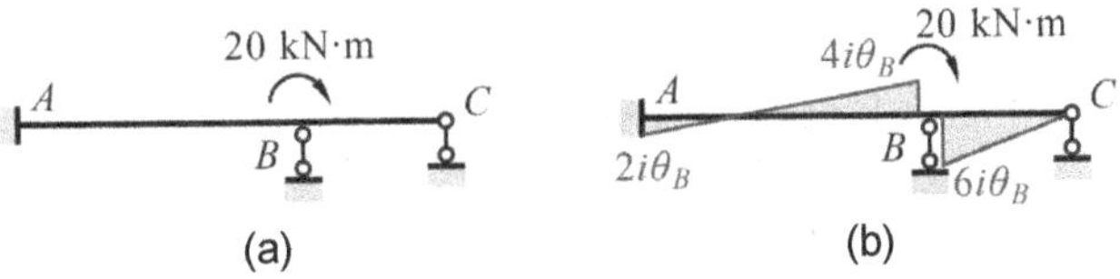

Figure 5.34. (a) The beam system. (b) The moment diagram.

where 0.4 and 0.6 are called the ***distribution factors*** (FD) of sections BA and BC, respectively, and may be denoted by μ_{BA} and μ_{BC}.

$$M_{AB}^{\Delta} = 2i\theta_B = 0.5 \times M_{BA} = 4 \tag{5.56}$$

$$M_{CB}^{\Delta} = 0 \times 20 = 0 \tag{5.57}$$

in which 0.5 and 0 are referred to as the ***transfer coefficients*** (TC) of members AB and BC, respectively, and may be denoted by λ_{AB} and λ_{CB}.

EXAMPLE 5.7

Draw the moment diagram of the beam shown in Figure 5.35(a).

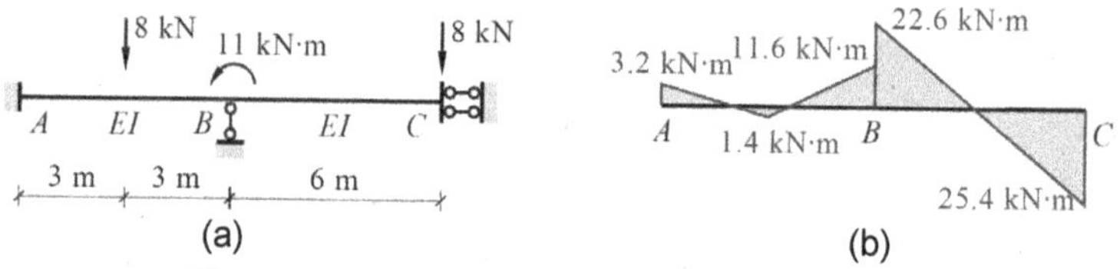

Figure 5.35. (a) The beam system. (b) Moment diagram.

Solution: The actual system shown in Figure (5.35a) is considered to be a superposition of systems (a) and (b) shown in Figure (5.35). The distribution factors μ_{BA} and μ_{BC} can be expressed as:

$$\mu_{BA} = \frac{4i}{4i+i} = 0.8 \tag{5.58}$$

$$\mu_{BC} = \frac{i}{4i+i} = 0.2 \tag{5.59}$$

where $i \equiv \frac{EI}{6}$. The transfer coefficients of members AB and BC are:

$$\lambda_{AB} = 0.5 \quad \text{and} \quad \lambda_{BC} = -1 \tag{5.60}$$

Thus, the end moments of the beam can be expressed as:

$$\begin{cases} M_{BA}^{\Delta} = \mu_{BA} \times (6-11) = 0.8 \times (-5) = -4 \\ M_{BC}^{\Delta} = \mu_{BC} \times (6-11) = 0.2 \times (-5) = -1 \end{cases} \tag{5.61}$$

and

$$\begin{cases} M_{AB}^{\Delta} = \lambda_{BA} \times M_{BA}^{\Delta} = 0.5 \times (-4) = -2 \\ M_{CB}^{\Delta} = \lambda_{BC} \times M_{BC}^{\Delta} = -1 \times (-1) = 1 \end{cases} \tag{5.62}$$

Thus, the end moments of the actual beam can be expressed as:

$$\begin{cases} M_{AB} = M_{AB}^{F} + M_{AB}^{\Delta} = -6 - 2 = -8 \\ M_{BA} = M_{BA}^{F} + M_{BA}^{\Delta} = 6 - 4 = 2 \\ M_{BC} = M_{BC}^{F} + M_{BC}^{\Delta} = -12 - 1 = -13 \\ M_{CB} = M_{CB}^{F} + M_{CB}^{\Delta} = -12 + 1 = -11 \end{cases} \tag{5.63}$$

Table 5.1. The moment distribution and transfer of the beam shown in Figure 5.35a.

Joint	A		B			C
Section	AB		BA	BC		CB
DF			0.8	0.2		
TC		0.5			−1	
M_{ij}^F	−6		6	−24		−24
M_i^E			18			
M_i			−11			
M_{ij}^{Δ}	2.8	←	5.6	1.4		−1.4
$M_{ij} = M_{ij}^F + M_{ij}^{\Delta}$	−3.2		11.6	−22.6		−25.4

The above procedure can be represented as in Table 5.1.

5.5.2 Moment Distribution Approach for MDOF Beams

EXAMPLE 5.8

Consider the beam shown in the upper figure of Figure 5.36. Use the moment distribution approach to draw the moment diagram.

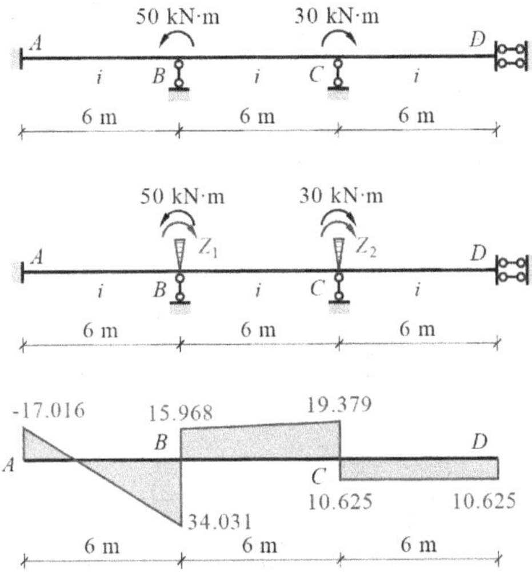

Figure 5.36. The beam system (upper), and the procedure (mid and lower) to solve the problem using the moment distribution approach.

***Solution*:**

$$i \equiv \frac{EI}{l} = \frac{1}{6}\left(3\times10^{10}\times\frac{1}{12}\times0.2\times0.6^3\right) = 1.8\times10^4 (kN\cdot m)$$

$$\mu_{BA} = \frac{4i}{4i+4i} = 0.5 \text{ and } \mu_{BC} = \frac{3i}{4i+4i} = 0.375$$

$$\mu_{CB} = \frac{4i}{4i+3i} = 0.571 \text{ and } \mu_{CD} = \frac{3i}{4i+3i} = 0.429$$

The moment diagram of the beam is given in Figure 5.37.

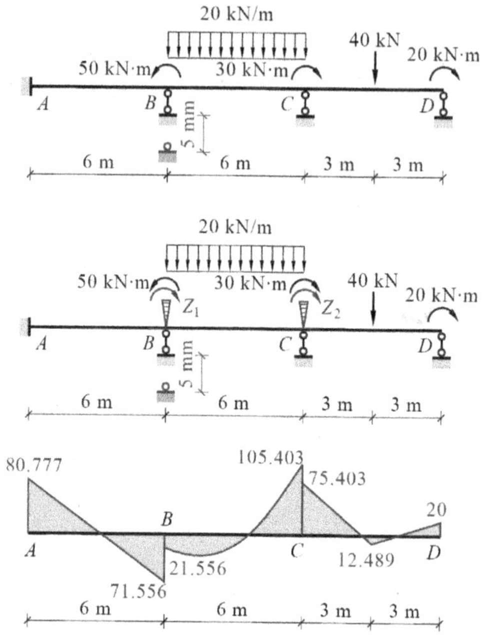

Figure 5.37. The moment diagram.

Table 5.2. The moment distribution and transfer of the beam shown in Figure 5.36.

Joint	*A*		*B*			*C*			*D*
Section	*AB*		*BA*	*BC*		*CB*	*CD*		*DC*
DF			0.5	0.5		0.8	0.2		
TC		0.5			0.5			–1	
M_i			–50			30			
	–12.500	←	–25.000	–25.000	→	–12.500			
				17.000	←	34.000	8.500		–8.500
	–4.250	←	–8.500	–8.500	→	–4.250			
				1.062	←	2.125	2.125		–2.125
	–0.266	←	–0.531	0.531					
M_{ij}	–17.016		–34.031	–15.968		19.379	10.625		–10.625

Table 5.3. The moment distribution and transfer of the beam shown in Figure 5.37.

Joint	*A*		*B*			*C*			*D*
Section	*AB*		*BA*	*BC*		*CB*	*CD*		*DC*
DF			0.5	0.5		0.571	0.429		
TC		0.5			0.5			0	
M_{ij}^F	−90		−90	30		150	−35		20
M_i			−50			30			
M_{ij}^{Δ}				−24.268	←	−48.535	−36.465	→	0
	8.567	←	17.134	17.134	→	8.567			
				−2.446	←	−4.892	−3.675	→	0
	0.612	←	1.223	1.223	→	0.612			
				−0.174	←	−0.349	−0.263	→	0
	0.044	←	0.087	0.087					
M_{ij}	−80.777		−71.556	21.556		105.403	−75.403		20

$$\begin{cases} M_{AB}^F = -6\times 1.8\times 10^4 \times \dfrac{0.005}{6} = -90kN\cdot m \\ M_{BA}^F = -6\times 1.8\times 10^4 \times \dfrac{0.005}{6} = -90kN\cdot m \\ M_{BC}^F = 6\times 1.8\times 10^4 \times \dfrac{0.005}{6} - \dfrac{1}{12}\times 20\times 6^2 = 90-60 = 30kN\cdot m \\ M_{CB}^F = 6\times 1.8\times 10^4 \times \dfrac{0.005}{6} + \dfrac{1}{12}\times 20\times 6^2 = 90+60 = 150kN\cdot m \\ M_{CD}^F = -\dfrac{3}{16}\times 40\times 6+10 = -35kN\cdot m \\ M_{DC}^F = 20kN\cdot m \end{cases}$$

EXERCISES

5.1 Use the moment distribution approach to draw the moment diagram of the beam shown in Figure 5.38.

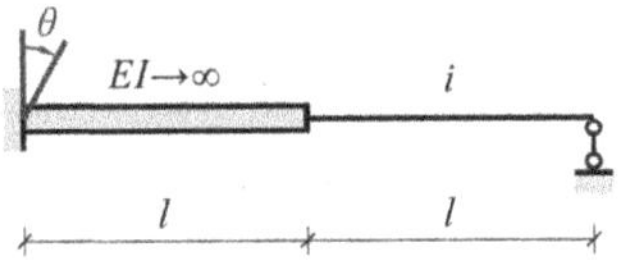

Figure 5.38. Exercise 5.1.

5.2 Use the moment distribution approach to draw the moment diagram of the beam shown in Figure 5.39.

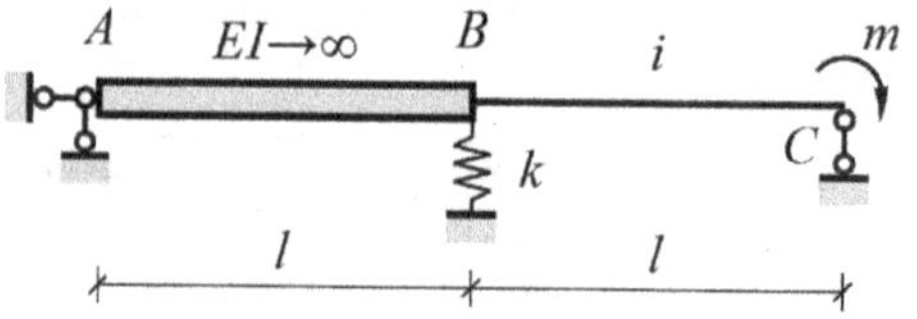

Figure 5.39. Exercise 5.2.

5.3 Use the moment distribution approach to draw the moment diagram of the beam shown in Figure 5.40.

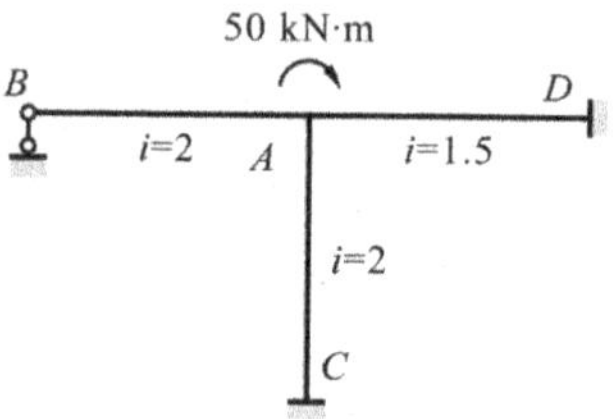

Figure 5.40. Exercise 5.3.

5.4 Use the moment distribution approach to draw the moment diagram of the beam shown in Figure 5.41.

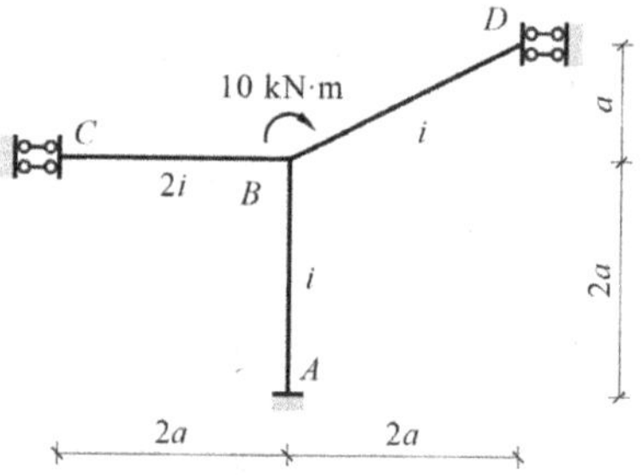

Figure 5.41. Exercise 5.4.

5.5 Use the moment distribution approach to draw the moment diagram of the beam shown in Figure 5.42.

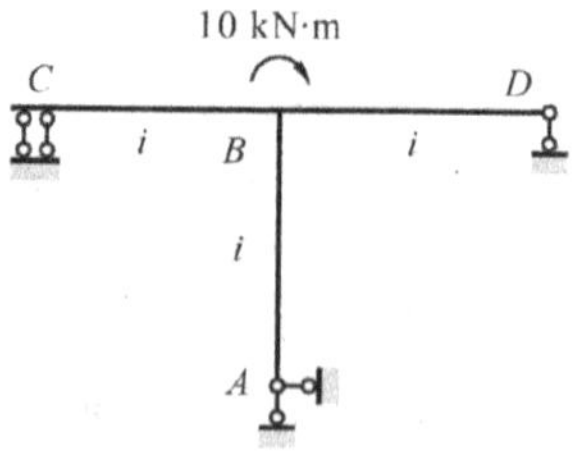

Figure 5.42. Exercise 5.5.

5.6 Use the moment distribution approach to draw the moment diagram of the beam shown in Figure 5.43.

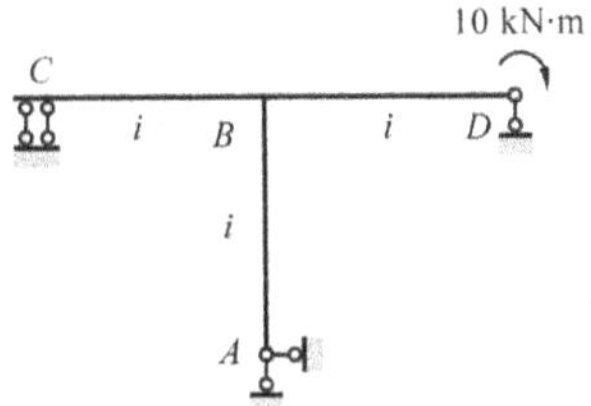

Figure 5.43. Exercise 5.6.

5.7 Use the moment distribution approach to draw the moment diagram of the beam shown in Figure 5.44.

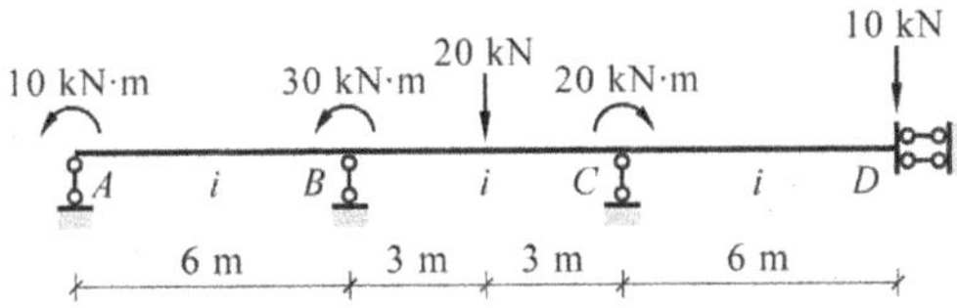

Figure 5.44. Exercise 5.7.

5.8 Use the moment distribution approach to draw the moment diagram of the beam shown in Figure 5.45.

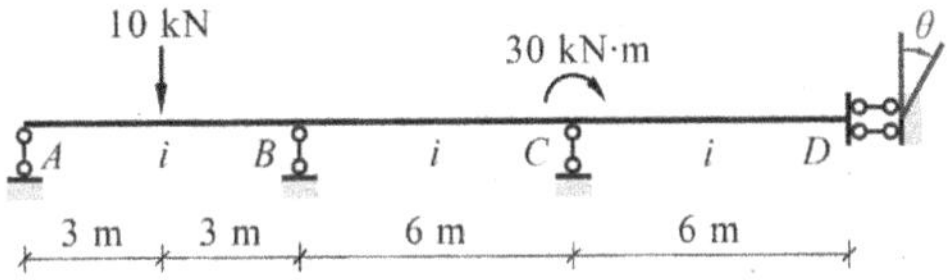

Figure 5.45. Exercise 5.8.

5.9 Apply the displacement method to construct the moment diagram of the structure shown in Figure 5.46, in which the rigidity of each member is *EI*.

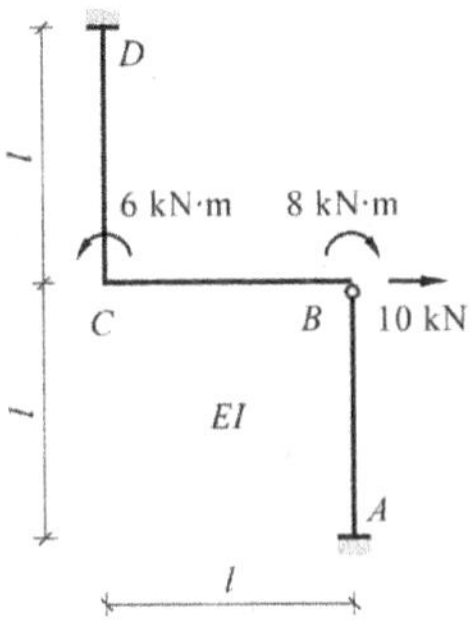

Figure 5.46. Exercise 5.9.

5.10 Apply the displacement method to construct the moment diagram of the structure shown in Figure 5.47, in which the rigidity of each member is EI.

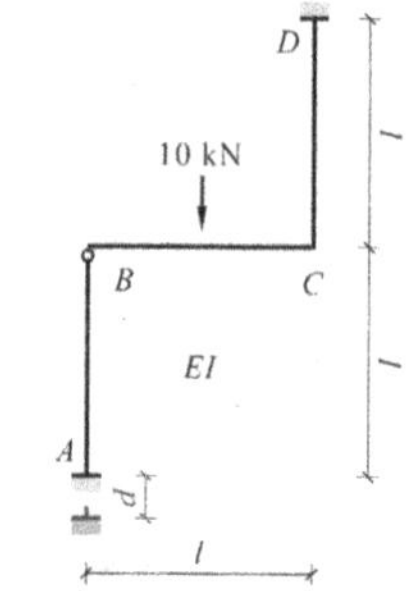

Figure 5.47. Exercise 5.10.

5.11 Use the moment distribution approach to draw the moment diagram of the beam shown in Figure 5.48.

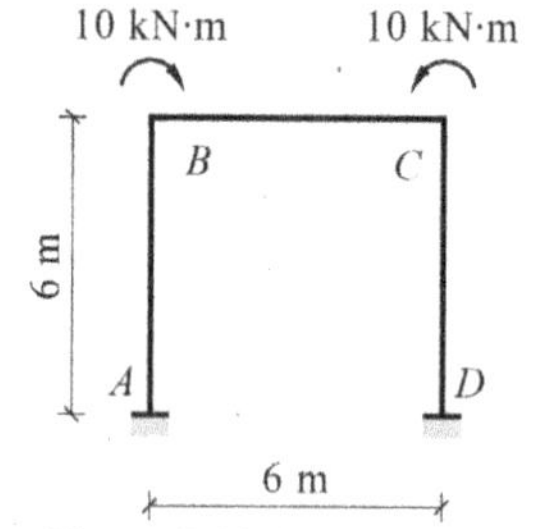

Figure 5.48. Exercise 5.11.

5.12 Use the moment distribution approach to draw the moment diagram of the beam shown in Figure 5.49.

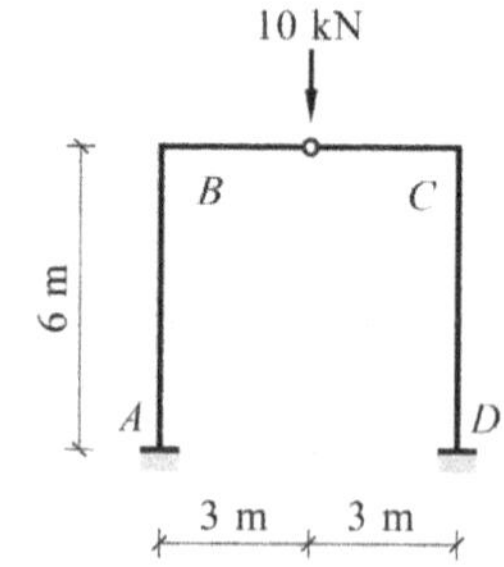

Figure 5.49. Exercise 5.12.

5.13 Which of the systems shown in Figure 5.50 can be used as a primary system to analyze the system shown in the upper left figure in Figure 5.9(a)?

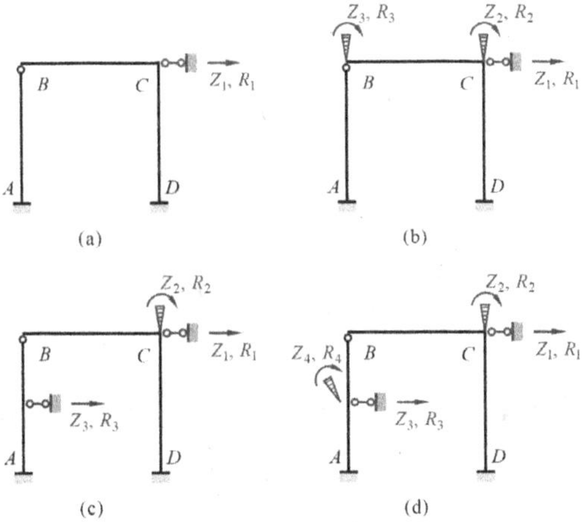

Figure 5.50. Exercise 5.13.

Chapter 6

Influence Lines for Statically Determinate Structures

6.1 Introduction

An ***influence line*** is a physical quantity of a function of a structure representing the variation of the quantity as a unit concentrated load moving over a specific member or members of the structure. Common functions studied with influence lines include reactions (forces that the structure supports must apply for the structure to remain static), shears, moments, and deflections. Influence lines are important in designing beams and trusses used in various types of engineering structures where loads move along their span. The influence lines show where a load will create the maximum effect for any of the functions studied.

6.2 Influence Lines for Beams

Figure 6.1 shows an example of the influence line for reaction forces at A.

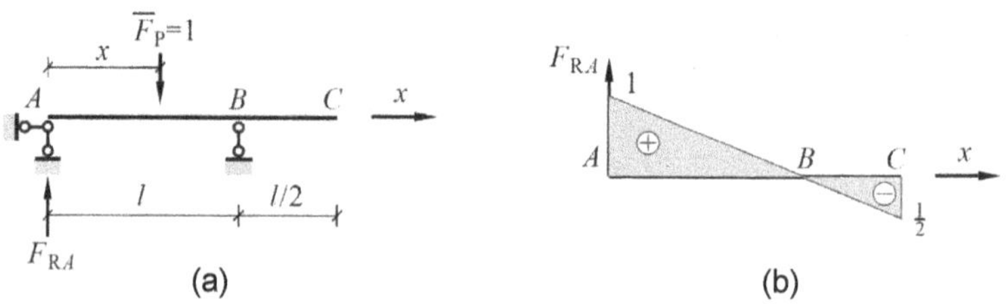

Figure 6.1. (a) A simply supported beam; (b) The influence line of the simply supported beam shown in (a).

EXAMPLE 6.1

Construct the influence lines for the moment and shear at point C of the beam shown in Figure 6.2a.

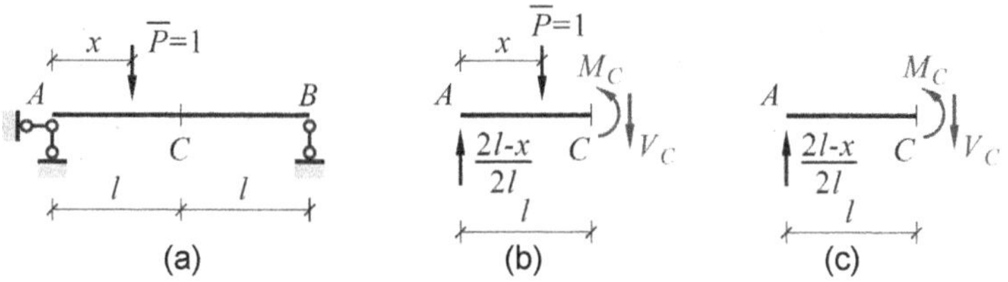

Figure 6.2. (a) A simply supported beam; (b) The free-body diagram of the beam segment *AC*.

***Solution*:** Draw the free-body diagram of the beam segment *AC*, shown in Figures 6.2b and c, which lead to:

$$\begin{cases} M_C = \dfrac{x}{2}, & x \le l \\ M_C = -\dfrac{x}{2} + l, & x > l \end{cases} \tag{6.1}$$

Equation (6.1) are the influence-line equations for the moment M_C, and are plotted in Figure (6.3a). The shear force V_C can be expressed as:

$$\begin{cases} V_C = -\dfrac{x}{2l}, & x \le l \\ V_C = -\dfrac{x}{2l} + 1, & x > l \end{cases} \tag{6.2}$$

These equations are the influence-line equations for shear V_C, and are plotted in Figure 6.3b.

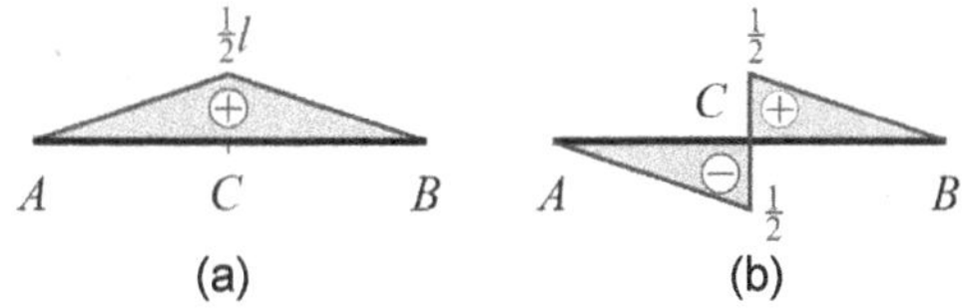

Figure 6.3. Influence line for bending moment M_C (a) and shear V_C (b).

6.2.1 Constructing Influence Lines by the Principle of Virtual Displacements

By first adopting the principle of virtual displacements to draw the influence line for the moment at point *C* of the simply supported beam shown in Figure 6.2a, and then constructing the mechanism shown in Figure 6.4a by inserting a pin at *C* of the simply supported beam, in which M_C is the released moment at *C*, one can obtain the results shown in Figure 6.4b, which gives the ***positive unit virtual displacement*** of the mechanism, namely, the virtual displacement due to $\theta_C = 1$. If $v(x)$ denotes the virtual-displacement component at point x in *a* direction opposite to that of the unit loading $\overline{P}$, then the principle of virtual displacements gives:

$$1 \times M_C(x) - 1 \times v(x) = 0 \tag{6.3}$$

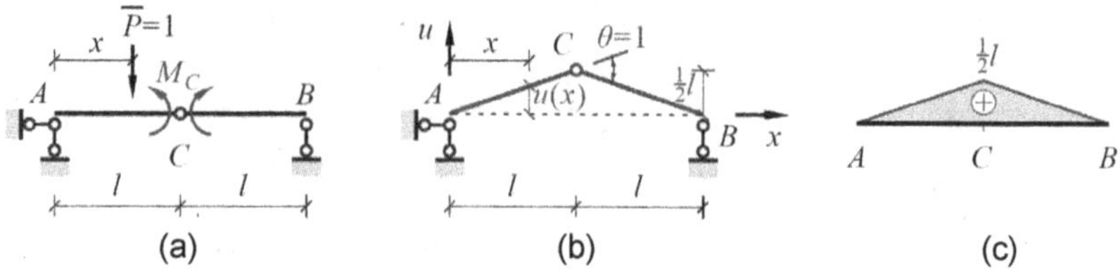

Figure 6.4. (a) Construct a mechanism by inserting a pin at point C; (b) the virtual displacement due to $\theta_C = 1$; (c) virtual displacement $y(x)$.

$$M_C(x) = y(x) \tag{6.4}$$

which implies that the influence line for M_C is identical with the corresponding virtual displacement $y(x)$ plotted in Figure 6.4c.

Roughly speaking, the influence line for an internal force is identical to the unit positive virtual displacement corresponding to the internal force.

EXAMPLE 6.2

Construct the influence line of a multiple span beam under a concentrated load $\overline{P}$, shown in Figure 6.5.

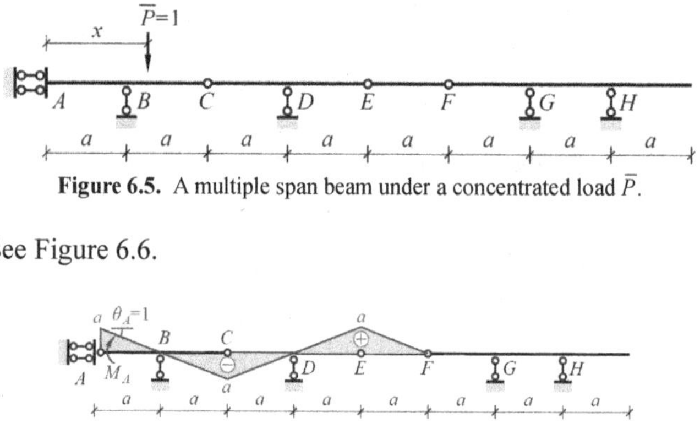

Figure 6.5. A multiple span beam under a concentrated load $\overline{P}$.

Solution: See Figure 6.6.

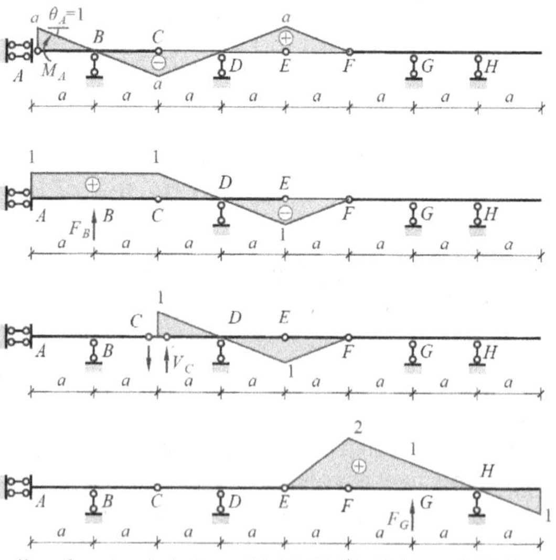

Figure 6.6. Influence lines from top to bottom: $M_A(x)$, $F_B(x)$, $V_C(x)$, and $F_G(x)$ under the concentrated load $\overline{P}$.

EXAMPLE 6.3

Construct the influence line of a simple frame under a concentrated load $\overline{P}$, shown in Figure 6.7.

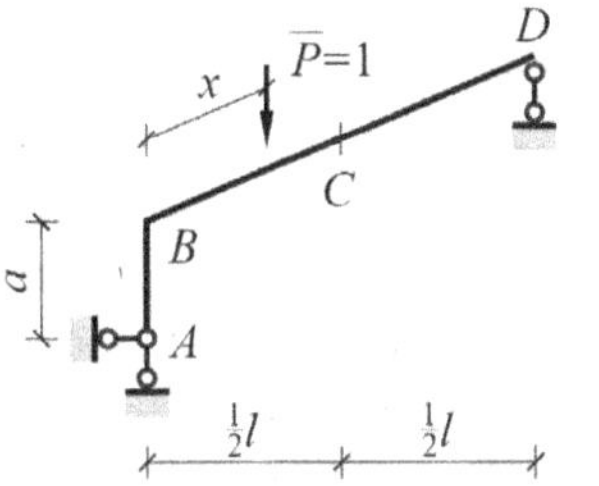

Figure 6.7. A simple frame subject to a concentrated load $\overline{P}$.

Solution: See Figure 6.8.

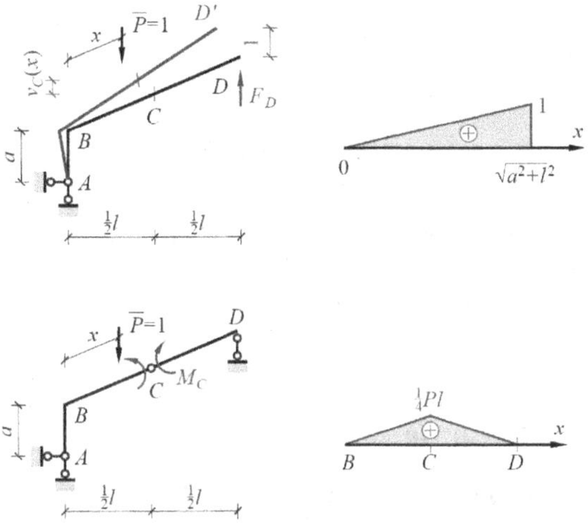

Figure 6.8. Influence lines for $F_D(x)$ (top) and $M_C(x)$ (bottom) under the concentrated load $\overline{P}$.

6.3 Influence Lines for Trusses

Construct the influence for section forces at 1, 2, 3, and 4 subject to a unit load F_p, shown in Figure 6.9.

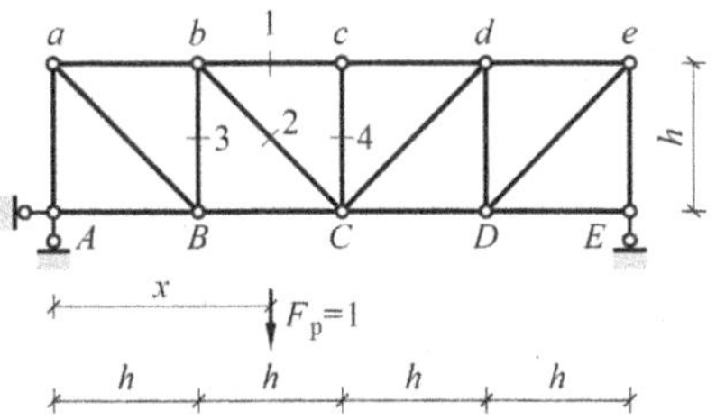

Figure 6.9. A truss structure subject to a unit load of F_p.

Solution: See Figure 6.10, Figure 6.11, Figure 6.12, and Figure 6.13.

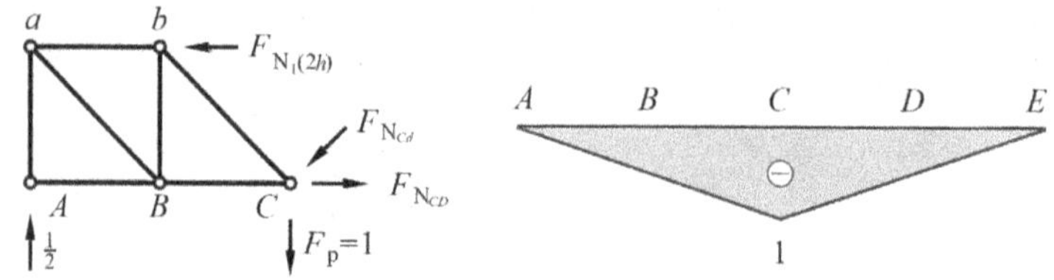

Figure 6.10. $N_{bc}(x)$ (section force at 1).

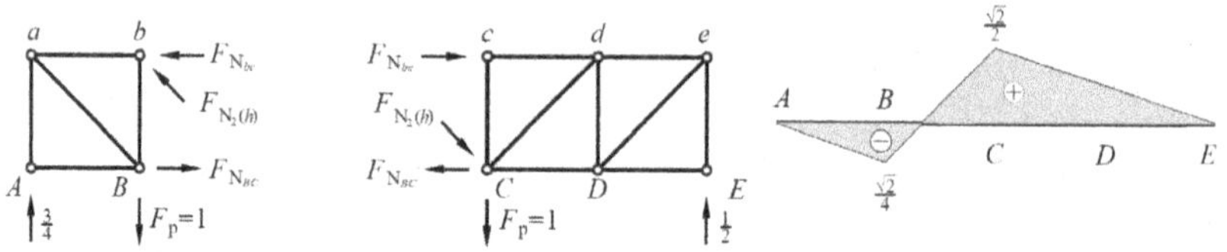

Figure 6.11. $N_{bc}(x)$ (section force at 2).

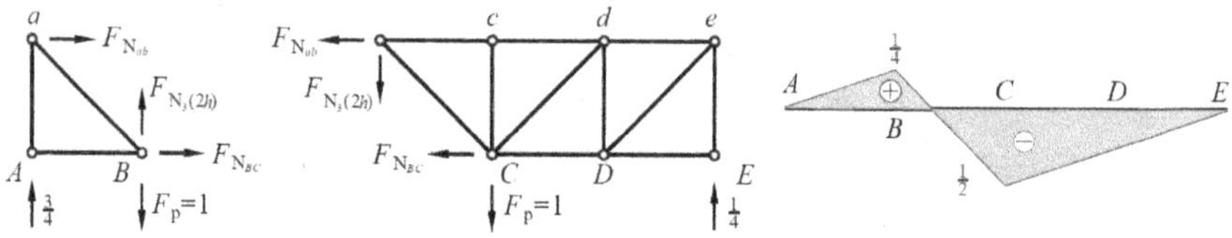

Figure 6.12. $N_{bB}(x)$ (section force at 3).

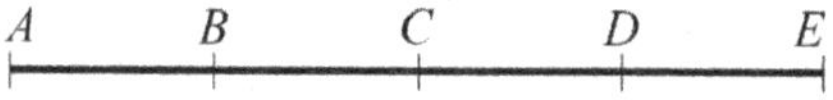

Figure 6.13. $N_{bB}(x)$ (section force at 4 is zero).

6.4 Maximum Response at a Specific Point

6.4.1 Maximum Response at a Point under Live Loads

One can determine the maximum and minimum moments (or the maximum positive and negative moments) at point C under the four loading cases of the beam shown in Figure 6.14.

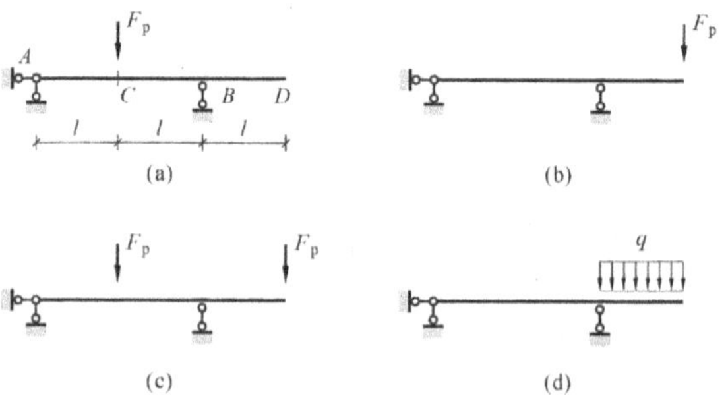

Figure 6.14. Four loading cases subject to concentrated load $P = \dfrac{ql}{4}$.

The influence line is shown in Figure 6.15, which gives the bending moment for four load cases as:

$$\begin{cases} M_C^I = P\times\dfrac{l}{2}=\dfrac{ql^2}{8} \\ M_C^{II} = P\times\left(-\dfrac{l}{2}\right)=-\dfrac{ql^2}{8} \\ M_C^{III} = P\times\dfrac{l}{2}+P\times\left(-\dfrac{l}{2}\right)=0 \\ M_C^{IV} = \int_B^D qM_C(x)\mathrm{d}x = q\int_B^D M_C(x)\mathrm{d}x = -\dfrac{ql^2}{4} \end{cases} \tag{6.5}$$

Thus, the maximum positive and negative moments at point C are $ql^2/8$ and $-ql^2/4$, respectively.

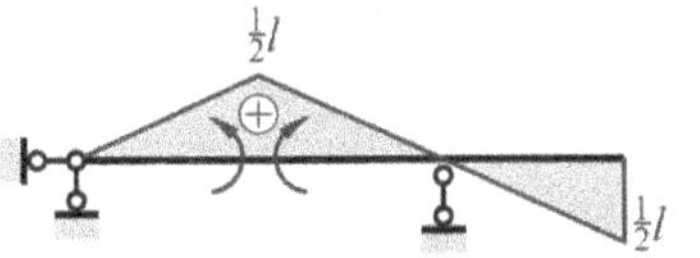

Figure 6.15. The influence line, $M_C(x)$, for the moment at point C.

6.4.2 Maximum Response at a Point under a Set of Concentrated Moving Loads

Theorem 6.1 *The influence caused by a set of concentrated moving loads is monotonic in a neighborhood of x at which none of the concentrated loads is located at an extreme point of the influence line.*

***Proof*:** Let $f(x)$ be the influence line for an effect at a point, and $F(x)$ denote the influence caused by a set of concentrated moving loads, P_i. The influence $F(x)$ can be expressed as:

$$F(x)=\sum_i P_i f(x_i) \tag{6.6}$$

where x_i, $i = 1, 2, ...,$ are the coordinates of loads P_i.

As none of the loads P_i is located at a local extreme point of the influence line, there exists a small increment Δx such that the change ΔF of the influence can be expressed as:

$$\begin{aligned} \Delta F &= F(x+\Delta x)-F(x) \\ &= \sum_i P_i f(x_i+\Delta x)-\sum_i P_i f(x_i) \\ &= \sum_i P_i\left[f(x_i)+k_i\Delta x\right]-\sum_i P_i f(x_i) \\ &= \sum_i k_i P_i \Delta x \end{aligned} \tag{6.7}$$

where k_i, $i = 1, 2, ...,$ are constants. Equation (6.7) shows that the influence ΔF is monotonic in the neighborhood $(x-\Delta x, x+\Delta x)$.

Theorem 6.2 *The influence caused by a set of concentrated moving loads attains a local maximum (or minimum) value if all the concentrated loads are located at local maximum (or minimum) points of the influence line.*

The proof of Theorem 7.2 is left to the reader to perform.

Theorem 6.3 *If the influence caused by a set of concentrated moving loads P_i has a local maximum (or minimum) at a point, then there is at least one concentrated load which is placed at a maximum (or minimum) point of the influence line.*

***Proof*:** There are four possible cases:

1. Case I: none of concentrated loads is located at a local extreme point;
2. Case II: all the loads are located at a local minimum point;
3. Case III: some, $P_{j,\min}$, of concentrated load are located at local minimum points, while none of the other loads, $P_{k,\text{non}}$, is located at a local maximum point;
4. Case IV: there is at least one concentrated load which is placed at a maximum point of the influence line.

Theorem 6.1 shows that in Case I the influence F has no local maximum at x.
Theorem 6.2 shows that in Case II the influence F attains a local minimum at x.

In Case III the influence F can be expressed as

$$F(x) = F_{\min}(x) + F_{\text{non}}(x) \tag{6.8}$$

where

$$F_{\min}(x) \equiv \sum_j P_{j,\min} f(x_j) \quad \text{and} \quad F_{\text{non}}(x) \equiv \sum_j P_{j,\text{non}} f(x_j) \tag{6.9}$$

Theorem 6.2 shows that the sub-influence $F_{\min}$ has a local minimum at x, while Theorem 6.1 shows that the sub-influence F_{non} is monotonic in a neighborhood of x. If F_{non} is increasing in the neighborhood of x, then:

$$F(x + \Delta x) > F(x) \tag{6.10}$$

If F_{non} is decreasing in the neighborhood of x, then:

$$F(x - \Delta x) > F(x) \tag{6.11}$$

Equations (6.10) and (6.11) shows that in Case III the influence F has no maximum at x.

All this shows that if the influence caused by a set of concentrated moving loads P_i attains a local maximum at a point, then there is at least one concentrated load which is placed at a maximum point of the influence line.

Similarly, we prove that if the influence caused by a set of concentrated moving loads P_i has a local minimum at a point, then there is at least one concentrated load which is placed at a minimum point of the influence line. This theorem is true.

Theorem 6.3 tells us the necessary condition that a local maximum (or minimum) influence will occur if atleast one concentrated load is placed at a maximum (or minimum) point of the influence line.

EXAMPLE 6.5

Determine the absolute maximum moment and shear at point C of the beam shown in Figure 6.16.

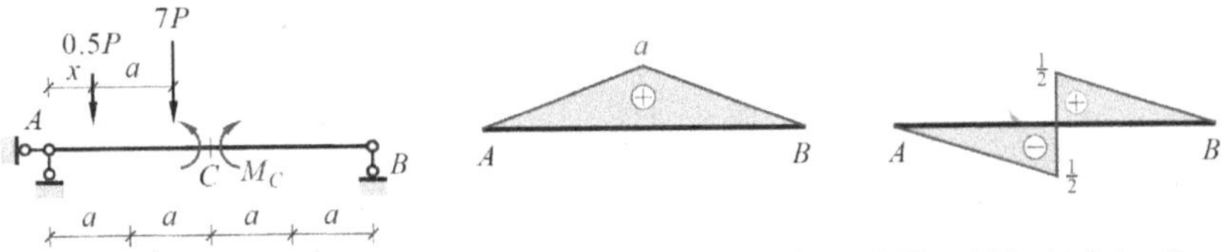

Figure 6.16. A simply supported beam subject to two concentrated forces 0.5P and 7P with $0 \leq x \leq 3a$, and the influence line of bending moment (middle) and shear force (right) at point C.

***Solution*:** (1) The absolute maximum moment $M_{C,\max}$ at point C.

$$M_{C,\max} = 0.5 \times P \times \frac{a}{2} + 7P \times a = 7.25Pa \tag{6.12}$$

(2) The absolute maximum shear $V_{C,\max}$ at point C.

Consider the cases, $x = 0$, a^-, a^+, $2a^-$, $2a^+$, $3a$, in which $0 \leq x \leq 3a$ are the ends of the domain of the shear $V_C(x)$, and $x = a^-$, a^+, $2a^-$, $2a^+$ are the possible extreme points of $V_C(x)$. The reason is given as follows:

$$\begin{aligned}
&\left|V_C\left(a^-\right)\right| > \left|V_C\left(a^+\right)\right| \\
&\left|V_C\left(a^-\right)\right| > \left|V_C\left(0\right)\right| \\
&\left|V_C\left(2a^+\right)\right| > \left|V_C\left(2a^-\right)\right| \\
&\left|V_C\left(2a^+\right)\right| > \left|V_C\left(3a\right)\right|
\end{aligned} \tag{6.13}$$

And,

$$V_C\left(a^-\right) = -0.5P \times \frac{1}{4} - 7P \times \frac{1}{2} = -3.625P$$

$$V_C\left(2a^+\right) = 0.5P \times \frac{1}{2} + 7P \times \frac{1}{4} = 2P$$

Thus, the absolute maximum shear at point C is: $V_C(a^-) = -3.625P$.

6.5 Moment Envelopes and Absolute Maximum Moments of Members

6.5.1 Definition of Moment Envelope

A simply supported beam shown in Figure 6.17 is subjected to the three live loads shown in Figure 6.18.

The ***moment envelope*** of a member is the boundary of the region filled by the moment diagrams of the member under all the possible loading cases, as given in Figure 6.19. From another aspect, the moment envelope of a member is the collection of all the points of the maximum (positive and negative) moments at each section.

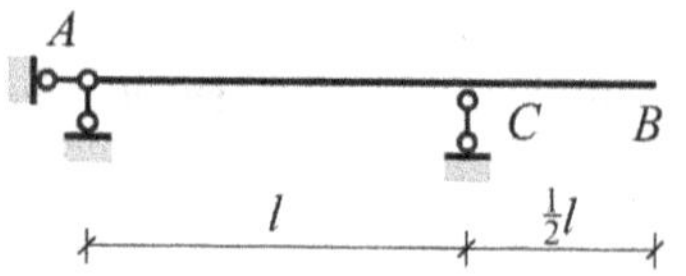

Figure 6.17. A simply supported beam.

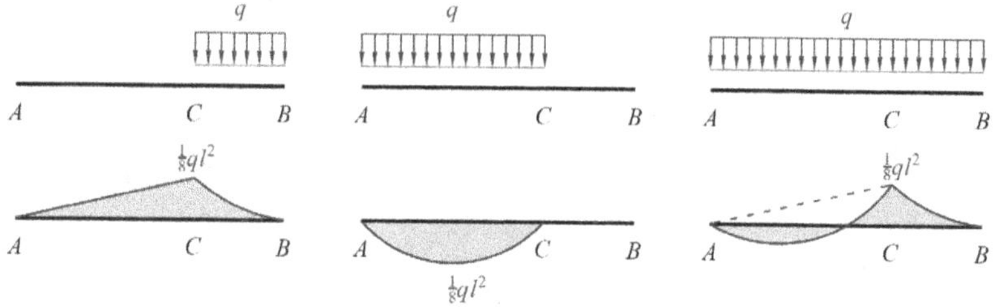

Figure 6.18. Three load scenarios (upper) and the corresponding moment diagrams (lower).

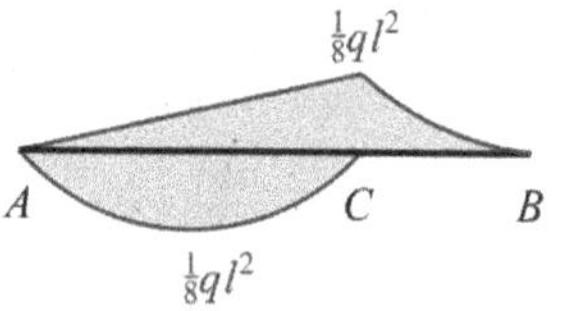

Figure 6.19. The moment envelope for all load scenarios in Figure 6.18.

6.5.2 *Moment Envelopes of Beams under Moving Loads*

Now let's construct the moment envelope of the simply supported beam subjected to a single moving load P, as shown in Figure 6.20a.

The influence line for the moment at point x, $0 \le x \le 2l$ is plotted in Figure 6.21a, expressed as:

$$\begin{cases} M^{+}_{x,\max} = \dfrac{Px(2l-x)}{2l} \\ M^{-}_{x,\max} = \dfrac{Px}{2} \end{cases} \tag{6.14}$$

for point $x(0 \le x \le 2l)$, The influence line for the moment at point x, $2l < x < 3l$ is plotted in Figure 6.21b, which shows that:

$$\begin{cases} M^{+}_{x,\max} = 0 \\ M^{-}_{x,\max} = P(3l-x) \end{cases} \tag{6.15}$$

Equations (6.14) and (6.15) lead to the moment envelop shown in Figure 6.20b, which shows that (i) the *absolute maximum positive moment of the beam* is $pl/2$; (ii) the *absolute maximum negative moment of the beam* is Pl; (iii) the *absolute maximum moment of the beam* is Pl.

Construct the moment envelop of the simply supported beam subjected to the two moving loads shown in Figure 6.22a. The beam is divided into n segments as

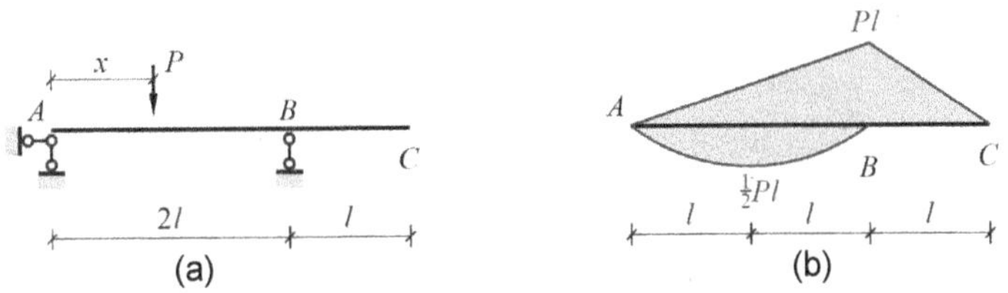

Figure 6.20. (a) $0 \leq x \leq 3l$. (b) The moment envelop of the beam under the moving load P.

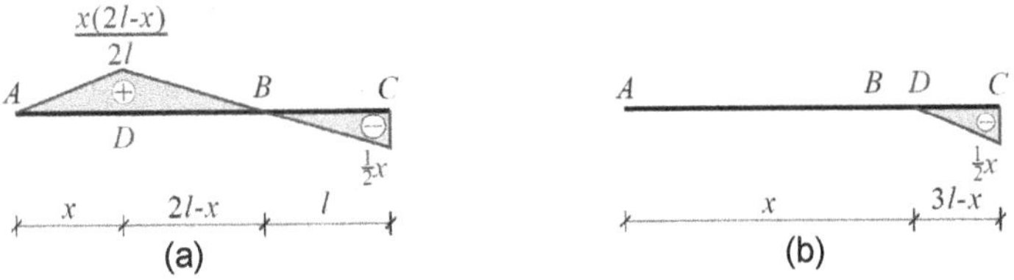

Figure 6.21. (a) The influence line for the moment at point x, $0 \leq x \leq 2l$. (b) The influence line for the moment at point x, $2l < x < 3l$.

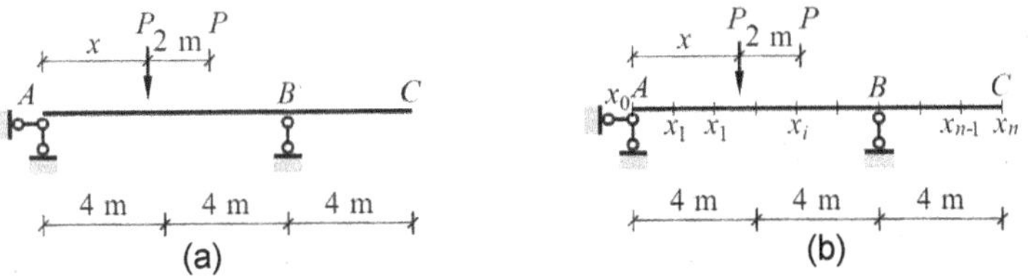

Figure 6.22. (a) A simply supported beam subjected to the two moving loads, where $0 \leq x \leq 10$m; (b) The beam in (a) is divided into n segments.

shown in Figure 6.22b. By using the influence line for the moment at x_i to determine the absolute maximum moments at x_i, one can obtain the moment envelope of the beam. Specifically, we draw the influence lines for the moment at x_i, shown in Figure 6.23. The maximum positive moment at x_i can be expressed as:

$$M^{+}_{x_i,\max} = \begin{cases} \dfrac{Px_i(14-2x_i)}{8}, & \text{when} \quad 0 \leq x_i \leq 4 \\ \dfrac{P(x_i-1)(8-x_i)}{4}, & \text{when} \quad 4 < x_i \leq 8 \\ 0, & \text{when} \quad 8 < x_i \leq 12 \end{cases} \tag{6.16}$$

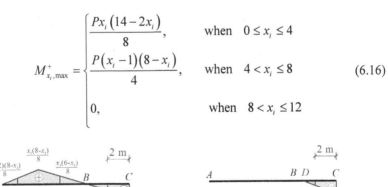

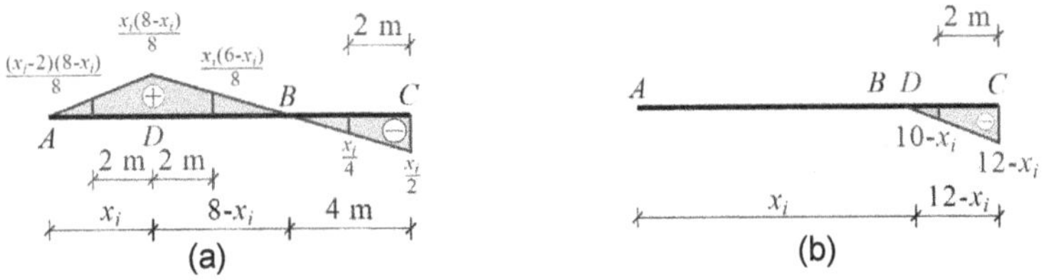

Figure 6.23. Influence lines for the bending moment at x_i, where (a) $0 \leq x_i \leq 8$. (b) $8 < x_i \leq 10$.

and the maximum negative moment at x_i can be expressed as:

$$M^{+}_{x_i,\max} = \begin{cases} \dfrac{3Px_i}{4}, & \text{when } 0 \le x_i \le 8 \\ -P(22-2x_i), & \text{when } 8 < x_i \le 10 \\ -P(12-x_i), & \text{when } 10 < x_i \le 12 \end{cases} \tag{6.17}$$

Then one obtains the moment envelope as shown in Figure 6.24.

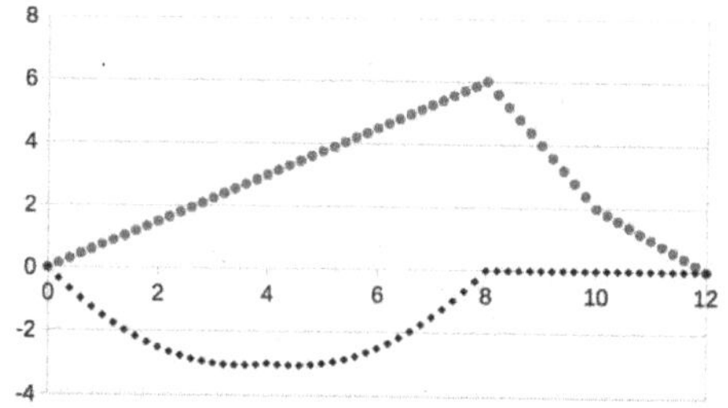

Figure 6.24. The bending moment envelope.

EXERCISES

6.1 Construct the influence lines for the moment at point A, the shear at joint C, the shear at the left of support E, and the reaction from support G of the beam shown in Figure 6.25a. Use the influence line for the moment at point A to determine the absolute maximum moment at point A of the beam under the three load cases shown in Figure 6.25b–d.

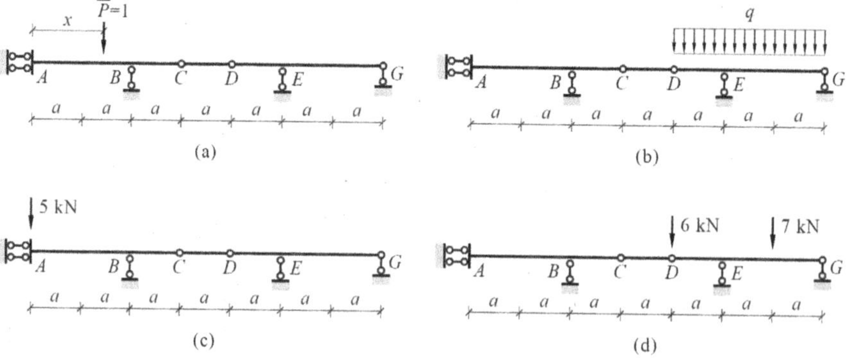

Figure 6.25. Exercise 6.1. (a) A beam is subjected to a unit moving load. (b) Load case I. (c) Load case II. (d) Load case III.

6.2 Draw the influence line for the beam shown in Figure 6.26.

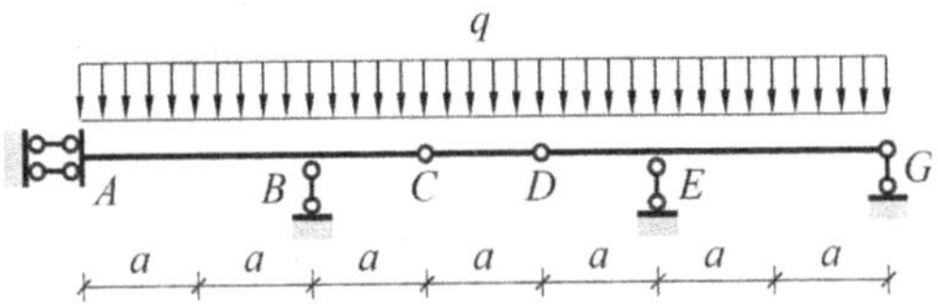

Figure 6.26. Exercise 6.2.

Chapter 7

Matrix Displacement Analysis

7.1 An Introductory Example

Consider the continuous beam shown in Figure 7.1.

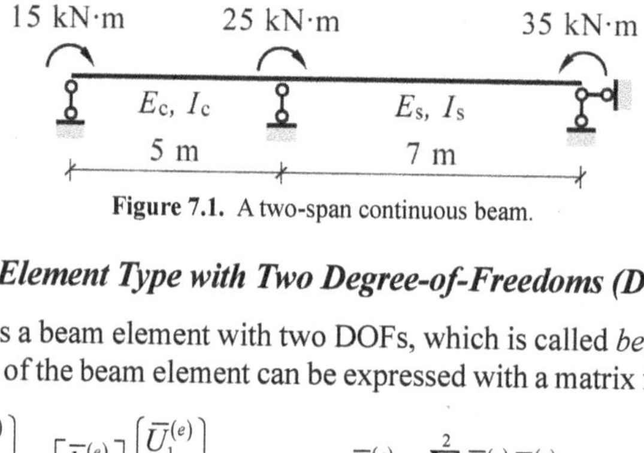

Figure 7.1. A two-span continuous beam.

7.1.1 A Beam Element Type with Two Degree-of-Freedoms (DOFs) – Beam1

Figure 7.2 shows a beam element with two DOFs, which is called *beam1*. The slope-deflect equation of the beam element can be expressed with a matrix form as follows:

$$\begin{Bmatrix} \overline{F}_1^{(e)} \\ \overline{F}_2^{(e)} \end{Bmatrix} = \left[\overline{k}^{(e)}\right] \begin{Bmatrix} \overline{U}_1^{(e)} \\ \overline{U}_2^{(e)} \end{Bmatrix} \quad or \quad \overline{F}_i^{(e)} = \sum_{j=1}^{2} \overline{k}_{ij}^{(e)} \overline{U}_j^{(e)}, \quad i = 1,2 \tag{7.1}$$

where,

$$\left[\overline{k}^{(e)}\right] = \begin{bmatrix} \overline{k}_{11}^{(e)} & \overline{k}_{12}^{(e)} \\ \overline{k}_{21}^{(e)} & \overline{k}_{22}^{(e)} \end{bmatrix} = \begin{bmatrix} 4i_e & 2i_e \\ 2i_e & 4i_e \end{bmatrix} \tag{7.2}$$

is called the ***element stiffness matrix***, and it is the digital representation of the slope-deflection equation of the beam element.

Table 7.1 is a digital representation, Beam1, of the beam element shown in Figure 7.2.

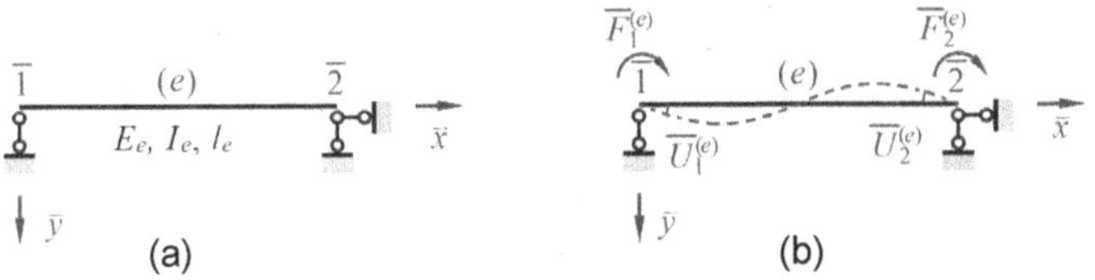

Figure 7.2. A digital beam element with two DOFs. (a) The geometry, node locations, and the element coordinate system for the beam element. (b) The end forces and displacements of the beam element.

Table 7.1. The digital definition of *Beam1*.

Nodal ID	Coordinates		Properties		End Forces	Element DOFs	Stiffness Eq.
	$\bar{x}$ (m)	$\bar{y}$ (m)	E (N/m²)	I_z (m²)	$\bar{F}_i^{(e)}$ (N · m)	$\bar{U}_i^{(e)}$ (rad)	
$\bar{1}$	0	0	E_e	I_e	$\bar{F}_1^{(e)}$	$\bar{U}_1^{(e)}$	Eq. (7.1)
$\bar{2}$	l_e	0			$\bar{F}_2^{(e)}$	$\bar{U}_2^{(e)}$	

Note: The end forces and end displacement, $\bar{F}_i^{(e)}$, $\bar{U}_i^{(e)}$, are positive if they are in the positive direction of $\bar{z}$ axis, which is toward the page.

7.1.2 Pre-processing – Discretizing and Digitizing of the Continuous Beam

The first step to analyze a structure by the matrix structural analysis is the ***pre-processing*** of the structure, whose goal is to develop a digital counterpart of the structure in the computer.

Before a structure can be digitized, it has to be discretized first. As shown in Figure 7.3a, the continuous beam can be discretized by three nodes 1, 2, 3, and can be described by three known nodal rotations, U_1, U_2, U_3. To digitize the structure, we define a ***global coordinate system*** to locate the nodes, and label all the nodes and elements with code-numbers, see Figure 7.3b.

Table 7.1 and Table 7.2 give a digital representation of the continuous beam shown in Figure 7.3.

In the next section, a method to formulate the stiffness equation of the continuous beam by using the data in Tables 7.1 and 7.2 will be presented.

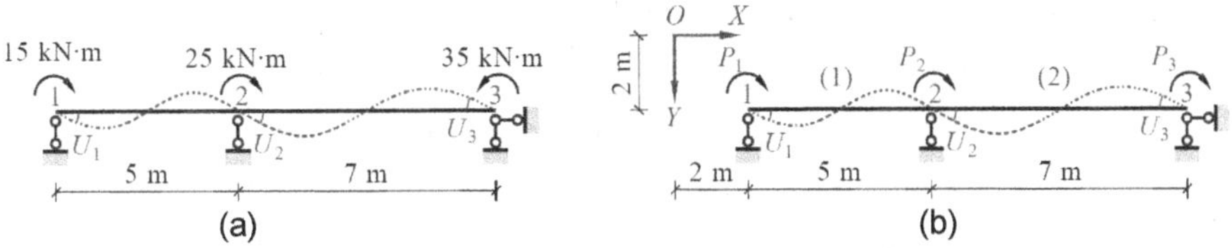

Figure 7.3. Pre-processing: (a) The continuous beam is discretized by three nodes. (b) The discretized structure is digitized by defining global coordinates *X*, *Y*, *Z*.

Table 7.2. The digital representation of the continuous beam in Figure 7.3b.

Nodal ID	Coordinates		Nodal Loads	Unknown DOFs	
	x(m)	y(m)	P_i (N · m)	(rad)	
1	2.0	2.0	15×10^3	U_1	
2	7.0	2.0	25×10^3	U_2	
3	14.0	2.0	-35×10^3	U_3	
Element ID	Nodes		Young's Modulus	Moment of Inertia	Element Type
	$\bar{1}$	$\bar{2}$	E (N²/m)	I_z (m⁴)	
(1)	1	2	3.00×10^{10}	1.440×10^{-5}	*Beam1*
(2)	2	3	2.06×10^{11}	1.152×10^{-5}	*Beam1*

Note: The nodal torques P_i and the nodal rotations U_i are positive if they are in the positive direction of the Z axis, which is toward the page.

7.1.3 Calculate the Element Stiffness Matrices of the Continuous Beam

Table 7.1 and Table 7.2 yield:

$$\boldsymbol{k}^{(1)} = 10^5 \times \begin{bmatrix} 3.456 & 1.728 \\ 1.728 & 3.456 \end{bmatrix} \tag{7.3}$$

and

$$\boldsymbol{k}^{(2)} = 10^5 \times \begin{bmatrix} 13.561 & 6.7803 \\ 6.7803 & 13.561 \end{bmatrix} \tag{7.4}$$

7.1.4 Assembling Stiffness Equation of the Structure by the Direct Stiffness Method

This section shall introduce a method, known as the ***direct stiffness method***, to formulate the stiffness equation of the continuous beam from the data listed in Tables 7.1 and 7.2.

Figure 7.4 shows the free-body diagrams of the members of the continuous beam. The moment equilibrium requirements at the three nodes give:

$$\begin{Bmatrix} P_1 \\ P_2 \\ P_3 \end{Bmatrix} = \begin{Bmatrix} F_{\bar{1}}^{(1)} \\ F_{\bar{2}}^{(1)} \\ 0 \end{Bmatrix} + \begin{Bmatrix} 0 \\ F_{\bar{1}}^{(2)} \\ F_{\bar{2}}^{(2)} \end{Bmatrix} \tag{7.5}$$

In general, one has:

$$\boldsymbol{P} = \sum_e \boldsymbol{F}^{(e)} \tag{7.6}$$

where $\boldsymbol{P}$ is the load vector, $\boldsymbol{F}^{(e)} = \left[F_1^{(e)}, \cdots, F_i^{(e)}, \cdots \right]^{\mathrm{T}}$, and $F_i^{(e)}$ is the contribution to the ith equilibrium equation from element (e) due to the structural nodal deformation. From Table 7.2, one can obtain the following ***element-node connectivity***:

$$(1) \begin{cases} \bar{1} \to 1 \\ \bar{2} \to 2 \end{cases}; \quad (2) \begin{cases} \bar{1} \to 2 \\ \bar{2} \to 3 \end{cases} \tag{7.7}$$

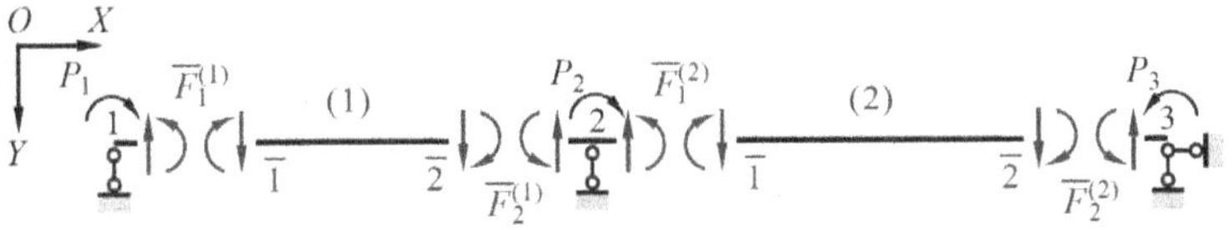

Figure 7.4. The free-body diagrams of the members of the continuous beam.

or the ***element connectivity matrix*** l as follows:

$$\ell = \begin{bmatrix} 1 & 2 \\ 2 & 3 \end{bmatrix} \tag{7.8}$$

where l_{ij} is the structural nodal ID of the jth node of the ith element.

The element-node connectivity immediately leads to the following ***element-DOF connectivity***:

$$\begin{cases} \bar{U}_1^{(1)} = U_1 \\ \bar{U}_2^{(1)} = U_2 \end{cases} \tag{7.9}$$

and

$$\begin{cases} \bar{U}_1^{(2)} = U_2 \\ \bar{U}_2^{(2)} = U_3 \end{cases} \tag{7.10}$$

By using Equation (7.9), the stiffness equation of element (1) can be rewritten as:

$$\begin{bmatrix} \bar{k}_{11}^{(1)} & \bar{k}_{12}^{(1)} & 0 \\ \bar{k}_{21}^{(1)} & \bar{k}_{22}^{(1)} & 0 \\ 0 & 0 & 0 \end{bmatrix} \begin{Bmatrix} U_1 \\ U_2 \\ U_3 \end{Bmatrix} = \begin{Bmatrix} \bar{F}_1^{(1)} \\ \bar{F}_2^{(1)} \\ 0 \end{Bmatrix} = \boldsymbol{F}^{(1)} \tag{7.11}$$

which is the stiffness equation of element (1) in terms of the *structural* DOFs. Similarly, by using Equation (7.10), the stiffness equation of element (2) in terms of the structural DOFs can be expressed as:

$$\begin{bmatrix} 0 & 0 & 0 \\ 0 & \bar{k}_{11}^{(2)} & \bar{k}_{12}^{(2)} \\ 0 & \bar{k}_{21}^{(2)} & \bar{k}_{22}^{(2)} \end{bmatrix} \begin{Bmatrix} U_1 \\ U_2 \\ U_3 \end{Bmatrix} = \begin{Bmatrix} 0 \\ \bar{F}_1^{(2)} \\ \bar{F}_2^{(2)} \end{Bmatrix} = \boldsymbol{F}^{(2)} \tag{7.12}$$

Substituting Equations (7.11) and (7.12) into Equation (7.5) gives:

$$\begin{bmatrix} \bar{k}_{11}^{(1)} & \bar{k}_{12}^{(1)} & 0 \\ \bar{k}_{21}^{(1)} & \bar{k}_{22}^{(1)} + \bar{k}_{11}^{(2)} & \bar{k}_{12}^{(2)} \\ 0 & \bar{k}_{21}^{(2)} & \bar{k}_{22}^{(2)} \end{bmatrix} \begin{Bmatrix} U_1 \\ U_2 \\ U_3 \end{Bmatrix} = \begin{Bmatrix} P_1 \\ P_2 \\ P_3 \end{Bmatrix} \tag{7.13}$$

which can be symbolically expressed as:

$$\boldsymbol{KU} = \boldsymbol{P} \tag{7.14}$$

where

$$\boldsymbol{U} = [U_1 \ \ U_2 \ \ U_3]^{\mathrm{T}} \tag{7.15}$$

$$\boldsymbol{P} = [P_1 \ \ P_2 \ \ P_3]^{\mathrm{T}} \tag{7.16}$$

are called the ***nodal displacement and load vectors*** of the structure respectively. Also,

$$\boldsymbol{K} = \begin{bmatrix} K_{11} & K_{12} & K_{13} \\ K_{21} & K_{22} & K_{23} \\ K_{31} & K_{32} & K_{33} \end{bmatrix} \tag{7.17}$$

is referred to as the ***global stiffness matrix*** or the stiffness matrix of the structure.

The procedure above to form the stiffness matrix $\boldsymbol{K}$ can be simplified into the following steps:

1. Form the element connectivity matrix from the element digital information.
2. Assemble the element stiffness matrices into the global stiffness matrix according to the element connectivity matrix. For example, the entries of $\overline{\boldsymbol{k}}^{(1)}$ will be added to rows 1, 2 and columns 1, 2 of $\boldsymbol{K}$, while the entries of $\overline{\boldsymbol{k}}^{(2)}$ will be added to rows 2, 3 and columns 2, 3 of $\boldsymbol{K}$. That is:

$$\begin{aligned} &\bar{k}_{11}^{(1)} \to K_{11}; \quad \bar{k}_{12}^{(1)} \to K_{12}; \\ &\bar{k}_{21}^{(1)} \to K_{21}; \quad \bar{k}_{22}^{(1)} \to K_{22}; \end{aligned} \tag{7.18}$$

and

$$\begin{aligned} &\bar{k}_{11}^{(2)} \to K_{22}; \quad \bar{k}_{12}^{(2)} \to K_{23}; \\ &\bar{k}_{21}^{(2)} \to K_{32}; \quad \bar{k}_{22}^{(2)} \to K_{33}; \end{aligned} \tag{7.19}$$

Then, one has Equation (7.13) or

$$10^5 \times \begin{bmatrix} 3.4560 & 1.7280 & 0 \\ 1.7280 & 17.017 & 6.7803 \\ 0 & 6.7803 & 13.5610 \end{bmatrix} \begin{Bmatrix} U_1 \\ U_2 \\ U_3 \end{Bmatrix} = \begin{Bmatrix} P_1 \\ P_2 \\ P_3 \end{Bmatrix} \tag{7.20}$$

This process shown above, called ***direct stiffness method***, is rather simple, and can be coded in a computer program very conveniently.

7.1.5 Solving and Post-processing

Solving Equation (7.20) yields the nodal displacements of the structure as follows.

$$\boldsymbol{U} = 10^{-2} \times \left[2.9691 \quad 2.7424 \quad -3.9522\right]^{\mathrm{T}} \ \ (\text{rad.}) \tag{7.21}$$

From $\boldsymbol{U}$, one can obtain the end displacements of element (e) by the following equations:

$$\begin{cases} \overline{U}_1^{(e)} = U_{\ell(e,1)} \\ \overline{U}_2^{(e)} = U_{\ell(e,2)} \end{cases}, \quad e = 1, 2 \tag{7.22}$$

where $l(e, i) \equiv l_{ei}$ is the structural node corresponding to the ith node of element (e). Then one has:

$$\overline{F}^{(1)} = 10^4 \times \begin{Bmatrix} 1.5000 \\ 1.4608 \end{Bmatrix}, \quad \text{and} \quad \overline{F}^{(2)} = 10^4 \times \begin{Bmatrix} 1.0392 \\ -3.5000 \end{Bmatrix} \tag{7.23}$$

by which one can draw the moment diagram of the continuous beam as shown in Figure 7.5.

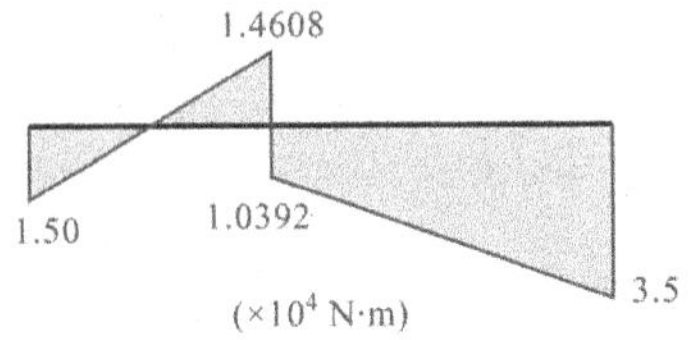

Figure 7.5. The moment diagram of the continuous beam shown in Figure 7.1.

7.1.6 MATLAB Codes – Beam1 Package

Input data for the continuous beam shown in Figure 7.1.

```
3                          % N node
2.0, 2.0, 15e3             % x, y, M z
7.0, 2.0, 25e3             % . . .
14.0, 2.0, -35e3
2                          % N e
1, 2, 3.00e10, 1.440e-5    % i, j, Ex, I z
2, 3, 2.06e11, 1.152e-5    % . . .
```

Result:

```
Nodal displacements =
2.9691e-02
2.7424e-02
–3.9522e-02
End forces =
1.5000e+04 1.0392e+04
1.4608e+04 –3.5000e+04
```

7.2 Boundary Conditions and the Beam Element with Six DOFs

7.2.1 Analyzing an Unrestrained Continuous Beam by Beam2 Element

An elastic unrestrained beam shown in Figure 7.6 is subjected to a set of external forces in equilibrium. The material's modulus of elasticity, the beam's moment of inertia and the cross-sectional area of each member are $E_1 = 3.0 \times 10^{10}$ N/m^2, $E_2 = 2.06 \times 10^{11}$ N/m^2, $A_1 = 0.020$ m^2, $A_2 = 0.015$ m^2, $I_1 = 6.6667 \times 10^{-5}$m^4, and $I_2 = 2.8125 \times 10^{-5}$m^4. This section develops a new type of beam element – *Beam2* – and applies the direct stiffness method to calculate the relative rotation between the two ends of the free continuous beam.

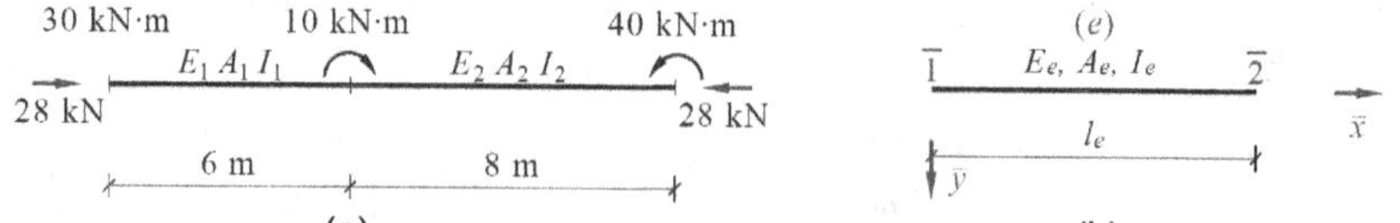

Figure 7.6. (a) An unconstrained continuous beam. (b) The Geometry of Beam2 element.

Developing a *Beam2* element type. To analyze this unrestrained continuous beam, we have to develop a new type of beam element with six DOFs, called ***Beam2***, see Figure 7.7(b). The end forces and end displacements of the beam are illustrated in Figure 7.7. The stiffness equation of such a beam element with six DOFs can be expressed as:

$$\begin{Bmatrix} \bar{F}_{1x}^{(e)} \\ \bar{F}_{1y}^{(e)} \\ \bar{M}_{1}^{(e)} \\ \bar{F}_{2x}^{(e)} \\ \bar{F}_{2y}^{(e)} \\ \bar{M}_{2}^{(e)} \end{Bmatrix} = \left[\begin{array}{ccc|ccc} \frac{E_e A_e}{l_e} & 0 & 0 & -\frac{E_e A_e}{l_e} & 0 & 0 \\ 0 & \frac{12 i_e}{l_e^2} & \frac{6 i_e}{l_e} & 0 & -\frac{12 i_e}{l_e^2} & \frac{6 i_e}{l_e} \\ 0 & \frac{6 i_e}{l_e} & 4 i_e & 0 & -\frac{6 i_e}{l_e} & 2 i_e \\ \hline -\frac{E_e A_e}{l_e} & 0 & 0 & \frac{E_e A_e}{l_e} & 0 & 0 \\ 0 & -\frac{12 i_e}{l_e^2} & -\frac{6 i_e}{l_e} & 0 & \frac{12 i_e}{l_e^2} & -\frac{6 i_e}{l_e} \\ 0 & \frac{6 i_e}{l_e} & 2 i_e & 0 & -\frac{6 i_e}{l_e} & 2 i_e \end{array}\right] \begin{Bmatrix} \bar{U}_{1x}^{(e)} \\ \bar{U}_{1y}^{(e)} \\ \bar{\theta}_{1}^{(e)} \\ \bar{U}_{2x}^{(e)} \\ \bar{U}_{2y}^{(e)} \\ \bar{\theta}_{2}^{(e)} \end{Bmatrix} \quad (7.24)$$

or symbolically written as,

$$\bar{\boldsymbol{k}}^{(e)}\bar{\boldsymbol{U}}^{(e)} = \bar{\boldsymbol{F}}^{(e)} \quad \text{or} \quad \begin{bmatrix} \bar{\boldsymbol{k}}_{11}^{(e)} & \bar{\boldsymbol{k}}_{12}^{(e)} \\ \bar{\boldsymbol{k}}_{21}^{(e)} & \bar{\boldsymbol{k}}_{22}^{(e)} \end{bmatrix} \begin{Bmatrix} \bar{\boldsymbol{U}}_{1}^{(e)} \\ \bar{\boldsymbol{U}}_{2}^{(e)} \end{Bmatrix} = \begin{Bmatrix} \bar{\boldsymbol{F}}_{1}^{(e)} \\ \bar{\boldsymbol{F}}_{2}^{(e)} \end{Bmatrix} \quad (7.25)$$

where $\bar{\boldsymbol{U}}^{(e)}$ and $\bar{\boldsymbol{F}}^{(e)}$ can be alternatively expressed as:

$$\bar{\boldsymbol{U}}^{(e)} = \left[\bar{U}_1^{(e)}, \cdots, \bar{U}_6^{(e)}\right]^{\mathrm{T}} \quad (7.26)$$

$$\bar{\boldsymbol{F}}^{(e)} = \left[\bar{F}_1^{(e)}, \cdots, \bar{F}_6^{(e)}\right]^{\mathrm{T}} \quad (7.27)$$

$\bar{F}_i^{(e)}$ is the end force at the ith DOF, $\bar{U}_i^{(e)}$, of *Beam2* element. Table 7.3 is the digital representation of *Beam2* element type.

Pre-processing of the unconstrained beam. As shown in Figure 7.8, three nodes, 1, 2, 3, are created to divide the continuous beam into two elements, thus global coordinates and a coding system to digitize the unconstrained continuous beam are established. Table 7.4 is the digital counterpart of the unconstrained continuous beam.

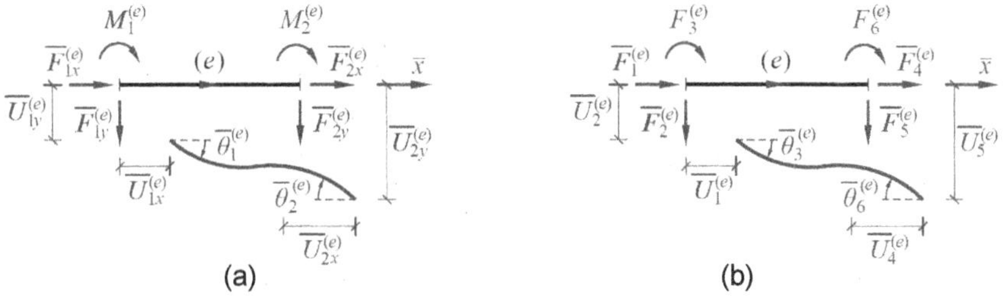

Figure 7.7. (a) The end forces and displacements of a Beam2 element. (b) The notations of end forces and end displacements used in computer programs.

Table 7.3. The digital definition of *Beam2*.

Nodes	Coordinates		Properties			End Forces	Elements DOFs	Stiffness Eq.
	$\bar{x}$ (m)	$\bar{y}$ (m)	E (N/m^2)	A_c (m^2)	I_z (m^4)	$\bar{F}_i^{(e)}$ (N · m)	$\bar{U}_i^{(e)}$ (rad)	
$\bar{1}$	0	0	E_e	A_e	I_e	$\bar{\boldsymbol{F}}_1^{(e)}$	$\bar{\boldsymbol{U}}_1^{(e)}$	Eq. (7.24)
$\bar{2}$	l_e	0				$\bar{\boldsymbol{F}}_2^{(e)}$	$\bar{\boldsymbol{U}}_2^{(e)}$	

Note: The end forces and end displacement, $\bar{F}_i^{(e)}$, $\bar{U}_i^{(e)}$, are positive if they are in the positive direction of the $\bar{z}$ axis, which is into the page.

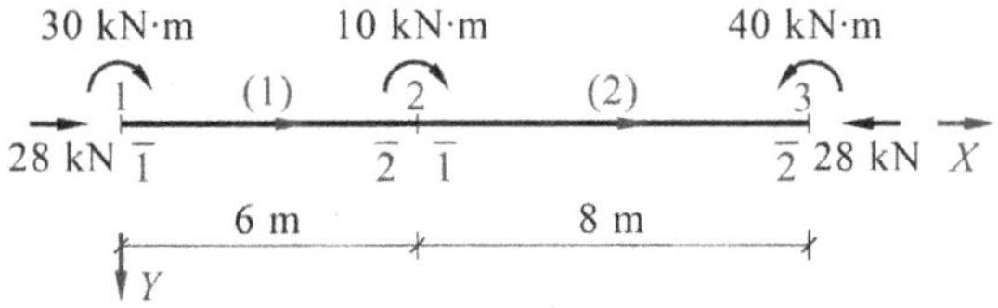

Figure 7.8. The coordinates and coding system used.

Table 7.4. The digital definition of the unconstrained continuous beam.

Nodal ID	Coordinates			Nodal Loads		DOFs
	x(m)	y(m)	P_{ix} (N)	P_{iy} (N · m)	M_i (N · m)	U_i
1	0.0	0.0	28×10^3	0.0	30×10^3	$U_1 = [U_1\ U_2\ U_3]^T$
2	6.0	0.0	0.0	0.0	10×10^3	$U_2 = [U_4\ U_5\ U_6]^T$
3	14.0	0.0	-28×10^3	0.0	-40×10^3	$U_3 = [U_7\ U_8\ U_9]^T$
Element ID	Nodes		Young's Modulus	Section Area	Moment of Inertia	Element Type
	$\bar{1}$	$\bar{2}$	E (N^2/m)	A_c (m^2)	I_z (m^4)	
(1)	1	2	3.00×10^{10}	0.020	6.6667×10^{-5}	*Beam2*
(2)	2	3	2.06×10^{11}	0.015	2.8125×10^{-5}	*Beam2*

Note: The nodal forces and displacements are positive if they are in the positive directions of the axes.

Calculating element stiffness matrices. By using Equation (7.24), one can obtain the stiffness matrices of elements (1) and (2) as follows:

$$\overline{\boldsymbol{k}}^{(1)} = 10^5 \times \begin{bmatrix} 1000 & 0 & 0 & -1000 & 0 & 0 \\ 0 & 1.1111 & 3.3334 & 0 & -1.1111 & 3.3334 \\ 0 & 3.3334 & 1.3333 & 0 & -3.3334 & 6.6667 \\ -1000 & 0 & 0 & 1000 & 0 & 0 \\ 0 & -1.1111 & -3.3334 & 0 & 1.1111 & -3.3334 \\ 0 & 3.3334 & 6.6667 & 0 & -3.3334 & 1.3333 \end{bmatrix} \quad (7.28)$$

and

$$\overline{\boldsymbol{k}}^{(2)} = 10^5 \times \begin{bmatrix} 3862.5 & 0 & 0 & -3862.5 & 0 & 0 \\ 0 & 1.3579 & 5.4316 & 0 & -1.3579 & 5.4316 \\ 0 & 5.4316 & 28.969 & 0 & -5.4316 & 14.484 \\ -3862.5 & 0 & 0 & 3862.5 & 0 & 0 \\ 0 & -1.3579 & -5.4316 & 0 & 1.3579 & -5.4316 \\ 0 & 5.4316 & 14.484 & 0 & -5.4316 & 28.969 \end{bmatrix} \quad (7.29)$$

Table 7.4 shows that the element connectivity is:

$$\ell = \begin{bmatrix} 1 & 2 \\ 3 & 4 \end{bmatrix} \quad \text{or} \quad (1)\begin{cases} \overline{1} \to 1 \\ \overline{2} \to 2 \end{cases}; \quad (2)\begin{cases} \overline{1} \to 2 \\ \overline{2} \to 3 \end{cases} \quad (7.30)$$

Forming the global stiffness matrix. Assembling the element stiffness matrices into the global stiffness matrix gives:

$$\boldsymbol{K}^{\text{fr}} \equiv \begin{bmatrix} \boldsymbol{K}_{11}^{\text{fr}} & \boldsymbol{K}_{12}^{\text{fr}} & \boldsymbol{K}_{13}^{\text{fr}} \\ \boldsymbol{K}_{21}^{\text{fr}} & \boldsymbol{K}_{22}^{\text{fr}} & \boldsymbol{K}_{23}^{\text{fr}} \\ \boldsymbol{K}_{31}^{\text{fr}} & \boldsymbol{K}_{32}^{\text{fr}} & \boldsymbol{K}_{33}^{\text{fr}} \end{bmatrix} = \begin{bmatrix} \overline{\boldsymbol{k}}_{11}^{(1)} & \overline{\boldsymbol{k}}_{12}^{(1)} & 0 \\ \overline{\boldsymbol{k}}_{21}^{(1)} & \overline{\boldsymbol{k}}_{22}^{(1)} + \overline{\boldsymbol{k}}_{11}^{(2)} & \overline{\boldsymbol{k}}_{12}^{(2)} \\ 0 & \overline{\boldsymbol{k}}_{21}^{(2)} & \overline{\boldsymbol{k}}_{22}^{(2)} \end{bmatrix} \quad (7.31)$$

Then the stiffness equation of the unconstrained continuous beam can be expressed as

$$\boldsymbol{K}^{\text{fr}}\boldsymbol{U}^{\text{fr}} = \boldsymbol{P}^{\text{fr}} \quad \text{or} \quad \sum_{j=1}^{9} K_{ij}^{\text{fr}} U_j = P_i, \quad i = 1, \cdots, 9 \quad (7.32)$$

or

$$\begin{bmatrix} \boldsymbol{K}_{11}^{\text{fr}} & \boldsymbol{K}_{12}^{\text{fr}} & \boldsymbol{K}_{13}^{\text{fr}} \\ \boldsymbol{K}_{21}^{\text{fr}} & \boldsymbol{K}_{22}^{\text{fr}} & \boldsymbol{K}_{23}^{\text{fr}} \\ \boldsymbol{K}_{31}^{\text{fr}} & \boldsymbol{K}_{32}^{\text{fr}} & \boldsymbol{K}_{33}^{\text{fr}} \end{bmatrix} \begin{Bmatrix} \boldsymbol{U}_1^{\text{fr}} \\ \boldsymbol{U}_2^{\text{fr}} \\ \boldsymbol{U}_3^{\text{fr}} \end{Bmatrix} = \begin{Bmatrix} \boldsymbol{P}_1^{\text{fr}} \\ \boldsymbol{P}_2^{\text{fr}} \\ \boldsymbol{P}_3^{\text{fr}} \end{Bmatrix} \quad (7.33)$$

in which

$$\sum_{j=1}^{3} \boldsymbol{K}_{ij}^{\text{fr}} \boldsymbol{U}_j^{\text{fr}} = \boldsymbol{P}_i^{\text{fr}}, \quad \text{for} \quad i = 1, 2, 3 \quad (7.34)$$

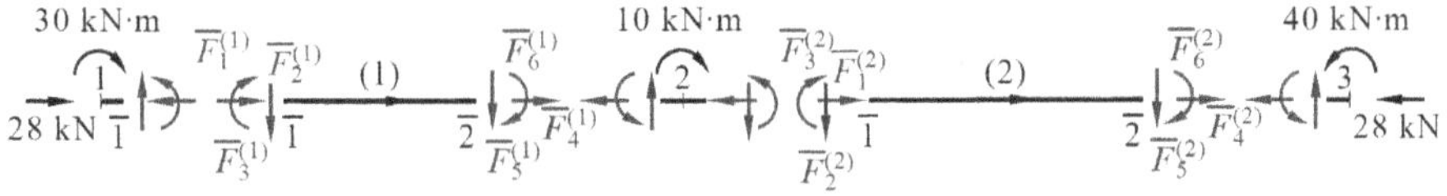

Figure 7.9. The free-body diagrams of the members of the unconstrained continuous beam.

are the equilibrium equations of node i as shown in Figure 7.9, and

$$\sum_{j=1}^{9} K_{ij}^{\mathrm{fr}} U_j^{\mathrm{fr}} = P_i^{\mathrm{fr}}, \qquad \text{for} \quad i = 1, \cdots, 9 \tag{7.35}$$

is the equilibrium equation at the ith DOF.

Imposing additional boundary conditions and solving the constrained stiffness equation. $\boldsymbol{K}^{\mathrm{fr}}$ is singular because the continuous beam is unrestrained and has numerous displacement solutions. It is necessary to introduce additional boundary conditions to find the relative rotation between the two ends. For example, one may define:

$$U_1 = U_2 = U_3 = 0, \quad \text{i.e.,} \quad U_{1x} = U_{1y} = \theta_1 = 0 \tag{7.36}$$

or

$$U_5 = U_7 = U_8 = 0, \quad \text{i.e.,} \quad U_{2y} = U_{3x} = U_{3y} = 0 \tag{7.37}$$

A DOF assigned with a value is called a ***prescribed DOF***, while an unknown DOF is called an ***active DOF***.

The combination of Equations (7.32) and (7.36) is equivalent to the following equation:

$$\begin{bmatrix} 1 & 0 & 0 & 0 & \cdots & 0 \\ 0 & 1 & 0 & 0 & \cdots & 0 \\ 0 & 0 & 1 & 0 & \cdots & 0 \\ 0 & 0 & 0 & K_{44} & \cdots & K_{49} \\ \cdot & \cdot & \cdot & \cdot & \cdot & \cdot \\ 0 & 0 & 0 & K_{94} & \cdots & K_{99} \end{bmatrix} \begin{Bmatrix} U_1 \\ U_2 \\ U_3 \\ U_4 \\ U_5 \\ U_6 \\ U_7 \\ U_8 \\ U_9 \end{Bmatrix} = \begin{Bmatrix} 0 \\ 0 \\ 0 \\ P_4 \\ P_5 \\ P_6 \\ P_7 \\ P_8 \\ P_9 \end{Bmatrix} \tag{7.38}$$

which can be rewritten as:

$$10^5 \times \begin{bmatrix} 4862.5 & 0 & 0 & -3862.5 & 0 & 0 \\ 0 & 2.4690 & 2.0983 & 0 & -1.3579 & 5.4316 \\ 0 & 2.0983 & 42.302 & 0 & -5.4316 & 14.484 \\ & 0 & 0 & 3862.5 & 0 & 0 \\ 0 & -1.3579 & -5.4316 & 0 & 1.3579 & -5.4316 \\ 0 & 5.4316 & 14.484 & 0 & -5.4316 & 28.969 \end{bmatrix} \begin{Bmatrix} U_4 \\ U_5 \\ U_6 \\ U_7 \\ U_8 \\ U_9 \end{Bmatrix} \tag{7.39}$$

$$= 10^3 \times \begin{bmatrix} 0 & 0 & 10 & -28 & 0 & -40 \end{bmatrix}$$

Solving Equation (7.39) yields:

$$\Delta\theta \equiv |U_3 - U_9| = 0.14523, \text{rad} \tag{7.40}$$

7.2.2 Post-Imposing of Boundary Conditions and Support Settlements

7.2.2 will illustrate a new way of analyzing structures using matrix displacement analysis. First, the stiffness equations of an unconstrained structure are formed. Then, one can extract the stiffness equations of the supported structure from the unconstrained stiffness equations by imposing the boundary conditions on them. If one imposes a simply-supported condition to a free beam, one can obtain the stiffness equation of a simply-supported beam. While if one imposes a fixed-end condition to the same unconstrained stiffness equation, then the stiffness equation of a cantilever is obtained. By such a post-imposing technique of boundary conditions, we can apply a single *Beam2* element type to solve various problems including the continuous beams shown in Figure 7.10.

In this section, we shall apply the direct stiffness method and the post-imposing technique of boundary conditions to find the end forces of each member in the continuous beam shown in Figure 7.10(a) subjected to nodal loading and a settlement. The material's modulus of elasticity, the beam's moment of inertia and the cross-sectional area of each member are $E_1 = 3.0 \times 10^{10}$ N/m², $E_2 = 2.06 \times 10^{11}$ N/m², $A_1 = 1.2 \times 10^{-2}$ m², $A_2 = 9.6 \times 10^{-3}$ m², $I_1 = 1.440 \times 10^{-5}$m⁴, and $I_2 = 1.152 \times 10^{-5}$m⁴, respectively.

Digitization of the beam. Table 7.5 is the digital counterpart of the supported continuous beam, induced by the coordinate and coding systems shown in Figure 7.10.

Decomposition of the structure. The supported continuous beam shown in Figure 7.10(a) is equivalent to the beam shown in Figure 7.10(b), with the following boundary conditions post-imposed to the beam:

$$\begin{cases} U_2 = U_5 = U_7 = 0 \\ U_8 = 0.01\left(\text{m}\right) \end{cases} \tag{7.41}$$

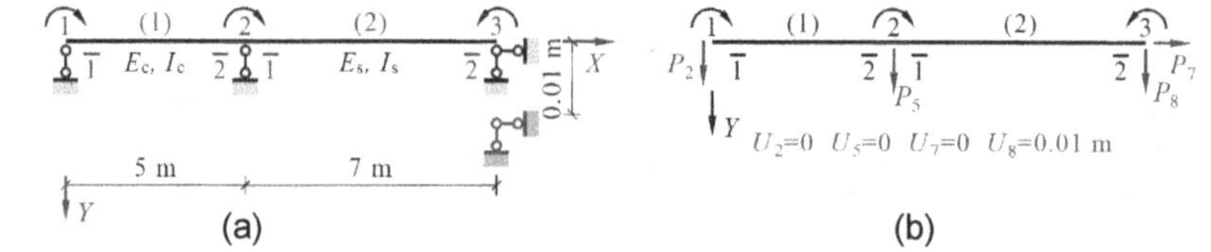

Figure 7.10. Post-processing of boundary conditions. (a) A supported beam subjected to nodal loads and a settlement. (b) The corresponding unconstrained beam with additional boundary conditions.

Table 7.5. The digital definition of the supported continuous beam shown in Figure 7.10(a).

Nodal ID	Coordinates			Nodal Loads		DOFs
	x(m)	y(m)	P_{ix} (N)	P_i (N . m)	M_i (N . m)	$\boldsymbol{U}_i$
1	0.0	0.0	0.0	0.0	15×10^3	$U_1 = [U_1\ U_2\ U_3]^{\mathrm{T}}$
2	5.0	0.0	0.0	0.0	25×10^3	$U_2 = [U_4\ U_5\ U_6]^{\mathrm{T}}$
3	12.0	2.0	0.0	0.0	-35×10^3	$U_3 = [U_7\ U_8\ U_9]^{\mathrm{T}}$
Element ID	Nodes		Young's Modulus	Section Area	Moment of Inertia	Element Type
	$\bar{1}$	$\bar{2}$	E (N²/m)	A_c (m²)	I_z (m⁴)	
(1)	1	2	3.00×10^{10}	0.0120	1.440×10^{-5}	*Beam2*
(2)	2	3	2.06×10^{11}	0.0096	1.152×10^{-5}	*Beam2*
Definition of Boundary Conditions						
ID of Prescribed DOFs			2	5	7	8
Values (m)			0.0	0.0	0.0	0.01

Forming the stiffness equation of the free beam. By using the procedure in Section 7.2.2, one can obtain the stiffness equation of the free continuous beam as follows:

$$\begin{bmatrix} K_{11}^{\mathrm{fr}} & \cdots & K_{19}^{\mathrm{fr}} \\ \vdots & \ddots & \vdots \\ K_{91}^{\mathrm{fr}} & \cdots & K_{99}^{\mathrm{fr}} \end{bmatrix} \begin{Bmatrix} U_1 \\ U_2 \\ U_3 \\ U_4 \\ U_5 \\ U_6 \\ U_7 \\ U_8 \\ U_9 \end{Bmatrix} = \begin{Bmatrix} P_1 \\ P_2 \\ P_3 \\ P_4 \\ P_5 \\ P_6 \\ P_7 \\ P_8 \\ P_9 \end{Bmatrix} = \begin{Bmatrix} 0 \\ P_2 \\ 15000 \\ 0 \\ P_5 \\ 25000 \\ P_7 \\ P_8 \\ -35000 \end{Bmatrix} \tag{7.42}$$

or

$$\sum_j K_{ij}^{\mathrm{fr}} U_j = P_i, \qquad i = 1, 2, \cdots, 9 \tag{7.43}$$

Forming the stiffness equation of the supported beam. Let

$$\boldsymbol{U}^{\mathrm{act}} \equiv [U_1\ \ U_3\ \ U_4\ \ U_6\ \ U_9]^{\mathrm{T}} \tag{7.44}$$

be the vector of ***active DOFs***, and

$$\boldsymbol{U}^{\mathrm{pre}} \equiv [U_2\ \ U_5\ \ U_7\ \ U_8]^{\mathrm{T}} = [0.0\ \ 0.0\ \ 0.0\ \ 0.01]^{\mathrm{T}} \tag{7.45}$$

Equation (7.43) can be rewritten as:

$$\sum_{j \in \boldsymbol{I}_{\mathrm{act}}} K_{ij}^{\mathrm{fr}} U_j + \sum_{j \in \boldsymbol{I}_{\mathrm{pre}}} K_{ij}^{\mathrm{fr}} U_j = P_i, \qquad i \in \boldsymbol{I}_{\mathrm{act}} \tag{7.46}$$

where $\boldsymbol{I}_{\mathrm{act}} = [1\ 3\ 4\ 6\ 9]^{\mathrm{T}}$ is the ID set of the active DOFs, and $\boldsymbol{I}_{\mathrm{pre}} = [2\ 5\ 7\ 8]^{\mathrm{T}}$ is the ID set of the prescribed DOFs.

Let

$$\boldsymbol{P}^{\text{act}} \equiv [P_1\ \ P_3\ \ P_4\ \ P_6\ \ P_9]^{\text{T}} = 10^3 \times [0.0\ \ 15\ \ 0.0\ \ 25\ \ -35]^{\text{T}} \tag{7.47}$$

and

$$\boldsymbol{P}^{\text{bc}} \equiv -\begin{Bmatrix} \sum_{j\in \boldsymbol{I}_{\text{pre}}} K_{1j}^{\text{fr}} U_j \\ \sum_{j\in \boldsymbol{I}_{\text{pre}}} K_{3j}^{\text{fr}} U_j \\ \sum_{j\in \boldsymbol{I}_{\text{pre}}} K_{4j}^{\text{fr}} U_j \\ \sum_{j\in \boldsymbol{I}_{\text{pre}}} K_{6j}^{\text{fr}} U_j \\ \sum_{j\in \boldsymbol{I}_{\text{pre}}} K_{9j}^{\text{fr}} U_j \end{Bmatrix} = 10^3 \times \begin{Bmatrix} 0 \\ 0 \\ 0 \\ 2.9059 \\ 2.9059 \end{Bmatrix} \tag{7.48}$$

which is the stiffness equation of the supported continuous beam, and where $\boldsymbol{P}^{\text{bc}}$ is the nodal loading vector induced by the prescribed DOFs.

Then Equation (7.46) can be rewritten as:

$$\boldsymbol{K}\,\boldsymbol{U}^{\text{act}} = \boldsymbol{P}^{\text{act}} + \boldsymbol{P}^{\text{bc}} \tag{7.49}$$

where

$$\boldsymbol{K} = \begin{bmatrix} K_{11}^{\text{fr}} & K_{13}^{\text{fr}} & K_{14}^{\text{fr}} & K_{16}^{\text{fr}} & K_{19}^{\text{fr}} \\ K_{31}^{\text{fr}} & K_{33}^{\text{fr}} & K_{34}^{\text{fr}} & K_{36}^{\text{fr}} & K_{39}^{\text{fr}} \\ K_{41}^{\text{fr}} & K_{43}^{\text{fr}} & K_{44}^{\text{fr}} & K_{46}^{\text{fr}} & K_{49}^{\text{fr}} \\ K_{61}^{\text{fr}} & K_{63}^{\text{fr}} & K_{64}^{\text{fr}} & K_{66}^{\text{fr}} & K_{69}^{\text{fr}} \\ K_{91}^{\text{fr}} & K_{93}^{\text{fr}} & K_{94}^{\text{fr}} & K_{96}^{\text{fr}} & K_{99}^{\text{fr}} \end{bmatrix} = 10^5 \times \begin{bmatrix} 720.00 & 0 & -720.00 & 0 & 0 \\ 0 & 3.4560 & 0 & 1.7280 & 0 \\ -720.00 & 0 & 3545.1 & 0 & 0 \\ 0 & 1.7280 & 0 & 17.017 & 6.7803 \\ 0 & 0 & 0 & 6.7803 & 13.561 \end{bmatrix} \tag{7.50}$$

is the stiffness matrix of the supported beam.

Find the values of the active DOFs. Solving Equation (7.49) yields:

$$\boldsymbol{U}^{\text{act}} = 10^{-2} \times [0\ \ 2.9122\ \ 0\ \ 2.8562\ \ -3.7948]^{\text{T}} \tag{7.51}$$

Calculating end forces of each member. Assembling $\boldsymbol{U}^{\text{act}}$ and $\boldsymbol{U}^{\text{pre}}$ together yields:

$$\boldsymbol{U} = 10^{-2} \times [0\ \ 0\ \ 2.9122\ \ 0\ \ 0\ \ 2.8562\ \ 0\ \ 1.0000\ \ -3.7948]^{\text{T}} \tag{7.52}$$

By using the element connectivity:

$$\ell = \begin{bmatrix} 1 & 2 \\ 3 & 4 \end{bmatrix}, \quad \text{i.e.,} \quad (1)\begin{cases} \bar{1} \to 1 \\ \bar{2} \to 2 \end{cases}; \quad (2)\begin{cases} \bar{1} \to 2 \\ \bar{2} \to 3 \end{cases} \tag{7.53}$$

one obtains the end-displacements:

$$\bar{U}^{(1)} = \begin{Bmatrix} U_1 \\ U_2 \end{Bmatrix} = 10^{-2} \times \begin{bmatrix} 0 & 0 & 2.9122 & 0 & 0 & 2.8562 \end{bmatrix}^{\mathrm{T}} \tag{7.54}$$

$$\bar{U}^{(2)} = \begin{Bmatrix} U_2 \\ U_3 \end{Bmatrix} = 10^{-2} \times \begin{bmatrix} 0 & 0 & 2.8562 & 0 & 1.0000 & -3.7948 \end{bmatrix}^{\mathrm{T}} \tag{7.55}$$

which lead to the following end-forces of elements (1) and (2):

$$\bar{F}^{(1)} = \bar{k}^{(1)}\bar{U}^{(1)} = 10^{3} \times \begin{bmatrix} 0 & 5.9807 & 15 & 0 & -5.9807 & 14.903 \end{bmatrix}^{\mathrm{T}} \tag{7.56}$$

$$\bar{F}^{(2)} = \bar{k}^{(2)}\bar{U}^{(2)} = 10^{3} \times \begin{bmatrix} 0 & -3.5576 & 10.097 & 0 & 3.5576 & -35.000 \end{bmatrix}^{\mathrm{T}} \tag{7.57}$$

EXAMPLES

EXAMPLE 7.1

Consider the continuous beam shown in Figure 7.11, whose material's modulus of elasticity, the beam's moment of inertia and cross-sectional area are $E = 3.0 \times 10^{10}$ N/m², $A_c = 1.2 \times 10^{-2}$m², $I_z = 1.440 \times 10^{-5}$m⁴, respectively. (i) Use the coordinate and label system shown in Figure 7.11(b) to find the end-forces of the members in the continuous beam, and draw its moment diagram; (ii) Calculate the end forces if the coordinate system shown in Figure 7.11(c) is used.

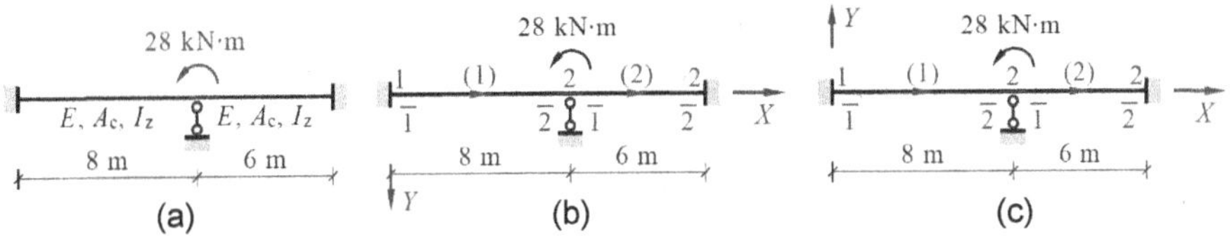

Figure 7.11. (a) A continuous beam. (b) Node and element label system. (c) Coordinate for determining end forces.

Solution: (1) Calculate the element stiffness matrices.

$$\bar{k}^{(1)} = 10^{5} \times \begin{bmatrix} 450 & 0 & 0 & -450 & 0 & 0 \\ 0 & 0.10125 & 0.405 & 0 & -0.10125 & 0.405 \\ 0 & 0.405 & 2.16 & 0 & -0.405 & 1.08 \\ -450 & 0 & 0 & 450 & 0 & 0 \\ 0 & -0.10125 & -0.405 & 0 & 0.10125 & -0.405 \\ 0 & 0.405 & 1.08 & 0 & -0.405 & 2.16 \end{bmatrix}$$

and

$$\bar{\boldsymbol{k}}^{(2)} = 10^5 \times \begin{bmatrix} 600 & 0 & 0 & -600 & 0 & 0 \\ 0 & 0.24 & 0.72 & 0 & -0.24 & 0.72 \\ 0 & 0.72 & 2.88 & 0 & -0.72 & 1.44 \\ -600 & 0 & 0 & 600 & 0 & 0 \\ 0 & -0.24 & 0.72 & 0 & 0.24 & -0.405 \\ 0 & -0.72 & 1.44 & 0 & -0.405 & 2.88 \end{bmatrix}$$

(2) Form the stiffness equation of the unconstrained structure. The element connectivity is:

$$(1)\begin{cases} \bar{1} \to 1 \\ \bar{2} \to 2 \end{cases} \quad \text{and} \quad (2)\begin{cases} \bar{1} \to 2 \\ \bar{2} \to 3 \end{cases}$$

i.e.,

$$(1)\begin{cases} \bar{U}_1^{(1)} = U_1 \\ \bar{U}_2^{(1)} = U_2 \\ \bar{U}_3^{(1)} = U_3 \\ \bar{U}_4^{(1)} = U_4 \\ \bar{U}_5^{(1)} = U_5 \\ \bar{U}_6^{(1)} = U_6 \end{cases} \quad \text{and} \quad (2)\begin{cases} \bar{U}_1^{(2)} = U_4 \\ \bar{U}_2^{(2)} = U_5 \\ \bar{U}_3^{(2)} = U_6 \\ \bar{U}_4^{(2)} = U_7 \\ \bar{U}_5^{(2)} = U_8 \\ \bar{U}_6^{(2)} = U_9 \end{cases}$$

The stiffness equation of the unconstrained structure is:

$$\sum_{j=1}^{3} \boldsymbol{K}_{ij}^{\text{fr}} \boldsymbol{U}_j = \boldsymbol{P}_i, \qquad i = 1, 2, 3$$

or

$$\sum_{j=1}^{9} K_{ij}^{\text{fr}} U_j = P_i, \qquad i = 1, \cdots, 9$$

where the stiffness coefficient K_{ij}^{fr} can be obtained by assembling the element matrices.

(3) Form the stiffness equation of the supported structure. After imposing the boundary conditions, one can obtain the following stiffness equation of the structure:

$$\sum_{j=4,6} K_{ij}^{\text{fr}} U_j = P_i, \qquad i = 4, 6$$

or

$$\boldsymbol{K}\begin{Bmatrix} U_4 \\ U_6 \end{Bmatrix} = \begin{Bmatrix} P_4 \\ P_6 \end{Bmatrix} = \begin{Bmatrix} 0 \\ -28000 \end{Bmatrix}$$

where

$$\boldsymbol{K} = \begin{bmatrix} \bar{k}_{44}^{(1)} + \bar{k}_{11}^{(2)} & \bar{k}_{46}^{(1)} + \bar{k}_{13}^{(2)} \\ \bar{k}_{64}^{(1)} + \bar{k}_{31}^{(2)} & \bar{k}_{66}^{(1)} + \bar{k}_{32}^{(2)} \end{bmatrix} = 10^6 \times \begin{bmatrix} 105 & 0 \\ 0 & 0.504 \end{bmatrix}$$

(4) Solving the stiffness equation yields:

$$\begin{bmatrix} U_4 & U_6 \end{bmatrix}^{\mathrm{T}} = \begin{bmatrix} 0 \\ -0.0556 \end{bmatrix}$$

(5) The end-forces of elements are:

$$\bar{\boldsymbol{F}}^{(1)} = 10^3 \times \begin{Bmatrix} 0 \\ -2.25 \\ -6 \\ 0 \\ 2.25 \\ -12 \end{Bmatrix} \quad \text{and} \quad \bar{\boldsymbol{F}}^{(2)} = 10^3 \times \begin{Bmatrix} 0 \\ -4 \\ -16 \\ 0 \\ 4 \\ -8 \end{Bmatrix}$$

which lead to the moment diagram shown in Figure 7.12.

(6) The end forces from the coordinates shown in Figure 7.11(c) are:

$$\bar{\boldsymbol{F}}^{(1)} = 10^3 \times \begin{Bmatrix} 0 \\ 2.25 \\ 6 \\ 0 \\ -2.25 \\ 12 \end{Bmatrix} \quad \text{and} \quad \bar{\boldsymbol{F}}^{(2)} = 10^3 \times \begin{Bmatrix} 0 \\ 4 \\ 16 \\ 0 \\ -4 \\ 8 \end{Bmatrix}$$

which also lead to the moment diagram shown in Figure 7.12 according to the sign convention that an end force is positive if it is in the positive direction of the corresponding axis.

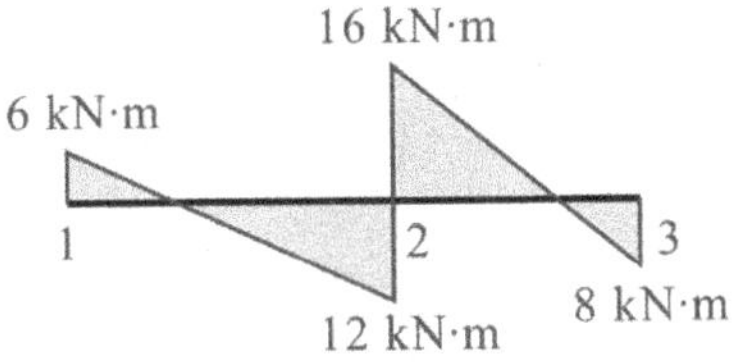

Figure 7.12. The moment diagram, in which a positive moment is on the tension side of the beam.

EXAMPLE 7.2

Consider a continuous beam, shown in Figure 7.13(a), subjected to loads and a rotation of 0.015 rad at the left support. Apply the direct stiffness method to find the end forces of the members of the structure. Young's modulus, the moment of inertia and the cross-sectional area are $E_1 = 3.0 \times 10^{10}$ N/m², $E_2 = 2.06 \times 10^{11}$ N/m², $A_1 = 1.2 \times 10^{-2}$ m², $A_2 = 9.6 \times 10^{-3}$ m², $I_1 = 1.440 \times 10^{-5}$m⁴, and $I_2 = 1.152 \times 10^{-5}$m⁴, respectively. Neglect the axial deformation of each member.

Solution: (1) Define the global coordinates and label according to elements and nodes shown in Figure 7.13(b).

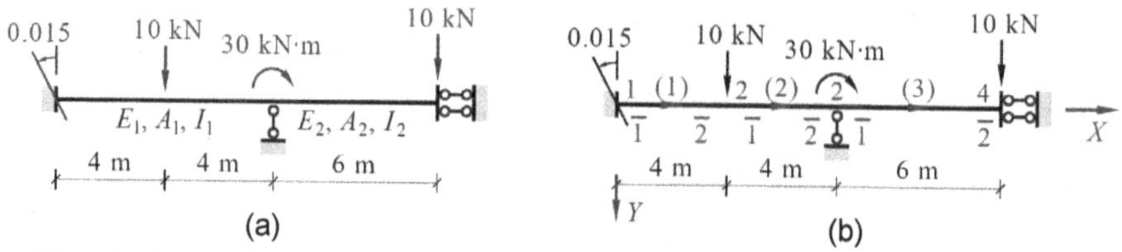

Figure 7.13. (a) A beam subjected to nodal loads and rotation. (b) Node and element label system.

(2) Calculate the element stiffness matrices. For $e = 1, 2$, one has:

$$\bar{\boldsymbol{k}}^{(e)} = 10^5 \times \begin{bmatrix} 900 & 0 & 0 & -900 & 0 & 0 \\ 0 & 0.81 & 1.62 & 0 & -0.81 & 1.62 \\ 0 & 1.62 & 4.32 & 0 & -1.62 & 2.16 \\ -900 & 0 & 0 & 900 & 0 & 0 \\ 0 & -0.81 & -1.62 & 0 & 0.81 & -1.62 \\ 0 & 1.62 & 2.16 & 0 & -1.62 & 4.32 \end{bmatrix}$$

and $e = 3$, one has:

$$\bar{\boldsymbol{k}}^{(3)} = 10^5 \times \begin{bmatrix} 3296 & 0 & 0 & -3296 & 0 & 0 \\ 0 & 1.3184 & 3.9552 & 0 & -1.3184 & 3.9552 \\ 0 & 3.9552 & 15.821 & 0 & -3.9552 & 7.9104 \\ -3296 & 0 & 0 & 3296 & 0 & 0 \\ 0 & -1.3184 & -3.9552 & 0 & 1.3184 & -0.405 \\ 0 & 3.9552 & 7.9104 & 0 & -0.405 & 15.821 \end{bmatrix}$$

(3) Forming the stiffness equation of the unconstrained structure: The element connectivity is:

$$(1)\begin{cases} \bar{1} \to 1 \\ \bar{2} \to 2 \end{cases}, \quad (2)\begin{cases} \bar{1} \to 2 \\ \bar{2} \to 3 \end{cases}, \quad (3)\begin{cases} \bar{1} \to 3 \\ \bar{2} \to 4 \end{cases}$$

i.e.,

$$(1)\begin{cases} \bar{U}_1^{(1)} = U_1 \\ \bar{U}_2^{(1)} = U_2 \\ \bar{U}_3^{(1)} = U_3 \\ \bar{U}_4^{(1)} = U_4 \\ \bar{U}_5^{(1)} = U_5 \\ \bar{U}_6^{(1)} = U_6 \end{cases}, \quad (2)\begin{cases} \bar{U}_1^{(2)} = U_4 \\ \bar{U}_2^{(2)} = U_5 \\ \bar{U}_3^{(2)} = U_6 \\ \bar{U}_4^{(2)} = U_7 \\ \bar{U}_5^{(2)} = U_8 \\ \bar{U}_6^{(2)} = U_9 \end{cases}, \quad (3)\begin{cases} \bar{U}_1^{(3)} = U_7 \\ \bar{U}_2^{(3)} = U_8 \\ \bar{U}_3^{(3)} = U_9 \\ \bar{U}_4^{(3)} = U_{10} \\ \bar{U}_5^{(3)} = U_{11} \\ \bar{U}_6^{(3)} = U_{12} \end{cases}$$

The stiffness equation of the unconstrained structure is:

$$\sum_{j=1}^{4} \boldsymbol{K}_{ij}^{\text{fr}} \boldsymbol{U}_j = \boldsymbol{P}_i, \qquad i = 1, 2, 3, 4$$

or

$$\sum_{j=1}^{12} K_{ij}^{\text{fr}} U_j = P_i, \qquad i = 1, \cdots, 12$$

in which the stiffness coefficients can be obtained by assembling the element stiffness matrices.

(4) Form the stiffness equation of the supported structure. The prescribed DOF vector is:

$$[U_1 \;\; U_2 \;\; U_3 \;\; U_4 \;\; U_7 \;\; U_8 \;\; U_{10} \;\; U_{12}] = [0 \;\; 0 \;\; -0.015 \;\; 0 \;\; 0 \;\; 0 \;\; 0 \;\; 0]$$

where $U_4 = U_7 = 0$ because we neglect the axial deformation.

The stiffness equation of the supported structure can be expressed as:

$$\boldsymbol{K}\,\boldsymbol{U}^{\text{act}} = \boldsymbol{P}^{\text{act}} + \boldsymbol{P}^{bc}$$

where

$$\boldsymbol{U}^{\text{act}} = [U_5 \;\; U_6 \;\; U_9 \;\; U_{11}] \tag{7.58}$$

is the active DOFs. The global stiffness matrix

$$\boldsymbol{K} = \begin{bmatrix} \bar{k}_{55}^{(1)} + \bar{k}_{22}^{(2)} & \bar{k}_{56}^{(1)} + \bar{k}_{23}^{(2)} & \bar{k}_{26}^{(2)} & 0 \\ \bar{k}_{65}^{(1)} + \bar{k}_{32}^{(2)} & \bar{k}_{66}^{(1)} + \bar{k}_{33}^{(2)} & \bar{k}_{36}^{(2)} & 0 \\ \bar{k}_{62}^{(2)} & \bar{k}_{63}^{(2)} & \bar{k}_{63}^{(2)} + \bar{k}_{33}^{(3)} & \bar{k}_{35}^{(3)} \\ 0 & 0 & \bar{k}_{53}^{(3)} & \bar{k}_{11}^{(3)} \end{bmatrix} = 10^5 \times \begin{bmatrix} 1.62 & 0 & 1.62 & 0 \\ 0 & 8.64 & 2.16 & 0 \\ 1.62 & 2.16 & 20.1408 & -3.9552 \\ 0 & 0 & -3.9552 & 1.3184 \end{bmatrix}$$

is assembled from the element matrices according to the element connectivity;

$$\boldsymbol{P}^{\text{act}} = \begin{Bmatrix} P_5 \\ P_6 \\ P_9 \\ P_{11} \end{Bmatrix} = \begin{Bmatrix} 10000 \\ 0 \\ 30000 \\ 15000 \end{Bmatrix} \tag{7.59}$$

and

$$\boldsymbol{P}^{\text{bc}} = -\begin{Bmatrix} \sum\limits_{j \in \boldsymbol{I}_{\text{pre}}} K_{5j}^{\text{fr}} U_j \\ \sum\limits_{j \in \boldsymbol{I}_{\text{pre}}} K_{6j}^{\text{fr}} U_j \\ \sum\limits_{j \in \boldsymbol{I}_{\text{pre}}} K_{9j}^{\text{fr}} U_j \\ \sum\limits_{j \in \boldsymbol{I}_{\text{pre}}} K_{11j}^{\text{fr}} U_j \end{Bmatrix} = -\begin{Bmatrix} -2430 \\ 3240 \\ 0 \\ 0 \end{Bmatrix} \tag{7.60}$$

(5) Solving the stiffness equation yields:

$$[U_5 \;\; U_6 \;\; U_9 \;\; U_{11}] = [-0.062 \;\; -0.023 \;\; 0.109 \;\; 0.4406]$$

(6) The end-forces of elements are:

$$\left[\overline{F}^{(1)} \quad \overline{F}^{(2)} \quad \overline{F}^{(3)}\right] = 10^3 \times \begin{bmatrix} 0.00 & 0.00 & 0.00 \\ -1.20 & 8.80 & -15.00 \\ -1.47 & 3.31 & -1.91 \\ 0.00 & 0.00 & 0.00 \\ 1.20 & -8.80 & 15.00 \\ -3.31 & 31.91 & -88.09 \end{bmatrix}$$

7.3 Frames Subjected to Nodal Loading – Change of Coordinates

7.3.1 Introduction

Consider a frame shown in Figure 7.14, whose Young's modulus, cross-sectional area and moment of inertia are $E = 2.06 \times 10^{11}$ N/m², $A = 9.6 \times 10^{-3}$m², and $I_z = 1.152 \times 10^{-5}$m⁴, respectively. It should be noticed that the assembling element stiffness matrices are essential for the analysis of the end forces of the members by using the direct stiffness method.

The derivation of the stiffness equation of the structure is to assemble the element stiffness matrices.

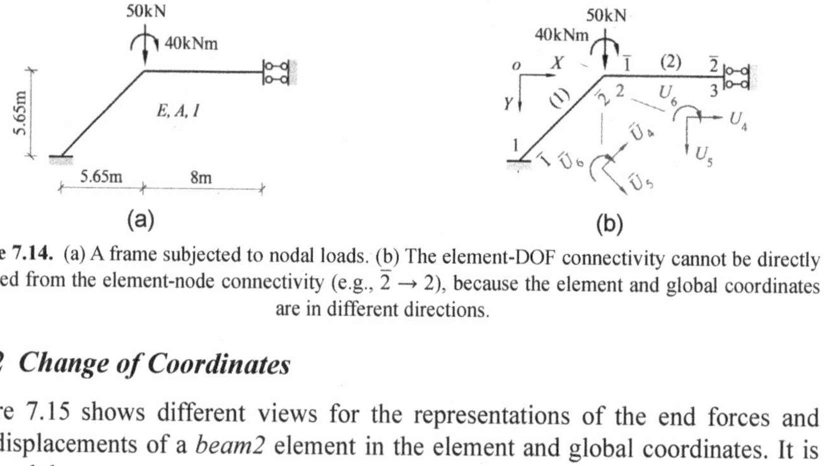

Figure 7.14. (a) A frame subjected to nodal loads. (b) The element-DOF connectivity cannot be directly obtained from the element-node connectivity (e.g., $\overline{2} \rightarrow 2$), because the element and global coordinates are in different directions.

7.3.2 Change of Coordinates

Figure 7.15 shows different views for the representations of the end forces and end displacements of a *beam2* element in the element and global coordinates. It is assumed that:

1. The Z and $\overline{z}$ axes are coincident, and into or out of the page.
2. All geometric or physical quantities are positive if they are in the positive direction of the correspondent. For example, the angle α is positive if it is in the positive direction of $\overline{z}$ according to the right-hand rule.

The component, $\overline{F}^{(e)}_{1x}$, of the end-force at node $\overline{1}$ in the $\overline{x}$-axis can be written in terms of the components of the end-force in the X, Y axes, i.e.,

$$\overline{F}^{(e)}_{1x} = F^{(e)}_{1x} \cos \alpha + F^{(e)}_{1y} \sin \alpha \tag{7.61}$$

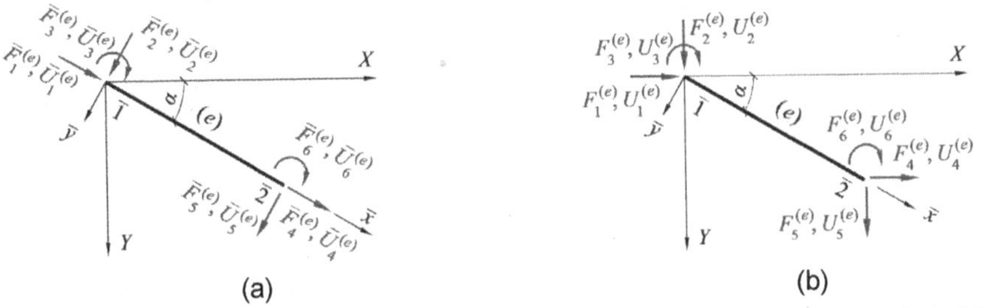

Figure 7.15. The end forces and end displacements of a *beam2* element . (a) *Y*-axis down view of the representation of the end forces and displacements in the element (local) coordinates. (b) *Y*-axis down view of the representation of the end forces and displacement in the global coordinates.

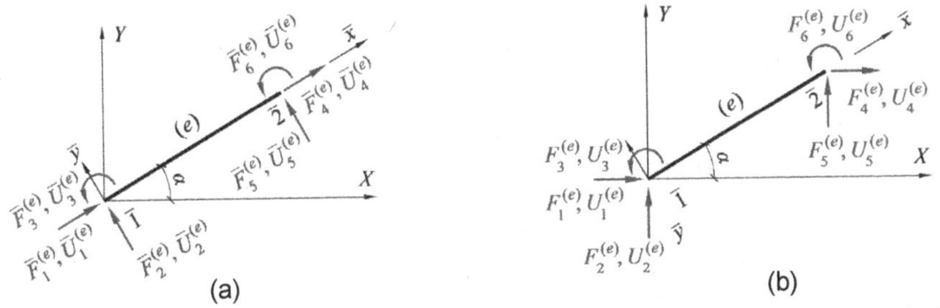

Figure 7.16. (a) The *Y*-axis up view of the representation of the end forces and displacements in the element (local) coordinates. (b) *Y*-axis up view of the representation of the end forces and displacement in the global coordinates.

Similarly, one has:

$$\begin{cases} \bar{F}_{1y}^{(e)} = -F_{1x}^{(e)} \sin\alpha + F_{1y}^{(e)} \cos\alpha \\ \bar{M}_1^{(e)} = M_1^{(e)} \\ \bar{F}_{2x}^{(e)} = F_{2x}^{(e)} \cos\alpha + F_{2y}^{(e)} \sin\alpha \\ \bar{F}_{2y}^{(e)} = -F_{2x}^{(e)} \sin\alpha + F_{2y}^{(e)} \cos\alpha \\ \bar{M}_2^{(e)} = M_2^{(e)} \end{cases} \tag{7.62}$$

Equations (7.61) and (7.62) can be expressed in the matrix form as follows:

$$\begin{Bmatrix} \bar{F}_{1x}^{(e)} \\ \bar{F}_{1y}^{(e)} \\ \bar{M}_1^{(e)} \\ \bar{F}_{2x}^{(e)} \\ \bar{F}_{2y}^{(e)} \\ \bar{M}_2^{(e)} \end{Bmatrix} = \begin{bmatrix} \cos\alpha & \sin\alpha & 0 & 0 & 0 & 0 \\ -\sin\alpha & \cos\alpha & 0 & 0 & 0 & 0 \\ 0 & 0 & 1 & 0 & 0 & 0 \\ 0 & 0 & 0 & \cos\alpha & \sin\alpha & 0 \\ 0 & 0 & 0 & -\sin\alpha & \cos\alpha & 0 \\ 0 & 0 & 0 & 0 & 0 & 1 \end{bmatrix} \begin{Bmatrix} F_{1x}^{(e)} \\ F_{1y}^{(e)} \\ M_1^{(e)} \\ F_{2x}^{(e)} \\ F_{2y}^{(e)} \\ M_2^{(e)} \end{Bmatrix} \tag{7.63}$$

which can be symbolically represented by:

$$\bar{\boldsymbol{F}}^{(e)} = \underset{\bar{e}\leftarrow g}{\boldsymbol{T}^{(e)}}\,\boldsymbol{F}^{(e)} \tag{7.64}$$

where $\underset{\bar{e}\leftarrow g}{\boldsymbol{T}^{(e)}}$ is called a ***transformation matrix of coordinates***, and is an orthogonal matrix, i.e.,

$$\left(\underset{\bar{e}\leftarrow g}{\boldsymbol{T}^{(e)}}\right)^{-1} = \left(\underset{\bar{e}\leftarrow g}{\boldsymbol{T}^{(e)}}\right)^{\mathrm{T}} \tag{7.65}$$

Analogously, the relationship between the coordinates of the end-displacement vectors of a beam, induced by local and global coordinate systems, can be expressed as:

$$\bar{\boldsymbol{U}}^{(e)} = \underset{\bar{e}\leftarrow g}{\boldsymbol{T}^{(e)}}\,\boldsymbol{U}^{(e)} \tag{7.66}$$

7.3.3 Element Stiffness Matrices in Global Coordinates and the Assembling Rules

We try to formulate the element stiffness equation in terms of global DOFs. The element stiffness equation in element local coordinates is:

$$\bar{\boldsymbol{k}}^{(e)}\bar{\boldsymbol{U}}^{(e)} = \bar{\boldsymbol{F}}^{(e)} \tag{7.67}$$

Inserting Equations (7.64) and (7.66) into Equation (7.67) gives:

$$\boldsymbol{k}^{(e)} = \left(\underset{\bar{e}\leftarrow g}{\boldsymbol{T}^{(e)}}\right)^{\mathrm{T}} \bar{\boldsymbol{k}}^{(e)}\,\underset{\bar{e}\leftarrow g}{\boldsymbol{T}^{(e)}} \tag{7.68}$$

where we used Equation (7.65). $\boldsymbol{k}^{(e)}$ is the element stiffness matrix in global coordinates, by which the stiffness equation in the global coordinates can be expressed as:

$$\boldsymbol{k}^{(e)}\,\boldsymbol{U}^{(e)} = \boldsymbol{F}^{(e)} \tag{7.69}$$

Here, $\boldsymbol{F}^{(e)}$ and $\boldsymbol{U}^{(e)}$ are respectively the end-force and end-displacement vectors in the global coordinates, defined as:

$$\boldsymbol{F}^{(e)} \equiv \begin{Bmatrix} F_{1x}^{(e)} \\ F_{1y}^{(e)} \\ M_1^{(e)} \\ F_{2x}^{(e)} \\ F_{2y}^{(e)} \\ M_2^{(e)} \end{Bmatrix} \equiv \begin{Bmatrix} F_1^{(e)} \\ F_1^{(e)} \\ M_3^{(e)} \\ F_4^{(e)} \\ F_5^{(e)} \\ M_6^{(e)} \end{Bmatrix}; \qquad \boldsymbol{U}^{(e)} \equiv \begin{Bmatrix} U_{1x}^{(e)} \\ U_{1y}^{(e)} \\ \theta_1^{(e)} \\ U_{2x}^{(e)} \\ U_{2y}^{(e)} \\ \theta_2^{(e)} \end{Bmatrix} \equiv \begin{Bmatrix} U_1^{(e)} \\ U_1^{(e)} \\ U_3^{(e)} \\ U_4^{(e)} \\ U_5^{(e)} \\ U_6^{(e)} \end{Bmatrix} \tag{7.70}$$

Because the element DOF $U_i^{(e)}$ must be in the direction of a structural DOF, the element-node connectivity leads to the element-DOF connectivity. For example, if an element-node connectivity yields:

$$(e)\begin{cases} \bar{1} \rightarrow i \\ \bar{2} \rightarrow j \end{cases} \tag{7.71}$$

the element-DOF connectivity is then:

$$(e):\begin{cases} U_1^{(e)} = U_{3i-2} \\ U_2^{(e)} = U_{3i-1} \\ U_3^{(e)} = U_{3i} \\ U_4^{(e)} = U_{3j-2} \\ U_5^{(e)} = U_{3j-1} \\ U_6^{(e)} = U_{3j} \end{cases} \tag{7.72}$$

Connectivity (7.72) shows that the element matrix $\boldsymbol{k}^{(e)}$ can be assembled into the global stiffness matrix K by the following rules:

$$\begin{cases} \boldsymbol{k}_{11}^{(e)} \to \boldsymbol{K}_{ii} \\ \boldsymbol{k}_{12}^{(e)} \to \boldsymbol{K}_{ij} \\ \boldsymbol{k}_{21}^{(e)} \to \boldsymbol{K}_{ji} \\ \boldsymbol{k}_{22}^{(e)} \to \boldsymbol{K}_{jj} \end{cases} \tag{7.73}$$

where $\boldsymbol{k}_{ij}^{(e)}$ and $\boldsymbol{K}_{ij}$ are the block matrices of the element and global stiffness matrices, respectively. For example, $\boldsymbol{k}_{26}^{(e)}$ will be added to row $3i - 1$ and column $3j$ of $\boldsymbol{K}$.

7.3.4 Analysis of the Frames Subjected to Nodal Loads

Consider the frame shown in Figure 7.14a.

1. Define the coordinates and labels shown in Figure 7.14b, to digitize the frame.
2. Calculate the element stiffness matrices. The element stiffness matrices in element local coordinates are:

$$\overline{\boldsymbol{k}}^{(e)} = 10^5 \times \begin{bmatrix} 2472 & 0 & 0 & -2472 & 0 & 0 \\ 0 & 0.55619 & 2.2248 & 0 & -0.55619 & 2.2248 \\ 0 & 2.2248 & 11.866 & 0 & -2.2248 & 5.9328 \\ -2472 & 0 & 0 & 2472 & 0 & 0 \\ 0 & -0.55619 & -2.2248 & 0 & 0.55619 & -2.2248 \\ 0 & 2.2248 & 5.9328 & 0 & -2.2248 & 11.866 \end{bmatrix} \tag{7.74}$$

for $e = 1, 2$. The element transformation matrices of coordinates are:

$$\underset{\bar{e}\leftarrow g}{\boldsymbol{T}^{(1)}} = \begin{bmatrix} 0.707 & -0.707 & 0 & 0 & 0 & 0 \\ 0.707 & 0.707 & 0 & 0 & 0 & 0 \\ 0 & 0 & 1 & 0 & 0 & 0 \\ 0 & 0 & 0 & 0.707 & -0.707 & 0 \\ 0 & 0 & 0 & 0.707 & 0.707 & 0 \\ 0 & 0 & 0 & 0 & 0 & 1 \end{bmatrix}, \quad \alpha = -\frac{\pi}{4} \tag{7.75}$$

and

$$\underset{\bar{e}\leftarrow g}{\boldsymbol{T}^{(2)}} = \boldsymbol{I}_{6\times 6}, \qquad \alpha = 0 \tag{7.76}$$

where $\boldsymbol{I}_{6\times 6}$ is the 6-by-6 identity matrix.

The element stiffness matrices in the global coordinates are:

$$\boldsymbol{k}^{(1)} = \left(\underset{\bar{e}\leftarrow g}{\boldsymbol{T}^{(1)}}\right)^{\mathrm{T}} \bar{\boldsymbol{k}}^{(1)} \underset{\bar{e}\leftarrow g}{\boldsymbol{T}^{(1)}}$$
$$= 10^5 \times \begin{bmatrix} 1236.3 & -1235.7 & 1.5731 & -1236.3 & 1235.7 & 1.5731 \\ -1235.7 & 1236.3 & 1.5731 & 1235.7 & -1236.3 & 1.5731 \\ 1.5731 & 1.5731 & 11.866 & -1.5731 & -1.5731 & 5.9328 \\ -1236.3 & 1235.7 & -1.5731 & 1236.3 & -1235.7 & -1.5731 \\ 1235.7 & -1236.3 & -1.5731 & -1235.7 & 1236.3 & -1.5731 \\ 1.5731 & 1.5731 & 5.9328 & -1.5731 & -1.5731 & 11.866 \end{bmatrix} \tag{7.77}$$

and

$$\boldsymbol{k}^{(2)} = \left(\underset{\bar{e}\leftarrow g}{\boldsymbol{T}^{(2)}}\right)^{\mathrm{T}} \bar{\boldsymbol{k}}^{(2)} \underset{\bar{e}\leftarrow g}{\boldsymbol{T}^{(2)}} = \bar{\boldsymbol{k}}^{(2)}$$
$$= 10^5 \times \begin{bmatrix} 24720 & 0 & 0 & -24720 & 0 & 0 \\ 0 & 5.5619 & 22.248 & 0 & -5.5619 & 22.248 \\ 0 & 22.248 & 118.66 & 0 & -22.248 & 59.328 \\ -24720 & 0 & 0 & 24720 & 0 & 0 \\ 0 & -5.5619 & -22.248 & 0 & 5.5619 & -1.5731 \\ 0 & 22.248 & 59.328 & 0 & -1.5731 & 118.66 \end{bmatrix} \tag{7.78}$$

3. Forming the stiffness equation of the unconstrained frame. The element connectivity is:

$$(1)\begin{cases} \bar{1} \rightarrow 1 \\ \bar{2} \rightarrow 2 \end{cases}, \qquad (2)\begin{cases} \bar{1} \rightarrow 2 \\ \bar{2} \rightarrow 3 \end{cases} \tag{7.79}$$

or

$$(1):\begin{cases} U_1^{(1)} = U_1 \\ U_2^{(1)} = U_2 \\ U_3^{(1)} = U_3 \\ U_4^{(1)} = U_4 \\ U_5^{(1)} = U_5 \\ U_6^{(1)} = U_6 \end{cases}, \qquad (2):\begin{cases} U_1^{(2)} = U_4 \\ U_2^{(2)} = U_5 \\ U_3^{(2)} = U_6 \\ U_4^{(2)} = U_7 \\ U_5^{(2)} = U_8 \\ U_6^{(2)} = U_9 \end{cases} \tag{7.80}$$

The stiffness equation of the unconstrained frame is:

$$\boldsymbol{K}^{\text{ft}}_{9\times 9}\,\boldsymbol{U} = \boldsymbol{P} \tag{7.81}$$

4. Form the stiffness equation of the supported structure. Using the boundary conditions:

$$[U_1\;\; U_2\;\; U_3\;\; U_7\;\; U_9] = [0\;\; 0\;\; 0\;\; 0\;\; 0] \tag{7.82}$$

in Equation (7.81) leads to the following stiffness equation of the supported frame:

$$\boldsymbol{K}\,\boldsymbol{U}^{\text{act}} = \boldsymbol{P}^{\text{act}} \tag{7.83}$$

where,

$$\boldsymbol{U}^{\text{act}} = [U_4\;\; U_5\;\; U_6\;\; U_8]^{\text{T}} \tag{7.84}$$

$$\boldsymbol{P}^{\text{act}} = [P_4\;\; P_5\;\; P_6\;\; P_8]^{\text{T}} = 10^3 \times [0\;\; 50\;\; 40\;\; 0] \tag{7.85}$$

and

$$\boldsymbol{K} = \begin{bmatrix} K^{\text{fr}}_{44} & K^{\text{fr}}_{45} & K^{\text{fr}}_{46} & K^{\text{fr}}_{48} \\ K^{\text{fr}}_{54} & K^{\text{fr}}_{55} & K^{\text{fr}}_{56} & K^{\text{fr}}_{58} \\ K^{\text{fr}}_{64} & K^{\text{fr}}_{65} & K^{\text{fr}}_{66} & K^{\text{fr}}_{68} \\ K^{\text{fr}}_{84} & K^{\text{fr}}_{84} & K^{\text{fr}}_{86} & K^{\text{fr}}_{88} \end{bmatrix} = \begin{bmatrix} 3.7083\times 10^8 & -1.2357\times 10^8 & -1.5731\times 10^8 & 0 \\ -1.2357\times 10^8 & 1.2368\times 10^8 & 6.5165\times 10^4 & -5.5620\times 10^4 \\ -1.5731\times 10^8 & 6.5165\times 10^4 & 2.3731\times 10^6 & -2.2248\times 10^5 \\ 0 & -5.5620\times 10^4 & -2.2248\times 10^5 & 5.5620\times 10^4 \end{bmatrix} \tag{7.86}$$

Here, we used connectivity (7.80) to assemble this stiffness matrix. For example,

$$K^{\text{fr}}_{45} = k^{(1)}_{45} + k^{(2)}_{12} = -1.2357 \times 10^8 + 0 = -1.2357 \times 10^8 \tag{7.87}$$

and

$$K^{\text{fr}}_{58} = k^{(2)}_{25} = -5.5620 \times 10^4 \tag{7.88}$$

5. Solving the stiffness equation of the supported structure yields:

$$\boldsymbol{U}^{\text{act}} = \begin{Bmatrix} U_4 \\ U_5 \\ U_6 \\ U_8 \end{Bmatrix} = \begin{Bmatrix} 23.651\times 10^{-4} \\ 6.7529\times 10^{-4} \\ 2.7066\times 10^{-2} \\ 1.0894\times 10^{-1} \end{Bmatrix} \tag{7.89}$$

6. Calculate the end forces of the structure.

The full-dimensional vector, $\boldsymbol{U}$, of the nodal displacements is:

$$\boldsymbol{U} = \begin{Bmatrix} U_1 \\ U_2 \\ U_3 \\ U_4 \\ U_5 \\ U_6 \\ U_7 \\ U_8 \\ U_9 \end{Bmatrix} = \begin{Bmatrix} 0 \\ 0 \\ 0 \\ 2.3651\times10^{-4} \\ 6.7529\times10^{-4} \\ 2.7066\times10^{-2} \\ 0 \\ 1.0894\times10^{-1} \\ 0 \end{Bmatrix} \tag{7.90}$$

One then has:

$$\overline{\boldsymbol{F}}^{(1)} = \overline{\boldsymbol{k}}^{(1)}\overline{\boldsymbol{U}}^{(1)} = \overline{\boldsymbol{k}}^{(1)} \underset{\bar{e}\leftarrow g}{\boldsymbol{T}^{(1)}}\, \boldsymbol{U}^{(1)} = 10^3 \times \begin{Bmatrix} -45.572 \\ -37.109 \\ 15.907 \\ 45.572 \\ 37.109 \\ 31.964 \end{Bmatrix}, \begin{pmatrix} \mathrm{N} \\ \mathrm{N} \\ \mathrm{N\cdot m} \\ \mathrm{N} \\ \mathrm{N} \\ \mathrm{N\cdot m} \end{pmatrix} \tag{7.91}$$

where $\boldsymbol{U}^{(1)}$ can be obtained from $\boldsymbol{U}$ and the element connectivity (7.80). Similarly, for element (2), one has:

$$\overline{\boldsymbol{F}}^{(2)} = \overline{\boldsymbol{k}}^{(2)}\overline{\boldsymbol{U}}^{(2)} = \overline{\boldsymbol{k}}^{(2)} \underset{\bar{e}\leftarrow g}{\boldsymbol{T}^{(2)}}\, \boldsymbol{U}^{(2)} = 10^3 \times \begin{Bmatrix} 58.465 \\ 0 \\ 8.0287 \\ -58.465 \\ 0 \\ -8.0287 \end{Bmatrix}, \begin{pmatrix} \mathrm{N} \\ \mathrm{N} \\ \mathrm{N\cdot m} \\ \mathrm{N} \\ \mathrm{N} \\ \mathrm{N\cdot m} \end{pmatrix} \tag{7.92}$$

EXAMPLE 7.3

Consider a frame, shown in Figure 7.17a, whose Young's modulus, cross-sectional area, and moment of inertia are $E = 2.06 \times 10^{11}$ N/m², $A = 1.5 \times 10^{-2}$m², $I_z = 2.813$

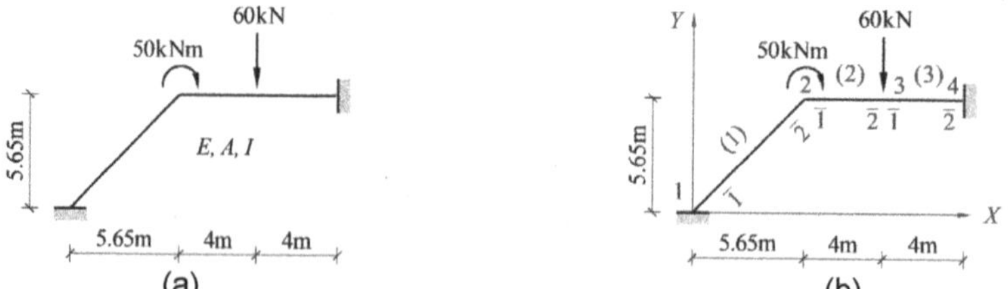

Figure 7.17. (a) A frame subjected to nodal loads. (b) Node and element system.

$\times$ $10^{-5}m^4$, respectively. Use the coordinate and label system shown in Figure 7.17b to find the end forces of the member of the frame. Neglect the axial deformation of each member.

***Solution*:**

1. Calculate the element matrices. The element stiffness matrices in the element local coordinates are:

$$\boldsymbol{k}^{(1)} = 10^5 \times \begin{bmatrix} 2742 & 0 & 0 & -2742 & 0 & 0 \\ 0 & 0.55619 & 2.2248 & 0 & -0.55619 & 2.2248 \\ 0 & 2.2248 & 11.866 & 0 & -2.2248 & 5.9328 \\ -2742 & 0 & 0 & 2742 & 0 & 0 \\ 0 & -0.55619 & -2.2248 & 0 & 0.55619 & -2.2248 \\ 0 & 2.2248 & 5.9328 & 0 & -2.2248 & 11.866 \end{bmatrix}$$

and

$$\boldsymbol{k}^{(e)} = 10^5 \times \begin{bmatrix} 4944 & 0 & 0 & -4944 & 0 & 0 \\ 0 & 4.4496 & 8.8992 & 0 & -4.4496 & 8.8992 \\ 0 & 8.8992 & 23.731 & 0 & -8.8992 & 11.866 \\ -4944 & 0 & 0 & 4944 & 0 & 0 \\ 0 & -4.4496 & -8.8992 & 0 & 4.4496 & -8.8992 \\ 0 & 8.8992 & 11.866 & 0 & -8.8992 & 23.731 \end{bmatrix}$$

for $e = 2, 3$. The angles of the elements are:

$$\alpha^{(1)} = \frac{\pi}{4}; \qquad \alpha^{(2)} = \alpha^{(3)} = 0$$

and the transformation matrices are:

$$\underset{\bar{e}\leftarrow g}{\boldsymbol{T}}^{(1)} = \begin{bmatrix} 0.707 & 0.707 & 0 & 0 & 0 & 0 \\ -0.707 & 0.707 & 0 & 0 & 0 & 0 \\ 0 & 0 & 1 & 0 & 0 & 0 \\ 0 & 0 & 0 & 0.707 & 0.707 & 0 \\ 0 & 0 & 0 & -0.707 & 0.707 & 0 \\ 0 & 0 & 0 & 0 & 0 & 1 \end{bmatrix}; \qquad \underset{\bar{e}\leftarrow g}{\boldsymbol{T}}^{(2)} = \boldsymbol{I}_{6\times6}$$

$$\underset{\bar{e}\leftarrow g}{\boldsymbol{T}}^{(3)} = \begin{bmatrix} -1 & 0 & 0 & 0 & 0 & 0 \\ 0 & -1 & 0 & 0 & 0 & 0 \\ 0 & 0 & 1 & 0 & 0 & 0 \\ 0 & 0 & 0 & -1 & 0 & 0 \\ 0 & 0 & 0 & 0 & -1 & 0 \\ 0 & 0 & 0 & 0 & 0 & 1 \end{bmatrix}$$

In addition, the element stiffness matrices in global coordinates are:

$$\boldsymbol{k}^{(1)} = 10^5 \times \begin{bmatrix} 1236.3 & 1235.7 & -1.5731 & -1236.3 & -1235.7 & -1.5731 \\ -1235.7 & 1236.3 & 1.5731 & -1235.7 & -1236.3 & 1.5731 \\ -1.5731 & 1.5731 & 11.866 & 1.5731 & -1.5731 & 5.9328 \\ -1236.3 & -1235.7 & 1.5731 & 1236.3 & 1235.7 & 1.5731 \\ -1235.7 & -1236.3 & -1.5731 & 1235.7 & 1236.3 & -1.5731 \\ -1.5731 & 1.5731 & 5.9328 & 1.5731 & -1.5731 & 11.866 \end{bmatrix}$$

$$\boldsymbol{k}^{(2)} = 10^5 \times \begin{bmatrix} 4944 & 0 & 0 & -4944 & 0 & 0 \\ 0 & 4.4496 & 8.8992 & 0 & -4.4496 & 8.8992 \\ 0 & 8.8992 & 23.731 & 0 & -8.8992 & 11.866 \\ -4944 & 0 & 0 & 4944 & 0 & 0 \\ 0 & -4.4496 & -8.8992 & 0 & 4.4496 & -8.8992 \\ 0 & 8.8992 & 11.866 & 0 & -8.8992 & 23.731 \end{bmatrix}$$

$$\boldsymbol{k}^{(3)} = 10^5 \times \begin{bmatrix} 4944 & 0 & 0 & -4944 & 0 & 0 \\ 0 & 4.4496 & -8.8992 & 0 & -4.4496 & -8.8992 \\ 0 & -8.8992 & 23.731 & 0 & 8.8992 & 11.866 \\ -4944 & 0 & 0 & 4944 & 0 & 0 \\ 0 & -4.4496 & -8.8992 & 0 & 4.4496 & 8.8992 \\ 0 & -8.8992 & 11.866 & 0 & 8.8992 & 23.731 \end{bmatrix}$$

2. Forming the stiffness equation of the unconstrained frame. The element connectivity is as under:

$$(1)\begin{cases} \bar{1} \to 1 \\ \bar{2} \to 2 \end{cases}; \quad (2)\begin{cases} \bar{1} \to 2 \\ \bar{2} \to 3 \end{cases}; \quad (3)\begin{cases} \bar{1} \to 3 \\ \bar{2} \to 2 \end{cases} \tag{7.93}$$

i.e.,

$$(1)\begin{cases} U_1^{(1)} = U_1 \\ U_2^{(1)} = U_2 \\ U_3^{(1)} = U_3 \\ U_4^{(1)} = U_4 \\ U_5^{(1)} = U_5 \\ U_6^{(1)} = U_6 \end{cases}; \quad (2)\begin{cases} U_1^{(2)} = U_4 \\ U_2^{(2)} = U_5 \\ U_3^{(2)} = U_6 \\ U_4^{(2)} = U_7 \\ U_5^{(2)} = U_8 \\ U_6^{(2)} = U_9 \end{cases}; \quad (3)\begin{cases} U_1^{(3)} = U_{10} \\ U_2^{(3)} = U_{11} \\ U_3^{(3)} = U_{12} \\ U_4^{(3)} = U_7 \\ U_5^{(3)} = U_8 \\ U_6^{(3)} = U_9 \end{cases} \tag{7.94}$$

The stiffness matrix of the unconstrained frame is:

$$\boldsymbol{K}^{\text{fr}}\,\boldsymbol{U} = \boldsymbol{P} \tag{7.95}$$

3. Form the stiffness equation of the supported structure. Imposing the prescribed DOFs

$$\boldsymbol{U}^{\text{pre}} = [U_1 \; U_2 \; U_3 \; U_4 \; U_5 \; U_7 \; U_{10} \; U_{11} \; U_{12}] = \mathbf{0}$$

into Equation (7.95) leads to the following stiffness equation of the supported frame:

$$\boldsymbol{K}\boldsymbol{U}^{\text{act}} = \boldsymbol{P}^{\text{act}}$$

where,

$$\boldsymbol{U}^{\text{act}} = [U_6 \; U_8 \; U_9]^{\mathrm{T}}$$

$$\boldsymbol{P}^{\text{act}} = [P_6 \; P_8 \; P_9]^{\mathrm{T}} = 10^5 \times [-60 \; -50 \; 0]^{\mathrm{T}} \tag{7.96}$$

and

$$\boldsymbol{K} = \begin{bmatrix} K_{66}^{\text{fr}} & K_{68}^{\text{fr}} & K_{69}^{\text{fr}} \\ K_{86}^{\text{fr}} & K_{88}^{\text{fr}} & K_{89}^{\text{fr}} \\ K_{96}^{\text{fr}} & K_{98}^{\text{fr}} & K_{99}^{\text{fr}} \end{bmatrix} = 10^5 \times \begin{bmatrix} 35.597 & -8.8992 & 11.866 \\ -8.8992 & 8.8992 & 0 \\ 11.866 & 0 & 47.462 \end{bmatrix}$$

Here, one obtains K_{ij}^{fr} by assembling the element stiffness matrices $\boldsymbol{k}^{(e)}$. For example, the element connectivity (7.94) leads to:

$$K_{69}^{\text{fr}} = k_{36}^{(2)} = 11.866$$

4. Solving the stiffness equation of the supported structure yields:

$$\boldsymbol{U}^{\text{act}} = \begin{Bmatrix} U_6 \\ U_8 \\ U_9 \end{Bmatrix} = \begin{Bmatrix} -0.0464 \\ -0.1025 \\ 0.0116 \end{Bmatrix}$$

5. Calculate the end forces of the structure.

$$\overline{\boldsymbol{F}}^{(1)} = \overline{\boldsymbol{k}}^{(1)}\overline{\boldsymbol{U}}^{(1)} = \overline{\boldsymbol{k}}^{(1)} \underset{\bar{e}\leftarrow g}{\boldsymbol{T}^{(1)}} \boldsymbol{U}^{(1)} = 10^3 \times \begin{Bmatrix} 0 \\ -10.312 \\ -27.5 \\ 0 \\ 10.312 \\ -55.0 \end{Bmatrix}, \begin{pmatrix} \text{N} \\ \text{N} \\ \text{N}\cdot\text{m} \\ \text{N} \\ \text{N} \\ \text{N}\cdot\text{m} \end{pmatrix} \tag{7.97}$$

Similarly, one has:

$$\overline{\boldsymbol{F}}^{(2)} = \overline{\boldsymbol{k}}^{(2)}\overline{\boldsymbol{U}}^{(2)} = \overline{\boldsymbol{k}}^{(2)} \underset{\bar{e}\leftarrow g}{\boldsymbol{T}^{(2)}} \boldsymbol{U}^{(2)} = 10^3 \times \begin{Bmatrix} 0 \\ 14.687 \\ -5.0002 \\ 0 \\ -14.687 \\ 63.75 \end{Bmatrix}, \begin{pmatrix} \text{N} \\ \text{N} \\ \text{N}\cdot\text{m} \\ \text{N} \\ \text{N} \\ \text{N}\cdot\text{m} \end{pmatrix} \tag{7.98}$$

$$\bar{F}^{(3)} = \bar{k}^{(3)}\bar{U}^{(3)} = \bar{k}^{(3)} \underset{\bar{e}\leftarrow g}{T^{(3)}} U^{(3)} = 10^3 \times \begin{Bmatrix} 0 \\ -35.313 \\ -77.5 \\ 0 \\ 35.313 \\ -63.75 \end{Bmatrix}, \begin{pmatrix} \mathrm{N} \\ \mathrm{N} \\ \mathrm{N\cdot m} \\ \mathrm{N} \\ \mathrm{N} \\ \mathrm{N\cdot m} \end{pmatrix} \quad (7.99)$$

7.4 Frames Subjected to In-span Loads and Equivalent Nodal Loads

The structures in the previous sections are subjected only to nodal loads. Here, we shall analyze structures subjected to axial loading. Consider the frame shown in Figure 7.18a, where the Young's modulus, the member's cross-sectional area and moments of inertia are $E = 2.06 \times 10^{11}$ N/m², $A = 9.6 \times 10^{-3}$ m², $I_z = 1.152 \times 10^{-5}$ m⁴, respectively. To digitize the frame, one can define a global coordinate system and label all the nodes and elements with code-numbers as shown in Figure 7.18b. This frame can then be represented by the data listed in Table 7.6 using the label system shown in Figure 7.18b.

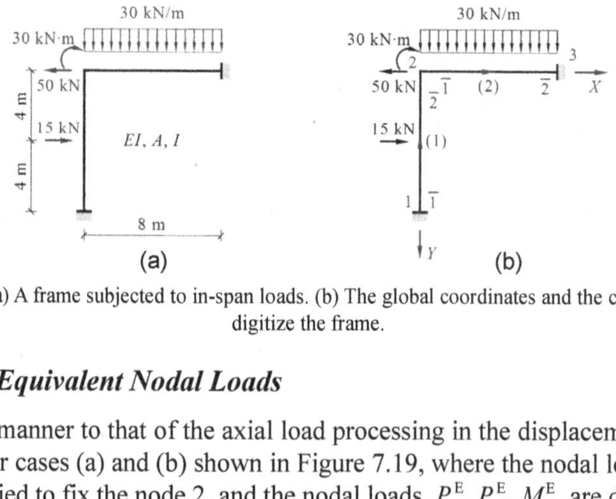

Figure 7.18. (a) A frame subjected to in-span loads. (b) The global coordinates and the code-numbers to digitize the frame.

7.4.1 The Equivalent Nodal Loads

In a similar manner to that of the axial load processing in the displacement method, let's consider cases (a) and (b) shown in Figure 7.19, where the nodal loads, P_4^{F}, P_5^{F}, M_6^{F}, are applied to fix the node 2, and the nodal loads, P_4^{E}, P_5^{E}, M_6^{E}, are defined as:

$$[P_4^{\mathrm{E}} \; P_5^{\mathrm{E}} \; M_6^{\mathrm{E}}] \equiv [P_4^{\mathrm{F}} \; P_5^{\mathrm{F}} \; M_6^{\mathrm{F}}] \quad (7.100)$$

Then, the case shown in Figure 7.18 is the superposition of cases (a) and (b) shown in Figure 7.19. Since the end forces of elements in case (a) are known, it is necessary to focus on solving it (b), which is subjected to nodal loading and can be solved by the method presented in the previous sections as long as P_4^{F}, P_5^{F}, M_6^{F} can be specified. Although P_4^{F}, P_5^{F}, M_6^{F} can be easily obtained by the way adopted in the displacement method, it is preferred to present a purely digital process to calculate them, which is convenient to be implemented into computer based numerical codes.

Table 7.6. Digital representation of the frame shown in Figure 7.18.

Definition of Nodes								
Nodal ID	Coordinates		Nodal Loads					
	x(m)	y(m)	P_x (N)	P_y (N)	M_z (kN · m)			
1	0.0	8.0	0.0	0.0	0.0			
2	0.0	0.0	-50×10^4	0.0	3.0×10^4			
3	8.0	0.0	0.0	0.0	0.0			
Definition of Elements								
ELT ID	Nodes		E_x	A_{cross}	I_z	q	P	$\bar{x}_p$
	$\bar{1}$	$\bar{2}$	(N/m²)	(m²)	(m⁴)	(q/m)	(N)	(m)
(1)	1	2	2.06×10^{11}	0.015	2.813×10^{-5}	0.0	15.0×10^3	4.0
(2)	2	3	2.06×10^{11}	0.015	2.813×10^{-5}	30.0×10^3	0.0	0.0
Definition of Boundary Conditions								
Prescribed DOFs			1	2	3	7	8	9
Values			0.0	0.0	0.0	0.0	0.0	0.0

Note: U_1 is the displacement of the first DOF of the structure, m^2 is the load on the second DOF of the structure.

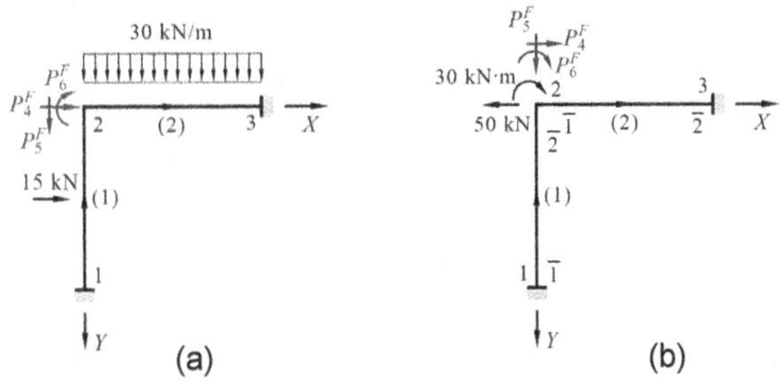

Figure 7.19. Superposition of the problem. (a) The frame subjected to the total nodal loads $\boldsymbol{P} + \boldsymbol{P}^{\text{E}}$. (b) The frame subjected to the fixed-node loads $\boldsymbol{P}^{\text{F}}$ and the axial loads.

First, the supported frames shown in Figures 7.19a and b are, respectively, equivalent to the unconstrained frames shown in Figures 7.20a and b plus the following prescribed nodal displacements:

$$\boldsymbol{U}^{\text{pre}} = [U_1 \;\; U_2 \;\; U_3 \;\; U_7 \;\; U_8 \;\; U_9]^{\text{T}} = \mathbf{0}^{\text{T}}_{6\times1} \tag{7.101}$$

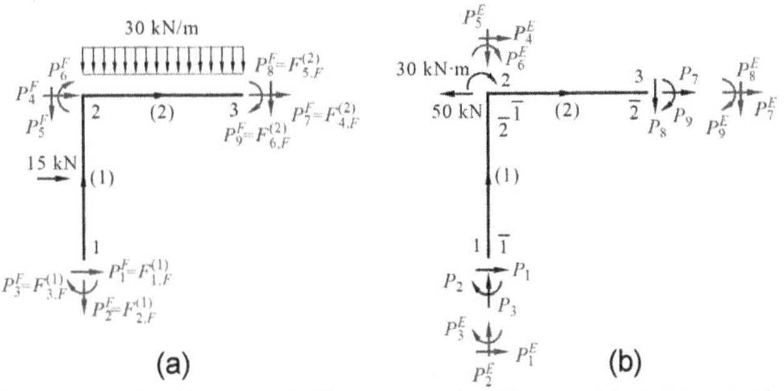

Figure 7.20. The unconstrained frames. (a) The unconstrained frame subjected to axial loading. (b) The unconstrained frame subjected to nodal loading.

A digital representation of the frame is illustrated in Table 7.6.

$$\boldsymbol{P} \equiv \begin{Bmatrix} \boldsymbol{P}_1 \\ \boldsymbol{P}_2 \\ \boldsymbol{P}_3 \end{Bmatrix} \equiv \begin{Bmatrix} P_1 \\ P_2 \\ P_3 \\ \hline P_4 \\ P_5 \\ P_6 \\ \hline P_7 \\ P_8 \\ P_9 \end{Bmatrix}; \quad \boldsymbol{P}^{\mathrm{F}} \equiv \begin{Bmatrix} \boldsymbol{P}_1^{\mathrm{F}} \\ \boldsymbol{P}_2^{\mathrm{F}} \\ \boldsymbol{P}_3^{\mathrm{F}} \end{Bmatrix} \equiv \begin{Bmatrix} P_1^{\mathrm{F}} \\ P_2^{\mathrm{F}} \\ P_3^{\mathrm{F}} \\ \hline P_4^{\mathrm{F}} \\ P_5^{\mathrm{F}} \\ P_6^{\mathrm{F}} \\ \hline P_7^{\mathrm{F}} \\ P_8^{\mathrm{F}} \\ P_9^{\mathrm{F}} \end{Bmatrix}; \quad \boldsymbol{P}^{\mathrm{E}} \equiv \begin{Bmatrix} \boldsymbol{P}_1^{\mathrm{E}} \\ \boldsymbol{P}_2^{\mathrm{E}} \\ \boldsymbol{P}_3^{\mathrm{E}} \end{Bmatrix} \equiv \begin{Bmatrix} P_1^{\mathrm{E}} \\ P_2^{\mathrm{E}} \\ P_3^{\mathrm{E}} \\ \hline P_4^{\mathrm{E}} \\ P_5^{\mathrm{E}} \\ P_6^{\mathrm{E}} \\ \hline P_7^{\mathrm{E}} \\ P_8^{\mathrm{E}} \\ P_9^{\mathrm{E}} \end{Bmatrix} \tag{7.102}$$

as the original nodal loading, the ***fixed-node nodal loading*** and the ***equivalent nodal loading***, in which:

$$\boldsymbol{P}^{\mathrm{E}} = -\boldsymbol{P}^{\mathrm{F}} \tag{7.103}$$

To determine $\boldsymbol{P}^{\mathrm{F}}$, one can draw the free-body diagrams of the frame's members, as shown in Figure 7.21, in which

$$\boldsymbol{F}_{\mathrm{F}}^{(e)} \equiv \begin{Bmatrix} \boldsymbol{F}_{1,\mathrm{F}}^{(e)} \\ \boldsymbol{F}_{2,\mathrm{F}}^{(e)} \end{Bmatrix} \equiv \begin{Bmatrix} F_{1,\mathrm{F}}^{(e)} \\ F_{2,\mathrm{F}}^{(e)} \\ F_{3,\mathrm{F}}^{(e)} \\ \hline F_{4,\mathrm{F}}^{(e)} \\ F_{5,\mathrm{F}}^{(e)} \\ F_{6,\mathrm{F}}^{(e)} \end{Bmatrix}, \qquad e = 1, 2 \tag{7.104}$$

is the fixed-end force of element (e) in the global coordinates, and can be obtained by the following trans-formation:

$$\boldsymbol{F}_{\mathrm{F}}^{(e)} \equiv \begin{Bmatrix} \boldsymbol{F}_{1,\mathrm{F}}^{(e)} \\ \boldsymbol{F}_{2,\mathrm{F}}^{(e)} \end{Bmatrix} \equiv \left(\underset{\bar{e} \leftarrow g}{\boldsymbol{T}}^{(e)} \right)^{\mathrm{T}} \begin{Bmatrix} \bar{\boldsymbol{F}}_{1,\mathrm{F}}^{(e)} \\ \bar{\boldsymbol{F}}_{2,\mathrm{F}}^{(e)} \end{Bmatrix} \equiv \left(\underset{\bar{e} \leftarrow g}{\boldsymbol{T}}^{(e)} \right)^{\mathrm{T}} \bar{\boldsymbol{F}}_{\mathrm{F}}^{(e)} \tag{7.105}$$

Specifically, one has:

$$\bar{\boldsymbol{F}}^{(1)} = \begin{Bmatrix} 0 \\ -7.5 \\ -15 \\ \hline 0 \\ -7.5 \\ 15 \end{Bmatrix} \times 10^3; \qquad \boldsymbol{F}^{(1)} = \begin{Bmatrix} -7.5 \\ 0 \\ -15 \\ \hline -7.5 \\ 0 \\ 15 \end{Bmatrix} \times 10^3 \tag{7.106}$$

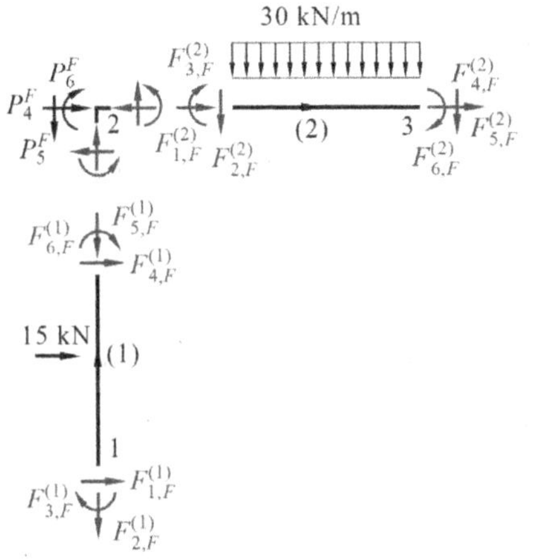

Figure 7.21. Free-body diagrams of the members of the frame.

and

$$\bar{\boldsymbol{F}}^{(2)} = \left\{\begin{array}{c} 0 \\ -120 \\ -160 \\ \hline 0 \\ -120 \\ 160 \end{array}\right\} \times 10^3; \qquad \boldsymbol{F}^{(2)} = \left\{\begin{array}{c} 0 \\ -120 \\ -160 \\ \hline 0 \\ -120 \\ 160 \end{array}\right\} \times 10^3 \tag{7.107}$$

Figure 7.19b and Figure 7.21 show that the fixed-node loading $\boldsymbol{P}^{\mathrm{F}}$ can be expressed as:

$$\boldsymbol{P}^{\mathrm{F}} = \left\{\begin{array}{c} P_1^{\mathrm{F}} \\ P_2^{\mathrm{F}} \\ P_3^{\mathrm{F}} \\ P_4^{\mathrm{F}} \\ P_5^{\mathrm{F}} \\ P_6^{\mathrm{F}} \\ P_7^{\mathrm{F}} \\ P_8^{\mathrm{F}} \\ P_9^{\mathrm{F}} \end{array}\right\} \equiv \left\{\begin{array}{c} F_{1,F}^{(1)} \\ F_{2,F}^{(1)} \\ F_{3,F}^{(1)} \\ \hline F_{4,F}^{(1)} \\ F_{5,F}^{(1)} \\ F_{6,F}^{(1)} \\ \hline 0 \\ 0 \\ 0 \end{array}\right\} + \left\{\begin{array}{c} 0 \\ 0 \\ 0 \\ \hline F_{1,F}^{(2)} \\ F_{2,F}^{(2)} \\ F_{3,F}^{(2)} \\ \hline F_{4,F}^{(2)} \\ F_{5,F}^{(2)} \\ F_{6,F}^{(2)} \end{array}\right\} = \left\{\begin{array}{c} -7.5 \\ 0 \\ -15 \\ \hline 7.5 \\ 120 \\ 145 \\ \hline 0 \\ -120 \\ 160 \end{array}\right\} \times 10^3 \tag{7.108}$$

For convenience in computer code implementation, Equation (7.108) can be also obtained by assembling $\boldsymbol{F}_{\mathrm{F}}^{(e)}$, $e = 1, 2$, into $\boldsymbol{P}^{\mathrm{F}}$ according to the element-DOF connectivity:

$$(1)\begin{cases} \bar{U}_1^{(1)} = U_1 \\ \bar{U}_2^{(1)} = U_2 \\ \bar{U}_3^{(1)} = U_3 \\ \bar{U}_4^{(1)} = U_4 \\ \bar{U}_5^{(1)} = U_5 \\ \bar{U}_6^{(1)} = U_6 \end{cases} \quad \text{and} \quad (2)\begin{cases} \bar{U}_1^{(2)} = U_4 \\ \bar{U}_2^{(2)} = U_5 \\ \bar{U}_3^{(2)} = U_6 \\ \bar{U}_4^{(2)} = U_7 \\ \bar{U}_5^{(2)} = U_8 \\ \bar{U}_6^{(2)} = U_9 \end{cases}$$

In general, for an element (e) with the connectivity

$$(e)\begin{cases} \bar{1} \to i \\ \bar{2} \to j \end{cases} \tag{7.109}$$

its end forces

$$\boldsymbol{F}_{\mathrm{F}}^{(e)} \equiv \begin{Bmatrix} \boldsymbol{F}_{1,\mathrm{F}}^{(e)} \\ \boldsymbol{F}_{2,\mathrm{F}}^{(e)} \end{Bmatrix} \tag{7.110}$$

should be assembled into $\boldsymbol{P}^{\mathrm{F}}$ according to the following rule:

$$\begin{cases} \boldsymbol{F}_{1,\mathrm{F}}^{(e)} \to \boldsymbol{F}_i^{\mathrm{F}} \\ \boldsymbol{F}_{2,\mathrm{F}}^{(e)} \to \boldsymbol{F}_j^{\mathrm{F}} \end{cases} \tag{7.111}$$

where

$$\boldsymbol{P}_i^{(e)} \equiv \begin{Bmatrix} P_{3i-2} \\ P_{3i-1} \\ P_{3i} \end{Bmatrix}; \quad \boldsymbol{P}_j^{(e)} \equiv \begin{Bmatrix} P_{3j-2} \\ P_{3j-1} \\ P_{3j} \end{Bmatrix} \tag{7.112}$$

7.4.2 The Stiffness Equation of the Structure

As soon as the fixed-end loading $\boldsymbol{P}^{\mathrm{F}}$ is assembled, one can obtain the equivalent nodal loading $\boldsymbol{P}^{\mathrm{F}}$ using Equation (7.103). Then the stiffness equation of the unconstrained frame subjected to the nodal loads $\boldsymbol{P}$ and $\boldsymbol{P}^{\mathrm{F}}$ is:

$$\boldsymbol{K}^{\mathrm{fr}}\boldsymbol{U} = \boldsymbol{P} + \boldsymbol{P}^{\mathrm{F}} \tag{7.113}$$

After imposing the boundary conditions (7.101), one obtains the stiffness equation of the supported frame as follows:

$$\boldsymbol{K}^{\mathrm{fr}}\boldsymbol{U}^{\mathrm{act}} = \boldsymbol{P}^{\mathrm{act}} + \boldsymbol{P}_{\mathrm{act}}^{\mathrm{F}} \tag{7.114}$$

where

$$\boldsymbol{K} = 10^5 \times \begin{bmatrix} 2472.6 & 0 & -2.2248 \\ 0 & 2472.6 & 2.2248 \\ -2.2248 & 2.2248 & 23.731 \end{bmatrix} \tag{7.115}$$

$$\boldsymbol{U}^{\text{act}} = [U_4 \;\; U_5 \;\; U_6] \tag{7.116}$$

$$\boldsymbol{P}^{\text{act}} = \begin{Bmatrix} P_4 \\ P_5 \\ P_6 \end{Bmatrix} = \begin{Bmatrix} -50 \\ 0 \\ 30 \end{Bmatrix} \times 10^3 \tag{7.117}$$

and

$$\boldsymbol{P}_{\text{act}}^{\text{E}} = \begin{Bmatrix} P_4^{\text{E}} \\ P_5^{\text{E}} \\ P_6^{\text{E}} \end{Bmatrix} = -\begin{Bmatrix} P_4^{\text{F}} \\ P_5^{\text{F}} \\ P_6^{\text{F}} \end{Bmatrix} = -\begin{Bmatrix} F_4^{(1)} + F_1^{(2)} \\ F_5^{(1)} + F_2^{(2)} \\ F_6^{(1)} + F_3^{(2)} \end{Bmatrix} = -\begin{Bmatrix} 7.5 \\ 120 \\ 145 \end{Bmatrix} \times 10^3 \tag{7.118}$$

7.4.3 The End Displacement and Forces of the Structure

Solving Equation (7.114) yields:

$$\boldsymbol{U}^{\text{act}} = [-1.0558 \times 10^{-4} \;\; 4.1902 \times 10^{-4} \;\; 7.3693 \times 10^{-4}]^{\text{T}} \tag{7.119}$$

As the original structure is the superposition of cases (a) and (b) shown in Figure 7.19, and since the nodal displacements of cases (a) in Figure 7.19 are zero, one can obtain the end displacements for each element from Figure 7.19(b) as follows:

$$\boldsymbol{U}^{(1)} = \begin{Bmatrix} 0 \\ 0 \\ 0 \\ \hline -1.0558 \times 10^{-4} \\ 4.1902 \times 10^{-4} \\ 7.3693 \times 10^{-2} \end{Bmatrix}; \qquad \boldsymbol{U}^{(2)} = \begin{Bmatrix} -1.0558 \times 10^{-4} \\ 4.1902 \times 10^{-4} \\ 7.3693 \times 10^{-2} \\ \hline 0 \\ 0 \\ 0 \end{Bmatrix} \tag{7.120}$$

It is noticed that the end forces, $\overline{\boldsymbol{F}}^{(e)}$, is the superposition of those in Figure 7.19(a) and (b), and should be expressed as:

$$\overline{\boldsymbol{F}}^{(e)} = \overline{\boldsymbol{k}}^{(e)} \underset{\bar{e} \leftarrow g}{\boldsymbol{T}}^{(e)} \boldsymbol{U}^{(e)} + \overline{\boldsymbol{F}}_{\text{F}}^{(e)}, \qquad e = 1, 2 \tag{7.121}$$

which yields:

$$\overline{\boldsymbol{F}}^{(1)} = \begin{Bmatrix} 1.0358 \times 10^5 \\ 8.9012 \times 10^3 \\ 2.8744 \times 10^4 \\ \hline -1.0358 \times 10^5 \\ -2.3901 \times 10^4 \\ 1.0247 \times 10^5 \end{Bmatrix}; \qquad \overline{\boldsymbol{F}}^{(2)} = \begin{Bmatrix} -2.6099 \times 10^4 \\ -1.0358 \times 10^5 \\ -7.2465 \times 10^4 \\ \hline 2.6099 \times 10^4 \\ -1.3642 \times 10^5 \\ 2.0381 \times 10^5 \end{Bmatrix} \tag{7.122}$$

In the discussion above, Equation (7.120) is obtained by using Figure 7.19b. For convenience in computer code implementation, they may be obtained by the following purely digital process:

1. Assemble the prescribed and active DOFs to form the full end-displacement vector $\boldsymbol{U}$.
2. Calculate $\boldsymbol{U}^{(e)}$ by using $\boldsymbol{U}$ and the element connectivity.

EXERCISES

7.1 Use matrix displacement method to calculate the rotations of node B of the frames shown in Figure 7.22a, in which support A undergoes a counterclockwise rotation of 0.01rad. The Young's modulus, the cross-sectional area, and the moment of inertia of each member are $E = 2.06 \times 10^{11}$ N/m², $A = 9.6 \times 10^{-3}$ m², $I_z = 1.152 \times 10^{-5}$ m⁴, respectively.

7.2 Use the matrix displacement method to calculate the rotations of node B of the frames shown in Figure 7.22b, in which support A undergoes a settlement of 5 cm. The Young's modulus, the cross-sectional area and the moment of inertia of each member are $E = 2.06 \times 10^{11}$ N/m², $A = 9.6 \times 10^{-3}$ m², $I_z = 1.152 \times 10^{-5}$ m⁴, respectively.

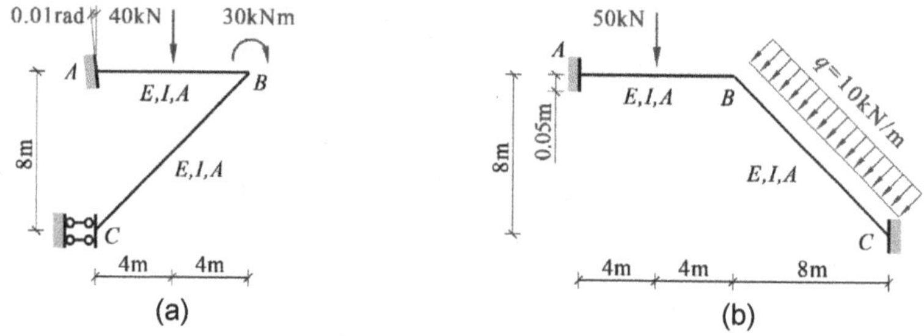

Figure 7.22. (a) Exercise 7.1. (b) Exercise 7.2.

Chapter 8
Dynamics of Structures

8.1 Introduction to Structural Dynamics

8.1.1 What is Structural Dynamics?

Structural Dynamics is a field of structural analysis which covers the responses of structures subjected to dynamic loading. Intuition may suggest that dynamic loads include wind, earthquakes and blasts impacting structures. However, the academic definition of dynamic loading is a bit complicated. *If the inertial forces are important for the analysis of the structural response to a load, then it is called a* ***dynamic loading***, and the problem involved is known as a ***structural dynamics problem***. In contrast, if the acceleration of the structure can be ignored in the analysis, then it is a static problem and the load involved is referred to as static loading. As might be expected intuitively, a time-varying excitation is usually a dynamic load. For example, the sinusoidal excitation shown in Figure 8.1a is a dynamic load applied to the cantilever. However, in some cases, a time-varying loading may be taken as a static load.

Figure 8.1b shows a car moving on a bridge. If the car moves slowly and steadily, then the loading applied on the bridge by the car may be regarded as a static load because the inertial force can be neglected in this case. Such a problem can be solved by the influence line method presented previously. Another example contrary to intuition is the ***step load***, P_a, which is suddenly applied to the bridge at time $t = \tau$, shown in Figure 8.2, and remains constant. This loading may be taken as a dynamic load because the inertial force cannot be ignored in the analysis.

Usually, some prior knowledge of a problem is required to distinguish between a static and a dynamic loading. This immediately raises a fundamental question: what

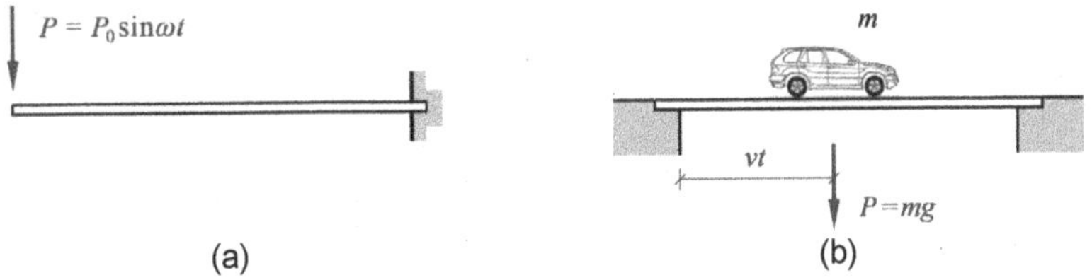

Figure 8.1. (a) A cantilever subjected to a sinusoidal loading. (b) A bridge subjected to a moving load.

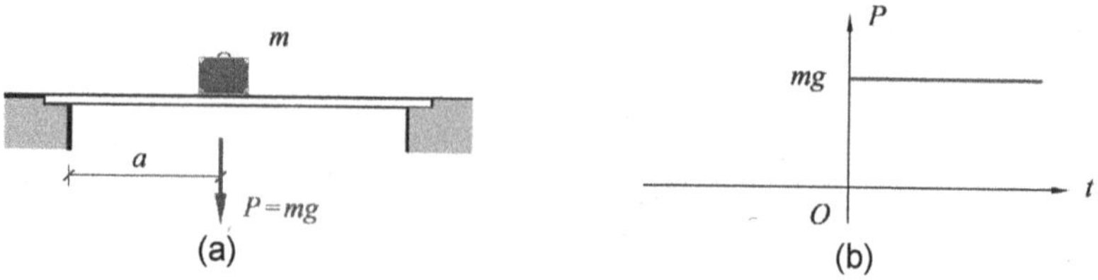

Figure 8.2. (a) A beam subjected to a step loading. (b) The time-history of the step loading.

should one do for a new problem without prior knowledge? In this case, one has to compare the results from static and dynamic analysis. If the difference between the results can be omitted, it means that the inertial effects are not significant, and one can treat the problem as a static one. Otherwise, it is a dynamic problem.

In the rest of this section, a simple example will be discussed to illustrate the general procedure of the dynamic analysis of structures. Some fundamental concepts will be introduced in the discussion.

8.1.2 Models for Dynamic Analysis

A 5-story building shown in Figure 8.3a is subjected to a wind drag force, which is regarded as a dynamic loading in the discussion that follows. The top displacement, the base shear, and the overturning moment (i.e., the base moment) will be examined, which are relevant for both structural engineering concept design and more detailed design of the building.

In structural analysis, a building is represented as a structural model by removing the walls, the windows, and the facilities in the building. For example, the 5-story building is usually modeled as a frame shown in Figure 8.3b, in which the wind action on the building is modeled as a time-varying distributed load on the frame. This is the first simplification. Since the structural dynamics covers only the dynamic properties of structures, to simplify the analysis, a further simplification is frequently introduced to model a simpler model for the dynamic analysis.

It is obvious that the structural model for the dynamic analysis of the building is not unique. For example, Figure 8.4 shows four possible structural models, which may be used in the dynamic analysis of the 5-story building. In Figure 8.4a, the multistorey is modeled as a cantilever beam with distributed mass, and subjected to a

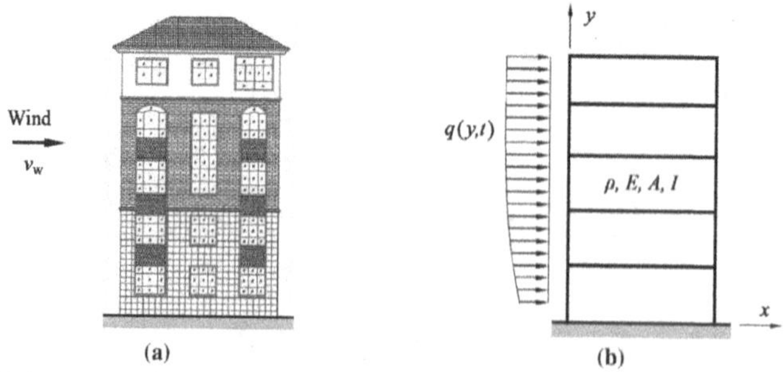

Figure 8.3. (a) A 5-story build is designed to resist wind. (b) The structural model for the analysis of wind resistance.

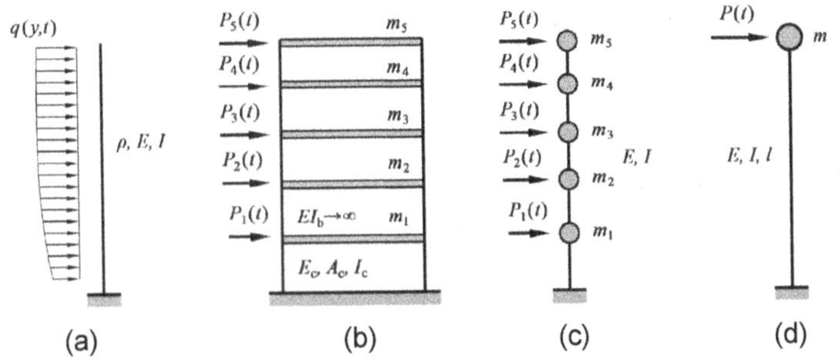

Figure 8.4. (a) A vertical cantilever beam with distributed mass for the dynamic analysis. (b) A frame with lumped masses for the dynamic analysis. (c) A vertical cantilever beam with 5 lumped masses for the dynamic analysis. (d) A vertical cantilever beam with a lumped mass.

distributed time-varying load $q(y, t)$. Figure 8.4b shows a frame model with 5 lumped masses at the floors. The building may also be modeled as a cantilever beam with lumped-masses and subjected to several concentrated time-varying loads $P_i(t)$, as described in Figure 8.4c. While the simplest model is shown in Figure 8.4d, in which a vertical cantilever beam with only one lumped mass at the beam end is subjected to a concentrated time-varying load $P(t)$ at the mass point.

In a dynamic analysis, a ***degree of freedom*** (DOF) of a dynamic system is defined as *an independent parameter required to locate the position of a mass point on the structural model*. For the simplest model shown in Figure 8.4d, by neglecting the axial deformation of the cantilever beam, the model has only one degree of freedom—the horizontal deflection at the mass point at time t because the axial deformation of the cantilever beam is neglected. Thus, it is a ***single-degree-of-freedom*** (SDOF) system. For the frame model shown in Figure 8.4b, the deflections at the floors are required to define the position of the masses on the system. It is a ***multi-degree-of-freedom*** (MDOF) system. Similarly, the model shown in Figure 8.4c is also a MDOF system. However, the cantilever model shown in Figure 8.4a has an infinite number of degrees of freedom (DOFs) because its mass is distributed rather than lumped. It is an ***infinite-degree-of-freedom*** (infinite-DOF) system. The model shown in Figure 8.4a is also known as a ***continuum model*** for dynamic analysis. The models shown in Figures 8.4b–c are referred to as the discrete models for dynamic systems.

The SDOF system shown in Figure 8.4d forms a simple structural model for the dynamic analysis of the wind-induced vibration of the 5-story building.

A complete model for dynamic analysis includes two parts: the structural model and the loading model. In the discrete models shown in Figures 8.4b–d, the wind actions are modeled as concentrated loading. The time history of a concentrated wind load may be modeled as a stochastic process shown in Figure 8.5a, a non-periodic deterministic load shown in Figure 8.5b, or a periodic deterministic load shown in Figure 8.5c. After selecting the SDOF system as the model for analysis, one needs to specify the mathematical description of the wind load on the system. For simplicity, it is assumed that the wind load on the SDOF system can be expressed as:

$$P(t) = P_s + P_d(t)$$

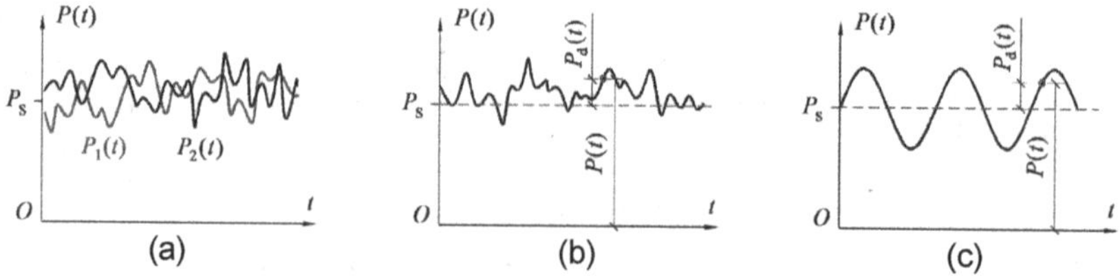

Figure 8.5. Mathematical presentations of load time history of a wind load may be modeled as: (a) a stochastic process, (b) a non-periodic deterministic load, (c) a periodic deterministic load.

where, P_s is the *equivalent static wind load*; P_d is the dynamic component of the wind load on the system and is called the *fluctuating wind load* associated with the wind turbulence. In this example, P_d is assumed to be a simple harmonic.

8.1.3 Equations of Motion and Initial Conditions

The state of a dynamic system is represented by a set of dynamic variables, normally the time-varying spatial coordinates of the masses on the system, which is governed by the ***equations of motion*** of the system. Thus, the first step to formulate equations of motions is to define a ***space-time coordinate system*** (i.e., a ***reference frame***) to describe the motion of the system.

As shown in Figure 8.6, we define a space-time coordinate system, whose origin is located at the fixed end of the cantilever, and whose time coordinate t starts at a time such that the fluctuating wind load $P_d(t)$ can be expressed as:

$$P_d(t) = P_0 \sin \tilde{\omega} t \tag{8.1}$$

where P_0 is the amplitude of the fluctuating wind loads.

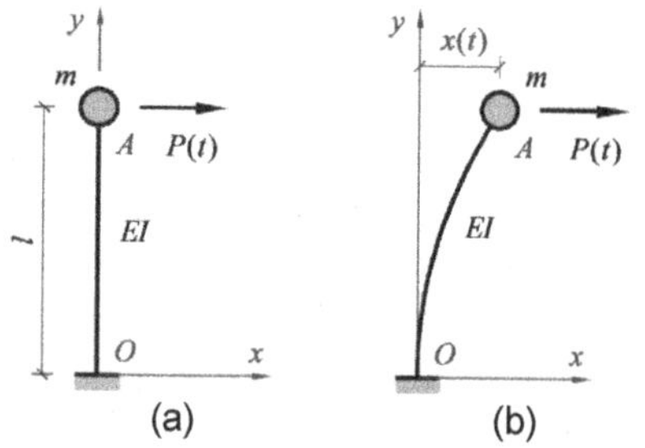

Figure 8.6. A space-time coordinate system: (a) the origin of the space coordinates is located at the fixed end of the cantilever, (b) deformed shape (illustrative) of the cantilever.

An equation of motion is the mathematical representation of the dynamic-equilibrium requirement in terms of the dynamic variables. To describe the dynamic equilibrium of the SDOF system, we draw the free-body diagram of the lumped mass m at time t, as shown in Figure 8.7a. F_s is the ***elastic restoring force*** exerted by the cantilever beam on the lumped mass at time t in a direction opposing the deformation. k denotes the ***lateral stiffness*** of the vertical cantilever, i.e., k is the

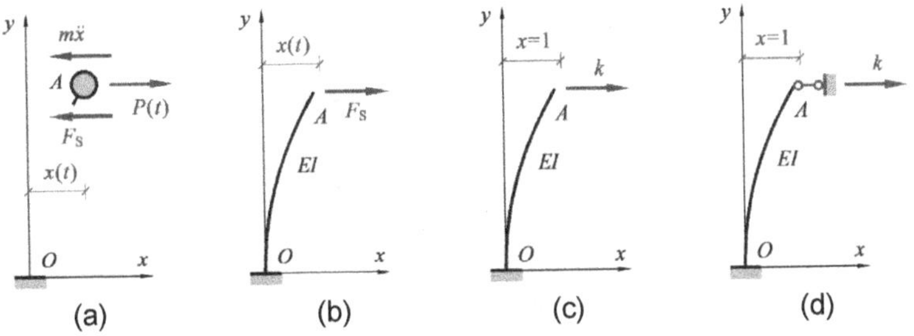

Figure 8.7. (a) Free-body diagram of the dynamic-equilibrium of the SDOF system, (b) F_s is the ***elastic restoring force*** exerted by the cantilever beam on the lumped mass at time t, (c) k denotes the lateral stiffness, (d) k is equal to the reaction from the roller.

force causing the massless cantilever to bend as shown in Figure 8.7(c). Then, the restoring force F_s can be expressed as:

$$F_s = kx(t) \tag{8.2}$$

The dynamic equilibrium in the direction of the unique DOF, i.e., the direction of the x-axis, yields:

$$m\ddot{x}(t) + F_s = P(t) \tag{8.3}$$

where a dot over x denotes differentiation with respect to time: thus $\dot{x}$ denotes the velocity of the lumped mass and $\ddot{x}$ its acceleration.

Substituting Equations (8.1) and (8.2) into Equation (8.3) gives:

$$m\ddot{x}(t) + kx(t) = P_s + P_d(t) \tag{8.4}$$

where $P_d(t)$ can be written as Equation (8.1).

Because k is equal to the reaction from the roller shown in Figure 8.7d, it can be expressed as:

$$k = \frac{3EI}{l^3} \tag{8.5}$$

where EI is the flexural rigidity of the cantilever.

Plugging Equation (8.5) into Equation (8.4) leads to:

$$m\ddot{x}(t) + \frac{3EI}{l^3}x(t) = P_s + P_d(t) \tag{8.6}$$

This is the forced-vibration equation governing the motion of the SDOF system. We have to specify the ***initial conditions*** of the system (i.e., the state of the system at time $t = 0$) when determining the dynamic response of the system to the wind loads. In this example, we seek the solution of Equation (8.6) subject to the following initial conditions:

$$\begin{cases} x(0) = \tilde{x}_0 \\ \dot{x}(0) = \tilde{v}_0 \end{cases} \tag{8.7}$$

where $\tilde{x}$ and $\tilde{v}$ are given constants.

Let x_s denote the ***static displacement*** caused by the equivalent static wind load P_s, i.e.,

$$x_s = \frac{P_s}{k} = \frac{P_s l^3}{3EI} \tag{8.8}$$

and $u(t)$ be the additional dynamic displacement, i.e.,

$$u(t) \equiv x(t) - x_s \quad \text{or} \quad x(t) \equiv x_s + u(t) \tag{8.9}$$

Substitution of Equations (8.8) and (8.9) into Equation (8.6) and the initial conditions (8.7) yields the following forced-vibration equation in terms of the additional dynamic displacement:

$$m\ddot{u}(t) + \frac{3EI}{l^3}u(t) = P_d(t) = P_0 \sin \tilde{\omega}t \tag{8.10}$$

where initial conditions are:

$$\begin{cases} u(0) = \tilde{x}_0 - \dfrac{P_s l^3}{3EI} \\ \dot{u}(0) = \tilde{v}_0 \end{cases} \tag{8.11}$$

Equation (8.10) is the mathematical representation of the dynamic system having a single degree of freedom, and is known as the mathematical model of the dynamic system.

8.1.4 Free Vibrations and Dynamic Properties

Before seeking the solution to the forced-vibration Equation (8.10), one needs to investigate the free vibration of the SDOF system. By omitting the contribution from damping (which will be discussed in Section 8.2.5 and 8.3), the free vibrations of the SDOF are governed by the following homogeneous differential equation:

$$m\ddot{u}(t) + ku(t) = 0 \tag{8.12}$$

Recall that the general solution of the homogeneous Equation (8.12) can be expressed as the linear combination of two linearly independent solutions. It is easy to verify that Equation (8.12) has the following linearly independent solutions:

$$u_1(t) = \cos \omega t \tag{8.13}$$

$$u_2(t) = \sin \omega t \tag{8.14}$$

where,

$$\omega = \sqrt{\frac{k}{m}} \tag{8.15}$$

Thus, the general solution, $u_c(t)$, of the free-vibration Equation (8.12) can be expressed as:

$$u_c(t) = A \cos \omega t + B \sin \omega t \tag{8.16}$$

where A and B are constants not yet determined and may be evaluated from specific initial conditions. For example,

$$u(t) = 0.05 \cos \omega t \tag{8.17}$$

(i.e., A = 0.05 and B = 0) is the solution to Equation (8.12) subject to the initial conditions

$$u(0) = 0.05 \quad \text{and} \quad \dot{u}(0) = 0 \tag{8.18}$$

The general solution (8.16) is called the ***complementary solution*** to the original forced-vibration equation, and it can proved that, for any given initial conditions, the SDOF system without any external excitation will vibrate at the same frequency ω. As ω totally depends on the natural properties of the system, it is called the ***natural frequency*** of the SDOF system.

8.1.5 Dynamic Responses to External Excitations

The general solution of the inhomogeneous Equation (8.10) can be expressed as the sum of the complementary solution $u_c(t)$ and an arbitrary particular solution $u_p(t)$, i.e.,

$$u(t) = u_c(t) + u_p(t) \tag{8.19}$$

where $u_c(t)$ is described by Equation (8.16).

To seek a particular solution to the inhomogeneous differential Equation (8.10), take a trial solution of the form:

$$u_p = C \sin \tilde{\omega} t \tag{8.20}$$

and one needs to find C such that the trial solution (8.20) satisfies Equation (8.10). Substituting the trial solution (8.20) into Equation (8.10) gives:

$$-mC\,\tilde{\omega}^2 \sin \tilde{\omega} t + kC \sin \tilde{\omega} t = P_0 \sin \tilde{\omega} t \tag{8.21}$$

which yields:

$$C = \frac{P_0}{k}\left[\frac{1}{1-(\tilde{\omega}/\omega)^2}\right] \tag{8.22}$$

This shows that the inhomogeneous Equation (8.10) has the following particular solution:

$$u_p = \frac{P_0}{k}\left[\frac{1}{1-(\tilde{\omega}/\omega)^2}\right]\sin \tilde{\omega} t \tag{8.23}$$

where $\tilde{\omega}$ and ω are the excitation and the natural frequencies of the SDOF system, respectively.

Plugging the complementary solution (8.16) and the particular solution (8.23) into the expression (8.19) gives the following general solution of the inhomogeneous Equation (8.10):

$$u(t) = A\cos \omega t + B\sin \omega t + \frac{P_0}{k}\left[\frac{1}{1-(\tilde{\omega}/\omega)^2}\right]\sin \tilde{\omega} t \tag{8.24}$$

where A and B are constants yet undetermined and can be evaluated by initial conditions. For example, if the system starts from the at-rest initial conditions, i.e.,

$$u(0) = 0 \quad \dot{u}(0) = 0 \tag{8.25}$$

one then has:

$$A = 0, \quad \text{and} \quad B = -\frac{P_0}{k}\left[\frac{\tilde{\omega}/\omega}{1-(\tilde{\omega}/\omega)^2}\right] \tag{8.26}$$

which yields:

$$u(t) = \frac{P_0}{k}\left[\frac{1}{1-(\tilde{\omega}/\omega)^2}\right]\left[\sin\tilde{\omega}t - (\tilde{\omega}/\omega)\sin\omega t\right] \tag{8.27}$$

This is the mathematical description of vibrations for the SDOF system induced by the fluctuating wind load. At time t, the top displacement $x(t)$ of the cantilever beam is the superposition of the static displacement and the additional dynamic displacement, i.e.,

$$\begin{aligned} x(t) &= \frac{P_s}{k} + \frac{P_0}{k}\left[\frac{1}{1-(\tilde{\omega}/\omega)^2}\right]\left[\sin\tilde{\omega}t - (\tilde{\omega}/\omega)\sin\omega t\right] \\ &= \frac{P_s l^3}{3EI} + \underbrace{\frac{P_0 l^3}{3EI}\left[\frac{1}{1-(\tilde{\omega}/\omega)^2}\right]\left[\sin\tilde{\omega}t - (\tilde{\omega}/\omega)\sin\omega t\right]}_{\text{the additional dynamic component}} \end{aligned} \tag{8.28}$$

where Equations (8.5), (8.8) and (8.27) are implemented.

Finally, one needs to calculate the overturning-moment and base-shear responses of the SDOF system, induced by the wind loading. Since the cantilever beam is assumed to be massless, the moment and shear at the fixed end at time t is equal to those of the clamped-roller beam subjected to a displacement $x(t)$ as shown in Figure 8.8. Thus, the overturning-moment, M_B, can be expressed as:

$$\begin{aligned} M_B &= \frac{3EI}{l^2}x(t) \\ &= P_s l + \underbrace{P_0 l\left[\frac{1}{1-(\tilde{\omega}/\omega)^2}\right]\left[\sin\tilde{\omega}t - (\tilde{\omega}/\omega)\sin\omega t\right]}_{\text{the additional dynamic component}} \end{aligned} \tag{8.29}$$

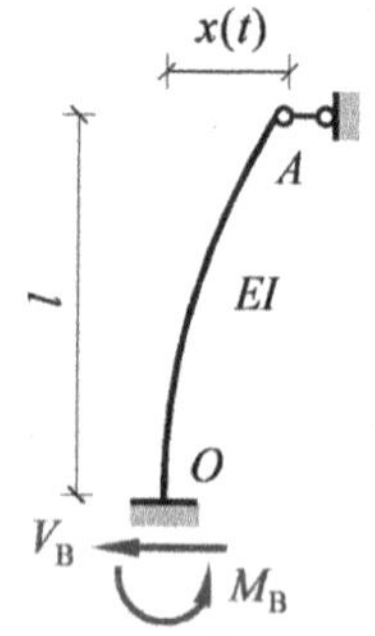

Figure 8.8. A clamped-roller beam subjected to a support displacement.

which gives the base shear V_B:

$$V_{\mathrm{B}} = P_{\mathrm{s}} + P_0 \underbrace{\left[\frac{1}{1-(\tilde{\omega}/\omega)^2}\right]\left[\sin\tilde{\omega}t - (\tilde{\omega}/\omega)\sin\omega t\right]}_{\text{the additional dynamic component}} \tag{8.30}$$

8.1.6 Summary

The essential steps in the dynamic analysis of a structure are summarized as below:

1) Construct a model for dynamic analysis by introducing some secondary assumptions, such as lumped masses and flexural rigid slabs. Select a model and determine its DOFs for the dynamic analysis.
2) Formulate the equation governing the motions of the structure.
3) Analyze the free vibrations of the system.
4) Calculate the displacement response caused by the dynamic loading.
5) Evaluate the internal-force responses of the system.

8.2 Equations of Motion

The formulation of the mathematical model (i.e., the equations of motion) is the most critical step in any dynamic analysis of structures, because the validity of the calculated results depends directly on how well the mathematical description can represent the behavior of the real physical system.

8.2.1 Stiffness Method: The Dynamic-Equilibrium Procedure

The method to formulate the equation governing the motions of the SDOF system in Section 8.1.3 is called ***the dynamic-equilibrium procedure of the stiffness method*** to establish the equations of motion. The procedure to formulate the equations of motion by the stiffness method is summarized with the following sequential steps:

1) Define a space-time coordinate system, formulate the external excitations, select the dynamic DOFs to locate masses, and select elastic Degrees Of Freedom to describe the deformation and internal forces of the structure.
2) Draw the deformed-configuration diagram of the dynamic system at time t, and label all the Degrees of Freedom for the analysis, including the dynamic Degree Of Freedom and the elastic Degree Of Freedom. For the system shown in Figure 8.6 the unique dynamic Degrees Of Freedom $x_1(t)$ are also the unique elastic Degrees Of Freedom of the system, in terms of which all the internal forces can be expressed.
3) Draw free-body diagrams of each lumped mass, and write the dynamic equilibrium requirements.
4) Specify all the forces exerted on the lumped masses, including the inertial forces induced by the motion of masses.
5) Substituting the formulations of all the forces into the dynamic-equilibrium conditions leads to the equations governing the motions of the system.

EXAMPLE 8.1

Consider a simple beam shown in Figure 8.9a. The flexural rigidity of the beam is EI and the length is l. For convenience, the mass of the beam is assumed to be lumped at the center of the beam. An external harmonic torque $\widetilde{M}_B(t)$ is applied at the right end.

Formulate the equation describing the motion of the simply supported beam with the single lumped mass at its center by using the stiffness method.

Solution:

(1) Define a space-time coordinate system shown in Figure 8.9(b), in which the origin of the spatial coordinates is located at the left end of the beam. The origin of the time coordinate is placed at a time such that the harmonic torque can be expressed as:

$$\widetilde{M}_B(t) = \widetilde{M}_{B0} \sin \tilde{\omega} t \tag{a}$$

where $\widetilde{M}_{B0}$ and $\tilde{\omega}$ are the amplitude and circular frequency of the harmonic excitation.

(2) Draw the deformed-configuration of the system at time t as shown in Figure 8.9(b), where the deflection, $y(t)$, at the center of the beam is the unique dynamic Degree Of Freedom. The deformation and internal forces of the system can be described by two elastic Degrees of Freedom, $y(t)$ and θ, in which θ is the rotation at the center of the beam.

(3) The free-body diagrams of each member of the system are shown in Figure 8.10. The equilibrium of the forces on the lumped mass in the y-direction requires the condition,

$$m\ddot{y} + V_{CA} - V_{CB} - mg = 0 \tag{b}$$

where V_{CA} and V_{CB} are the end shears of the beam segments CA and CB, respectively. See Figures 8.10(a) and 8.10(c).

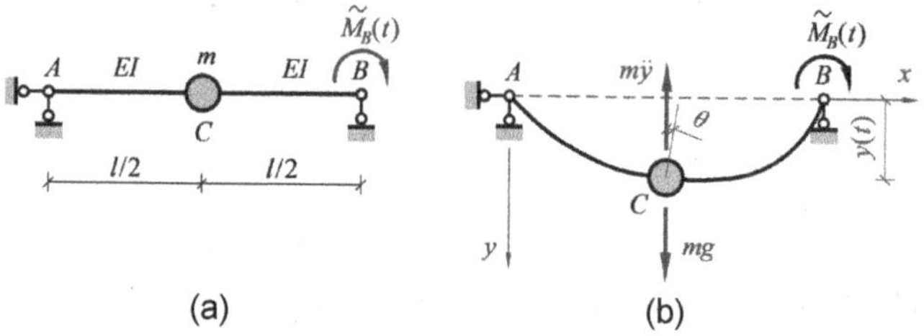

Figure 8.9. (a) A simply supported beam with a lumped mass. (b) The configuration diagram of the dynamic system at time t and the space-time coordinates for the analysis.

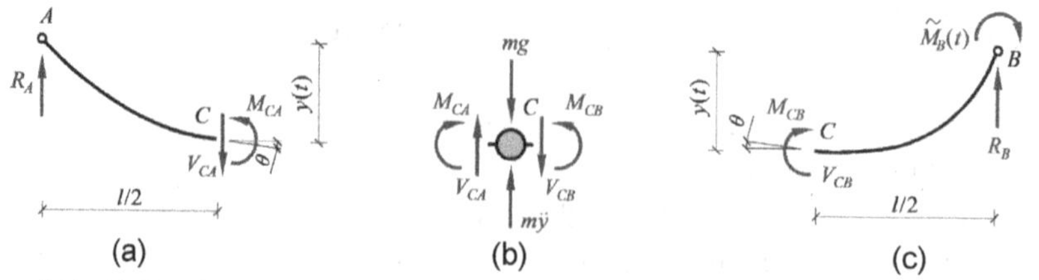

Figure 8.10. (a) The free-body diagram of the beam segment AC. (b) The free-body diagram of the lumped mass. (c) The free-body diagram of the beam segment CB.

(4) Specify all the forces exerted on lumped masses. To formulate the shear V_{CA}, consider the fixed-roller beam subjected to the support displacements, $y(t)$ and θ, as shown in Figure 8.11a. The shear V_{CA} is equal to that at end C of the fixed-roller beam, and can be expressed as:

$$V_{CA} = \frac{24EI}{l^3} y(t) - \frac{12EI}{l^2}\theta \tag{c}$$

Similarly, to formulate the shear V_{CB}, one can consider the fixed-roller beam subjected to support displacements, $y(t)$ and θ, and the harmonic torque $\tilde{M}_B(t)$ at node 3 as shown in Figure 8.11b. The shear V_{CB} is equal to that at end 2 of this fixed-roller beam, and can be expressed as

$$V_{CB} = -\frac{24EI}{l^3} y(t) - \frac{12EI}{l^2}\theta - \frac{3\tilde{M}_B(t)}{l} \tag{d}$$

(5) Formulate the equation governing the motions of the simply supported beam. Substitution of Equations (c) and (d) into Equation (b) gives

$$m\ddot{y} + \frac{48EI}{l^3} y = mg - \frac{3\tilde{M}_B(t)}{l} \tag{e}$$

Let

$$y_s \equiv \frac{mgl^3}{48EI} \tag{f}$$

where, y_s is the static displacement at the center of the beam caused by its weight because $48EI/l^3$ is the stiffness factor corresponding to the elastic Degree Of Freedom, y.

Let $v(t)$ be the additional dynamic displacement of the lumped mass, i.e.,

$$v(t) \equiv y(t) - y_s \tag{g}$$

Plugging Equation (g) into Equation (e) yields:

$$m\ddot{v} + \frac{48EI}{l^3} v = -\frac{3\tilde{M}_{B0}(t)}{l} \tag{h}$$

Equation (h) above is the vibration equation governing the additional dynamic displacement, $v(t)$, of the simply supported beam with a lumped mass.

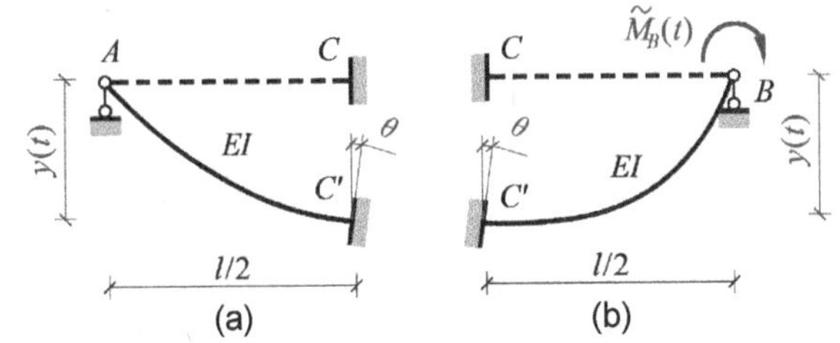

Figure 8.11. (a) The fixed-roller beam subjected to support displacements, whose support reaction is equal to the shear V_{CA}. (b) The fixed-roller beam subjected to support displacements and the harmonic torque, whose support reaction is equal to the shear V_{CB}.

8.2.2 Stiffness Method: The Virtual Constraint Approach

Another process to formulate equations of motions by the stiffness method is the so-called ***virtual-constraint procedure***. This section will illustrate how to establish equations describing the motion of a structure by the virtual-constraint approach.

Consider again the simply supported beam in Example 8.1, which is re-plotted in Figure 8.12a for convenience. Here, let's make a detour to avoid using elastic Degrees Of Freedom. First, consider the ***primary system*** shown in Figure 8.12(b), which is constructed by adding a vibrating *virtual roller 1* to the original dynamic system shown in Figure 8.12a. The primary system is excited by the vertical vibrations of the virtual roller support and the harmonic torque at point *B*.

Let $R_1(t)$ be the reaction from the virtual roller 1. If we can find a vertical motion, $y_1(t)$, of the virtual roller support such that:

$$R_1(t) = 0 \tag{8.31}$$

then the primary system is equivalent to the original dynamic system, i.e., $y_1(t)$ is equal to the deflection at the center of the simple beam shown in Figure 8.12a. Now the original dynamic problem has been transformed to obtain the virtual-support motion, $y_1(t)$, such that the reaction from the virtual support vanishes.

To formulate the reaction, $R_1(t)$, from the virtual roller, one can draw the deformed-configuration diagram of the primary system at time t as shown in Figure 8.13(a) and the free-body diagram of the lumped mass shown in Figure 8.13b. The reaction $R_1(t)$ is the superposition of the reactions from the virtual roller, induced

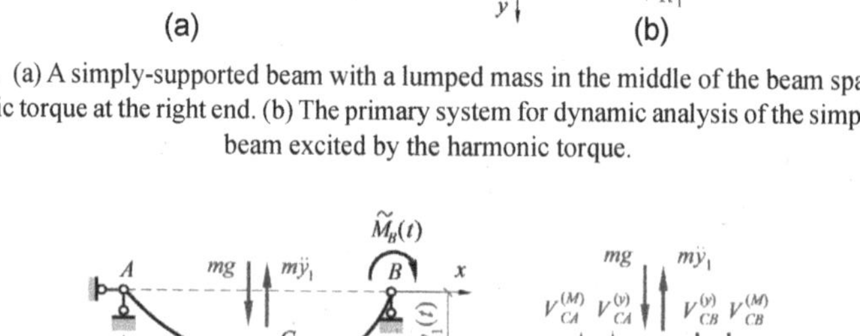

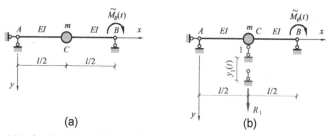

Figure 8.12. (a) A simply-supported beam with a lumped mass in the middle of the beam span, subjected by a harmonic torque at the right end. (b) The primary system for dynamic analysis of the simply supported beam excited by the harmonic torque.

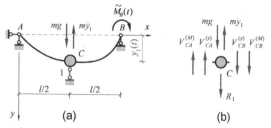

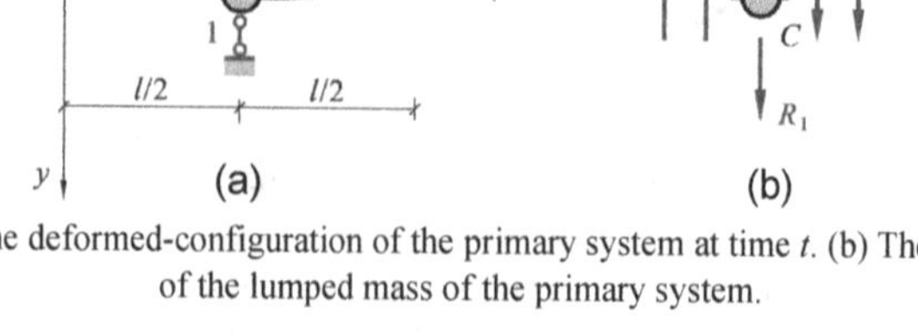

Figure 8.13. (a) The deformed-configuration of the primary system at time t. (b) The free-body diagram of the lumped mass of the primary system.

by the dynamic torque, the inertial force, the weight, and the settlement of the virtual support, i.e.,

$$R_1 = m\ddot{y} - mg + R_1^{(y)} + R_1^{(M)} \tag{8.32}$$

where,

$$R_1^{(y)} = V_{CA}^{(y)} - V_{CB}^{(y)} \tag{8.33}$$

is the reaction caused by the support displacement $y_1(t)$, $V_{ij}^{(y)}$ is the shear force caused by the support displacement; and

$$R_1^{(M)} = V_{CA}^{(M)} - V_{CB}^{(M)} \tag{8.34}$$

is the reaction caused by the dynamic torque and $V_{ij}^{(M)}$ is the shear force caused by the dynamic torque.

The reaction $R_1^{(M)}$ can be expressed as:

$$R_1^{(M)} = f_{1M}\,\widetilde{M}(t) \tag{8.35}$$

where the factor f_{1M} is the reaction from the virtual support 1 induced by a unit torque $\widetilde{M} = 1$ at node B.

To calculate the reaction factor f_{1M}, one can consider the statically indeterminate beam, shown in Figure 8.14a, subjected to a unit torque at point B. The moment diagram of the beam induced by the unit torque can be obtained using the moment distribution approach and is plotted in Figure 8.14b. From this moment diagram one can compute the end shear of each beam segment, which immediately yields the following reaction, f_{1M}, from the roller support:

$$f_{1M} = \overline{V}_{CA}^{(y)} - \overline{V}_{CB}^{(y)} = \frac{1}{2l} - \left(-\frac{5}{2l}\right) = \frac{3}{l} \tag{8.36}$$

where $\overline{V}_{CA}^{(y)}$, $\overline{V}_{CB}^{(y)}$ are the shear forces at points to the left and right of C caused by the unit torque. Plugging Equation (8.36) into Equation (8.35) gives:

$$R_1^{(M)} = \frac{3\widetilde{M}_B(t)}{l} \tag{8.37}$$

On the other hand, the reaction $R_1^{(y)}$ may be written as

$$R_1^{(y)} = k_{11} \cdot y_1(t) \tag{8.38}$$

where k_{11} is the reaction from the virtual support 1, caused by its unit settlement $y_1(t) = 1$.

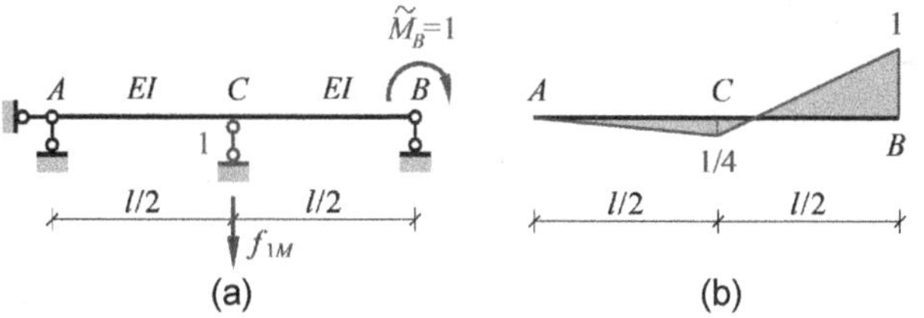

Figure 8.14. (a) A statically indeterminate beam subjected to a unit torque at the right end. (b) The moment diagram of the beam caused by the unit torque.

To compute the stiffness factor k_{11}, consider the statically indeterminate beam, shown in Figure 8.15a, subjected to a unit support settlement. Figure 8.15b shows its moment diagram, which yields:

$$k_{11} = 2\times\frac{12EI/l^2}{l/2} = \frac{48EI}{l^3} \tag{8.39}$$

Substitution of Equation (8.39) into Equation (8.38) gives:

$$R_1^{(y)} = \frac{48EI}{l^3} y_1(t) \tag{8.40}$$

Substituting Equations (8.37) and (8.40) leads to:

$$R_1(t) = m\ddot{y}_1 - mg + \frac{48EI}{l^3} y_1(t) + \frac{3\tilde{M}_B(t)}{l} \tag{8.41}$$

Plugging Equation (8.41) into Equation (8.31) gives:

$$m\ddot{y}_1 + \frac{48EI}{l^3} y_1(t) = mg - \frac{3\tilde{M}_B(t)}{l} \tag{8.42}$$

Equation (8.42) is the vibration equation governing the displacement response $y_1(t)$ at the center of the beam shown in Figure 8.12a.

Let

$$y_{1s} \equiv \frac{mgl^3}{48EI} \tag{8.43}$$

and

$$v_1(t) \equiv y_1(t) - y_{1s} \tag{8.44}$$

where y_{1s} is the static displacement at the center of the beam caused by its weight, and $v_1(t)$ is the additional dynamic displacement of the lumped mass.

Plugging Equation (8.44) into Equation (8.42) yields:

$$m\ddot{v}_1 + \frac{48EI}{l^3} v_1 = -\frac{3\tilde{M}_{B0}}{l} \sin\tilde{\omega}t \tag{8.45}$$

where the Equation (8.43) is used.

Equation (8.45) is the equation describing the additional dynamic deflection, $v_1(t)$, of the simply supported beam shown in Figure 8.12a.

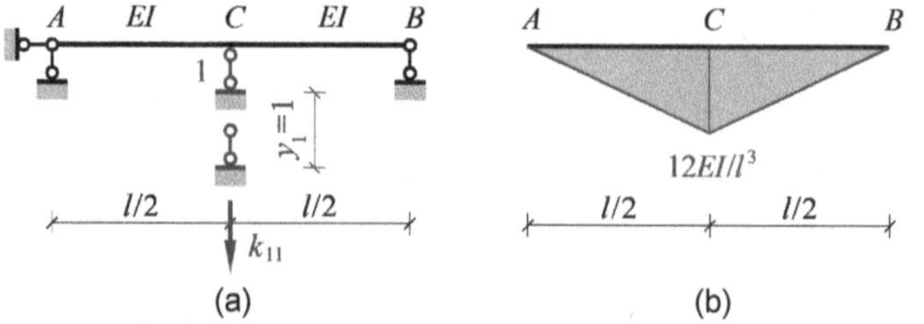

Figure 8.15. (a) A statically indeterminate beam subjected to a unit support settlement. (b) The moment diagram of the beam caused by the unit support settlement.

The virtual-constraint procedure to formulate the equations of the motions is summarized next in a step-by-step form:

1) Define a space-time coordinate system, formulate the external excitations, and label the dynamic DOFs for the analysis.
2) Construct the primary system by adding some virtual vibrating supports to constrain all the dynamic Degrees Of Freedom. The primary system is subjected to the virtual-support motion and the original dynamic loading.
3) Draw the deformed-configuration diagram of the primary system at time t, and exert all the forces on the primary system, including the inertial forces induced by the motions of masses.
4) Formulate the reactions from the virtual constraints caused by the weights, inertial forces, the virtual-support displacements, and the original dynamic loading.
5) Establish the equations of motion according to the zero-reaction condition that all the reactions from virtual constraints should be zero.
6) Formulate the equations governing the additional dynamic displacements by eliminating the static displacements from the equations describing the total displacements of the structure.

The following example illustrates how the virtual-constraint approach will be applied to formulate the equations of the motions of a multi-degree-of-freedom (MDOF) system.

EXAMPLE 8.2

Consider the two-story frame subjected to the lateral dynamic loads, P_g, P_1 and P_2, as shown in Figure 8.16. The flexural rigidity of each column in the first story is EI_1, while the flexural rigidity of each column in the second story is EI_2. The height of each story is h. The mass of this frame is lumped in the girders, with values shown in Figure 8.16, and all the columns are assumed to be weightless. The girders are assumed to be rigid, and the axial deformation of each member can then be neglected. Formulate the equations governing the story displacements of the frame.

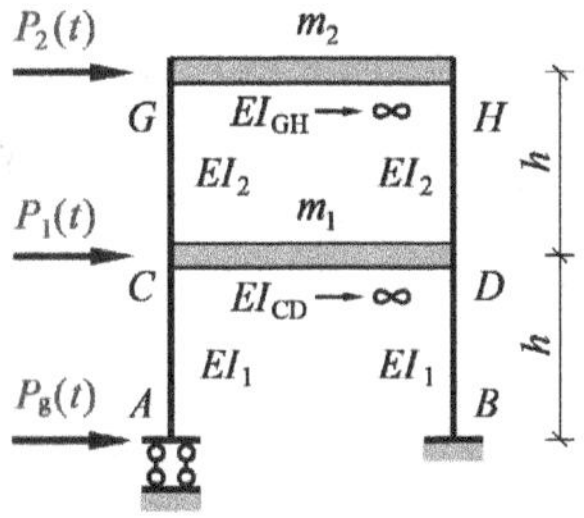

Figure 8.16. A two-story frame subjected to lateral dynamic loading.

***Solution*:**

(1) Construct the primary system by adding two lateral virtual rollers, 1, 2, at points H and D, as shown in Figure 8.17a. The origin of the spatial coordinates is located at the bottom, A, of the left column of the undeformed frame.

(2) Draw the deformed configuration of the primary system at time t, see Figure 8.17b, where R_i is the reaction on the frame from the virtual roller i. (R_i is denoted by a dashed line because the reaction and the support should not be plotted in the same figure.) The zero-reaction rule to ensure that the primary system is equivalent to the original frame is that:

$$R_i = 0, \quad i = 1, 2 \tag{a}$$

(3) Formulate the reactions, R_1 and R_2, which can be expressed as:

$$\begin{cases} R_1 = m\ddot{x}_1 - P_1(t) + R_{1g} + R_{1X} \\ R_2 = m\ddot{x}_2 - P_2(t) + R_{2g} + R_{2X} \end{cases} \tag{b}$$

Here, R_{ig} is the reaction from the virtual roller i caused by the loading $P_g(t)$; and R_{iX} is the reaction from the virtual roller i caused by the virtual-support displacements, $x_1(t)$ and $x_2(t)$.

R_{ig}, $i = 1, 2$, are indicated in Figure 8.18(a). To specify R_{ig}, draw the moment diagram of the primary system due the load P_g (Figure 8.18b), which leads to:

$$\begin{cases} R_{1g} = -P_g(t) \\ R_{2g} = 0 \end{cases} \tag{c}$$

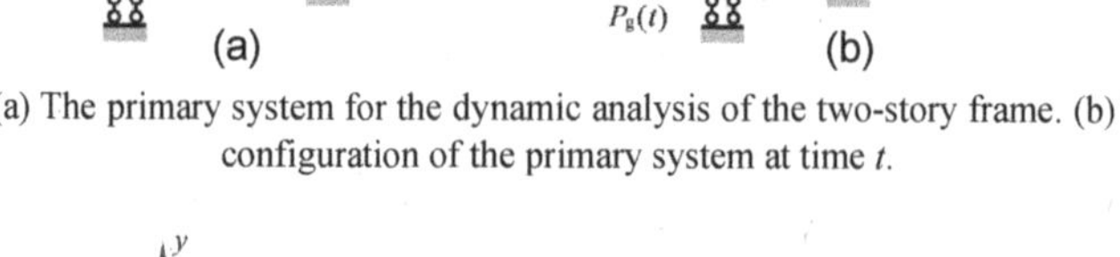

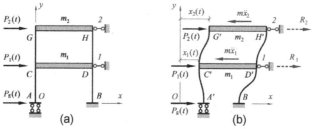

Figure 8.17. (a) The primary system for the dynamic analysis of the two-story frame. (b) The deformed configuration of the primary system at time t.

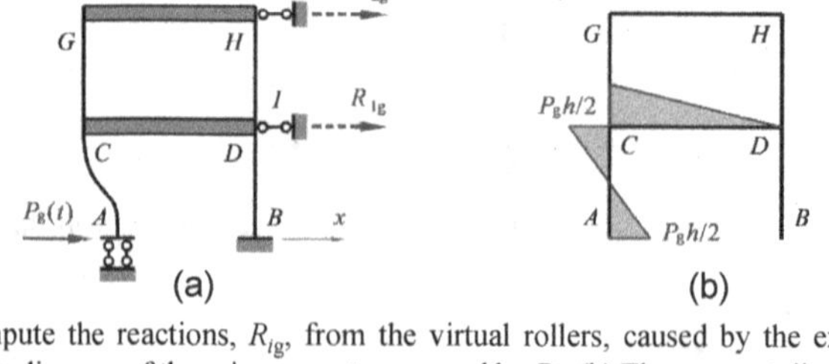

Figure 8.18. Compute the reactions, R_{ig}, from the virtual rollers, caused by the external loading P_g. (a) The deformation diagram of the primary system caused by P_g. (b) The moment diagram of the primary system, caused by P_g.

To specify R_{1X} and R_{2X}, the equations above can be rewritten as:

$$\begin{cases} R_{1X} = k_{11}x_1(t) + k_{12}x_2(t) \\ R_{2X} = k_{21}x_1(t) + k_{22}x_2(t) \end{cases} \tag{d}$$

where k_{ij} is the reaction from the virtual support i due to the unit support displacement $x_j = 1$, see Figure 8.19a and Figure 8.20a. The moment diagram of the primary system induced by the unit support displacements $x_1 = 1$ is shown in Figure 8.19b. What is caused by the unit support displacement $x_2 = 1$ is shown in Figure 8.20b. All this yields:

$$\begin{cases} k_{11} = \dfrac{1}{h}\left(2\times\dfrac{6EI_1}{h^2}\right) + 2\times\dfrac{1}{h}\left(2\times\dfrac{6EI_2}{h^2}\right) = \dfrac{12EI_1 + 24EI_2}{h^3} \\ k_{22} = 2\times\dfrac{1}{h}\left(2\times\dfrac{6EI_2}{h^2}\right) = \dfrac{24EI_2}{h^3} \\ k_{12} = k_{21} = -2\times\dfrac{1}{h}\left(2\times\dfrac{6EI_2}{h^2}\right) = -\dfrac{24EI_2}{h^3} \end{cases} \tag{e}$$

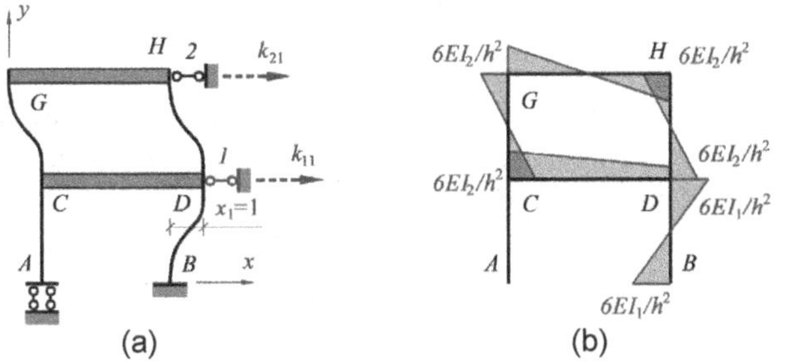

Figure 8.19. Computing the reactions, k_{i1}, from the virtual rollers, caused by the unit support displacement, $x_1 = 1$. (a) The deformation diagram of the primary system caused by $x_1 = 1$. (b) The moment diagram of the primary system caused by the unit support where, displacement, $x_1 = 1$.

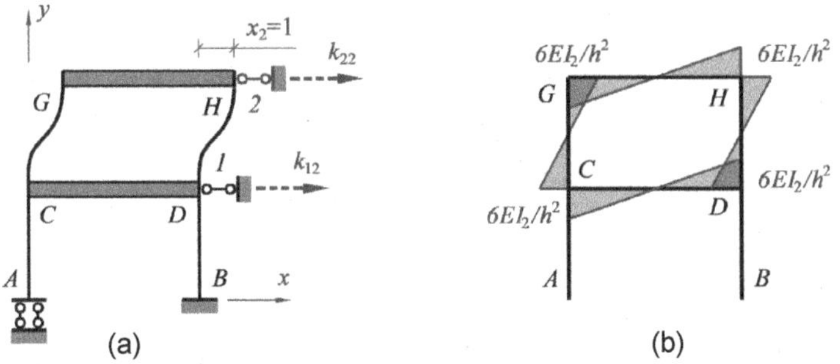

Figure 8.20. Computing the reactions, k_{i2}, from the virtual rollers, caused by the unit support displacement, $x_2 = 1$. (a) The deformation diagram of the primary system caused by $x_2 = 1$. (b) The moment diagram of the primary system caused by the unit support displacement, $x_2 = 1$.

Substitution of Equations (b)–(e) into Equation (a) yields:

$$\begin{bmatrix} m_1 & 0 \\ 0 & m_2 \end{bmatrix} \begin{Bmatrix} \ddot{x}_1 \\ \ddot{x}_2 \end{Bmatrix} + \frac{12}{h^3} \begin{bmatrix} EI_1 + 2EI_2 & -2EI_2 \\ -2EI_2 & 2EI_2 \end{bmatrix} \begin{Bmatrix} x_1 \\ x_2 \end{Bmatrix} = \begin{Bmatrix} P_1(t) + P_g(t) \\ P_2(t) \end{Bmatrix} \tag{f}$$

This is the forced vibration equation of the two-story frame shown in Figure 8.16.

8.2.3 Flexibility Method to Formulate Equations of Motion

The key step of the stiffness approach is to formulate the stiffness matrix. However, for some structures, it is easier to calculate the flexibility matrix and invert it to the stiffness matrix. For example, in dynamic analysis, a high-rise building is usually modeled as a vertical cantilever beam with n lumped masses as shown in Figure 8.21a. The stiffness equations of motion can be expressed as:

$$\begin{bmatrix} m_1 & \cdots & 0 \\ \vdots & \ddots & \vdots \\ 0 & \cdots & m_n \end{bmatrix} \begin{Bmatrix} \ddot{x}_1 \\ \vdots \\ \ddot{x}_n \end{Bmatrix} + \begin{bmatrix} k_{11} & \cdots & k_{1n} \\ \vdots & \ddots & \vdots \\ k_{n1} & \cdots & k_{nn} \end{bmatrix} \begin{Bmatrix} x_1 \\ \vdots \\ x_n \end{Bmatrix} = \begin{Bmatrix} P_1(t) \\ \vdots \\ P_n(t) \end{Bmatrix} \tag{8.46}$$

or

$$\boldsymbol{M}\ddot{\boldsymbol{x}} + \boldsymbol{K}\boldsymbol{x} = \boldsymbol{P} \tag{8.47}$$

where the stiffness influence coefficients are identified in Figure 8.21b. When $n \le 2$, it is easy to compute the stiffness influence coefficients. However, when $n \ge 3$, it is inconvenient to calculate the stiffness influence coefficients directly. Next, let's apply the flexibility method to solve this problem.

Now that the inertial forces have been applied on the system, as shown in Figure 8.21b, the problem has been viewed as a static one. In this case, the translational

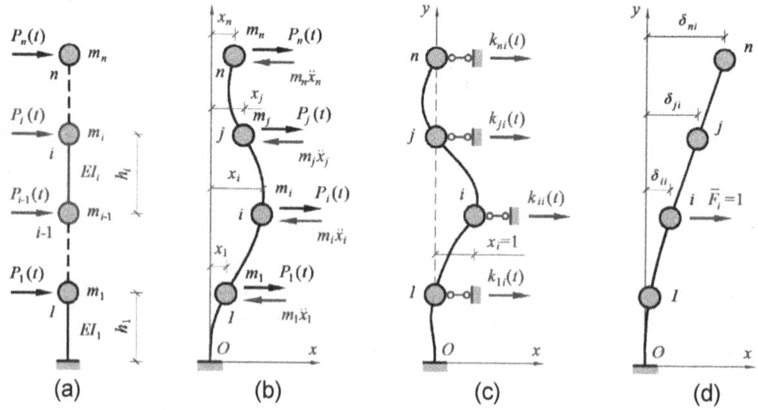

Figure 8.21. (a) A vertical cantilever beam with n lumped masses. (b) The dynamic Degrees Of Freedom of the cantilever beam with lumped masses. (c) Definition of stiffness influence coefficients. (d) Definition of flexibility influence coefficients.

Degree Of Freedom, x_j, is the displacement at point of the cantilever caused by the external loads, $P_i - m\ddot{x}_i$, $i = 1,\ldots, n$, and can be expressed as:

$$x_j = \sum_{i=1}^{n} \delta_{ji}\left(P_i - m\ddot{x}_i\right), \quad j = 1,\cdots,n \tag{8.48}$$

or

$$\boldsymbol{x} = \boldsymbol{\delta P} - \boldsymbol{\delta M\ddot{x}} \tag{8.49}$$

where the flexibility influence coefficient δ_{ji} is the horizontal deflection at point j due to unit force applied at point i (Figure 8.21b), and $\boldsymbol{\delta}$ is called the ***flexibility matrix*** of the system. Equation (8.49) is the flexibility form of the equations of motion, while Equation (8.46) is referred to as their stiffness form.

Determining the flexibility influence coefficient δ_{ji} is a standard problem in Chapter 4, and we draw the moment diagrams, $\bar{M}_i$ and $\bar{M}_j$, of the cantilever due to the unit forces at points i and j, respectively, see Figure 8.22. The graphic multiplication of $\bar{M}_i$ and $\bar{M}_j$ shows that, for $i \le j$, the flexibility influence coefficient, δ_{ji}, can be expressed as:

$$\delta_{ji} = \sum_{s=1}^{i}\frac{1}{EI_s}\left[\left(h_s\sum_{r=s+1}^{i}h_r\right)\left(h_s+\sum_{r=s+1}^{j}h_r\right)+\frac{h_s^2}{2}\left(\frac{2h_s}{3}+\sum_{r=s+1}^{j}h_r\right)\right] \tag{8.50}$$

For $i > j$, δ_{ji} can be calculated by the symmetry of a flexibility matrix, i.e.,

$$\delta_{ji} = \delta_{ij} \tag{8.51}$$

If the flexural rigidity and height of each story are EI and h, respectively, then Equation (8.50) can be rewritten as:

$$\delta_{ji} = \frac{h^3}{EI}\sum_{s=1}^{i}\left[\left(i-s\right)\left(\frac{1}{2}+j-s\right)+\frac{1}{2}\left(\frac{2}{3}+j-s\right)\right] \tag{8.52}$$

One can obtain the stiffness matrix, $\boldsymbol{K}$ by inverting the flexibility matrix, $\boldsymbol{\delta}$, i.e.,

$$\boldsymbol{K} = \boldsymbol{\delta}^{-1} \tag{8.53}$$

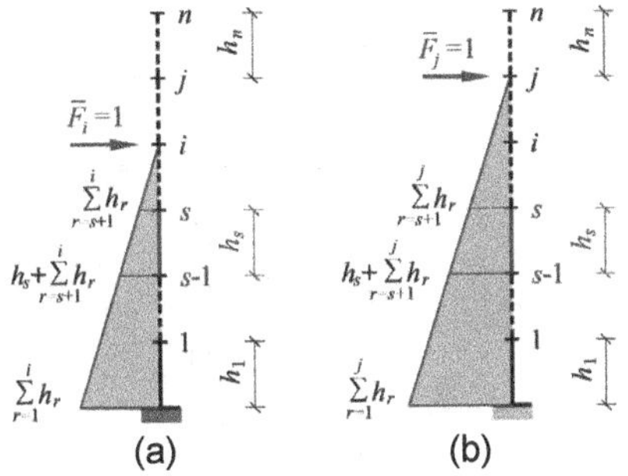

Figure 8.22. (a) The moment diagram, $\bar{M}_i$, of the cantilever due to the unit force at point i. (b) The moment diagram, $\bar{M}_j$, of the cantilever due to the unit force at point j.

The following example will illustrate the application of the flexibility method to formulate the equations of motion of a dynamic model frequently used in seismic analysis.

EXAMPLE 8.3

The earthquake response of a building may be investigated by applying a horizontal base motion $u_g(t)$ on the bottom end of the cantilever beam with two lumped masses, as shown in Figure 8.23a. The flexural rigidity and height of each story are EI and h, respectively. Formulate the equations governing the elastic deformation produced by the earthquake.

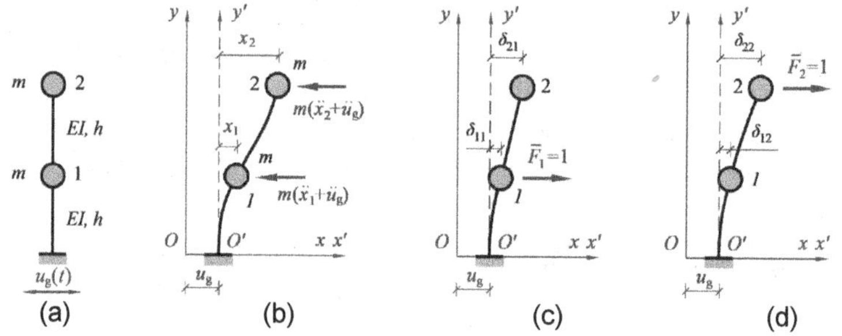

Figure 8.23. (a) A vertical cantilever beam with two lumped masses, subjected to base movement. (b) The dynamic Degrees of Freedom of the cantilever beam with lumped masses. (c) Definition of flexibility influence coefficients, δ_{11} and δ_{21}. (d) Definition of flexibility influence coefficients, δ_{22} and δ_{12}.

***Solution*:**

Define a space-time coordinate system Oxy as in Figure 8.23b such that the origin of the spatial coordinates is placed at the point, at which the bottom of the vertical cantilever is located at time $t = 0$. In addition, one needs to define a spatial coordinate $O'x'y'$ whose origin is fixed at the base of the cantilever beam as shown in Figure 8.23b.

The flexibility influence coefficients δ_{ij}, $i, j = 1, 2$, are identified in Figure 8.23c and Figure 8.23b, which shows that the dynamic DOF vector, $\boldsymbol{x} = [x_1, x_2]^{\mathrm{T}}$, can be expressed as:

$$\boldsymbol{x} = -m\boldsymbol{\delta}\ddot{\boldsymbol{x}} - m\boldsymbol{\delta}\begin{Bmatrix} \ddot{u}_g \\ \ddot{u}_g \end{Bmatrix} \tag{8.54}$$

which is the flexibility form of the equations of motion.

Inverting the flexibility matrix δ and pre-multiplying both sides of Equation (8.54) by the inverse yields the stiffness form of the equations of motion as follows:

$$\boldsymbol{M}\ddot{\boldsymbol{x}} + \boldsymbol{K}\boldsymbol{x} = -\begin{Bmatrix} m\ddot{u}_g \\ m\ddot{u}_g \end{Bmatrix} \tag{8.55}$$

where:

$$\boldsymbol{K} = \boldsymbol{\delta}^{-1} \tag{8.56}$$

$$\boldsymbol{M} = m\boldsymbol{I}_{2\times 2} \tag{8.57}$$

For two-Degrees Of Freedom shown in Figure 8.23(a), Equation (8.52) gives:

$$\boldsymbol{K}=\delta^{-1}=\left(\frac{h^3}{EI}\begin{bmatrix}1/3 & 5/6\\ 5/6 & 8/3\end{bmatrix}\right)^{-1}=\frac{EI}{h^3}\begin{bmatrix}96/7 & -30/7\\ -30/7 & 12/7\end{bmatrix} \tag{8.58}$$

Substituting Equation (8.58) into Equation (8.55) yields:

$$\begin{bmatrix}m & 0\\ 0 & m\end{bmatrix}\begin{Bmatrix}\ddot{x}_1\\ \ddot{x}_2\end{Bmatrix}+\frac{EI}{h^3}\begin{bmatrix}96/7 & -30/7\\ -30/7 & 12/7\end{bmatrix}\begin{Bmatrix}x_1\\ x_2\end{Bmatrix}=-\begin{Bmatrix}m\ddot{u}_g\\ m\ddot{u}_g\end{Bmatrix} \tag{8.59}$$

This is the equation of motion of the vertical cantilever with two lumped masses.

8.2.4 *Stiffness Method: The Matrix Displacement Approach and Static Condensation*

This section shall apply the direct stiffness method (i.e., the matrix displacement method presented in Chapter 8) to formulate the equations of motion of a structure. Consider the frame subjected to a vertical harmonic load at the lumped mass of Figure 8.24a, whose static stiffness equations have been studied in the previous section. The Young's modulus, the cross-sectional area and the moment of inertia of each member are 2.06×10^{11} N/m^2, $A=0.0096$ m^2, $I=1.152\times10^{-5}$ m^4, respectively.

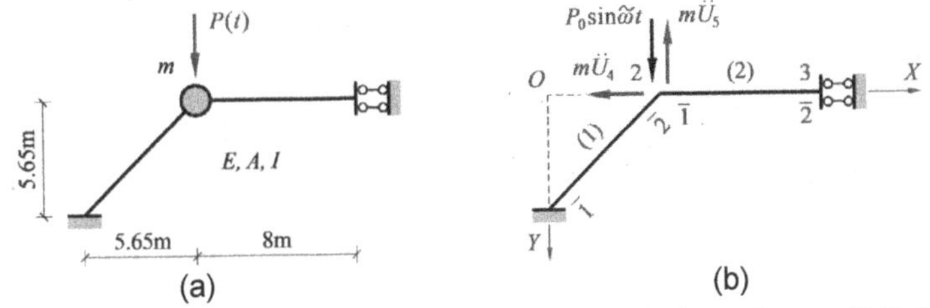

Figure 8.24. (a) A frame subjected to a vertical harmonic load at the lumped mass. (b) Definition of a reference system *OXY*, labels for nodes and elements, and force equilibrium at the lumped mass.

As shown in Figure 8.24b, we define a reference system *OXY* and label all the nodes and elements. The harmonic load can be expressed as:

$$P(t)=P_0\sin\tilde{\omega}t \tag{8.60}$$

Let's set this system with 9 elastic Degrees Of Freedoms, in which U_1, U_2, U_3, U_7 and U_9 are prescribed, U_4, U_5 and U_6 are active, and U_4, U_5 are dynamic Degrees Of Freedom. Now that the inertial forces, $m\ddot{U}_4$ and $m\ddot{U}_5$, have been exerted on the frame, the problem has been transformed into a static one, which is a frame subjected to the nodal loads. Repeating the process in the previous section gives:

$$\begin{bmatrix}m & 0 & 0 & 0\\ 0 & m & 0 & 0\\ 0 & 0 & 0 & 0\\ 0 & 0 & 0 & 0\end{bmatrix}\begin{Bmatrix}\ddot{U}_4\\ \ddot{U}_5\\ \ddot{U}_6\\ \ddot{U}_8\end{Bmatrix}+\boldsymbol{K}_{\mathrm{ee}}\begin{Bmatrix}U_4\\ U_5\\ U_6\\ U_8\end{Bmatrix}=\begin{Bmatrix}0\\ P_0\sin\tilde{\omega}t\\ 0\\ 0\end{Bmatrix} \tag{8.61}$$

where:

$$\boldsymbol{K}_{ee} = \begin{bmatrix} 3.7083\times10^8 & -1.2357\times10^8 & -1.5731\times10^5 & 0 \\ -1.2357\times10^8 & 1.2368\times10^8 & 6.5165\times10^4 & -5.5620\times10^4 \\ -1.5731\times10^5 & 6.5165\times10^4 & 2.3731\times10^6 & -2.2248\times10^5 \\ 0 & -5.5620\times10^4 & -2.2248\times10^5 & 5.5620\times10^4 \end{bmatrix} \tag{8.62}$$

is the stiffness matrix of the frame in static analysis, which is the stiffness matrix between the elastic Degrees Of Freedom. The subscript (or superscript) "ee" is used to form the stiffness matrix, $\boldsymbol{K}$, in the dynamic analysis.

By solving the third and fourth equations in (8.61) for U_6 and U_8 and substituting the result into the first and second equations, one obtains:

$$\begin{bmatrix} m & 0 \\ 0 & m \end{bmatrix}\begin{Bmatrix} \ddot{U}_4 \\ \ddot{U}_5 \end{Bmatrix} + \begin{bmatrix} 3.7081 & -1.2359 \\ -1.2359 & 1.2361 \end{bmatrix}\begin{Bmatrix} U_4 \\ U_5 \end{Bmatrix} = \begin{Bmatrix} 0 \\ P_0 \sin\tilde{\omega}t \end{Bmatrix} \tag{8.63}$$

This is the equation of the motion of the frame.

The process of eliminating those Degrees Of Freedom of a structure to which a zero mass is assigned is called ***static condensation***. Next, let's illustrate the general procedure of the static condensation, which is easy to implement in computer software. To this end, it is assumed that zero-mass and nonzero mass Degrees Of Freedom have been segregated, so that the equations governing the elastic Degrees Of Freedom can be written in partitioned form:

$$\begin{bmatrix} \boldsymbol{M} & \boldsymbol{0} \\ \boldsymbol{0} & \boldsymbol{0} \end{bmatrix}\begin{Bmatrix} \ddot{\boldsymbol{U}}_m \\ \ddot{\boldsymbol{U}}_0 \end{Bmatrix} + \begin{bmatrix} \boldsymbol{K}_{mm} & \boldsymbol{K}_{m0} \\ \boldsymbol{K}_{0m} & \boldsymbol{K}_{00} \end{bmatrix}\begin{Bmatrix} \boldsymbol{U}_m \\ \boldsymbol{U}_0 \end{Bmatrix} = \begin{Bmatrix} \boldsymbol{P}_m \\ \boldsymbol{P}_0 \end{Bmatrix} \tag{8.64}$$

where,

$$\boldsymbol{M} = \begin{bmatrix} m & 0 \\ 0 & m \end{bmatrix} \tag{8.65}$$

is the mass matrix of the system; $\boldsymbol{U}_0$ denotes the Degrees Of Freedom with zero mass and $\boldsymbol{U}_m$ is the Degree Of Freedom with mass; $\boldsymbol{K}_{m0}$ is the stiffness matrix between nonzero-mass and zero-mass Degrees Of Freedom (i.e., $K_{ij}^{(m0)}$ is the reaction from the ith nonzero-mass virtual constraint due to the unit displacement at the jth zero-mass constraint); $\boldsymbol{K}_{mm}$ is the stiffness matrix between nonzero-mass Degrees Of Freedom and $\boldsymbol{K}_{00}$ is the stiffness matrix between zero-mass Degrees Of Freedom; $\boldsymbol{P}_m$ and $\boldsymbol{P}_0$ are the loads exerted on nonzero-mass and zero-mass Degrees Of Freedom, respectively. The two partitioned equations are:

$$\boldsymbol{M}\ddot{\boldsymbol{U}}_m + \boldsymbol{K}_{mm}\boldsymbol{U}_m + \boldsymbol{K}_{m0}\boldsymbol{U}_0 = \boldsymbol{P}_m \tag{8.66}$$

and

$$\boldsymbol{K}_{0m}\boldsymbol{U}_m + \boldsymbol{K}_{00}\boldsymbol{U}_0 = \boldsymbol{P}_0 \tag{8.67}$$

Even though $\boldsymbol{P}_0$ may be a dynamic load, Equation (8.67) permits a static relation between $\boldsymbol{U}_0$ and $\boldsymbol{U}_m$:

$$\boldsymbol{U}_0 = -\boldsymbol{K}_{00}^{-1}\boldsymbol{K}_{0m}\boldsymbol{U}_m + \boldsymbol{K}_{00}^{-1}\boldsymbol{P}_0 \tag{8.68}$$

Inserting Equation (8.68) into Equation (8.66) yields:

$$\boldsymbol{M}\ddot{\boldsymbol{U}}_{\mathrm{m}} + \boldsymbol{K}\boldsymbol{U}_{\mathrm{m}} = \boldsymbol{P}_{\mathrm{m}} + \boldsymbol{C}^{\mathrm{P}}_{\mathrm{m0}}\boldsymbol{P}_{0} \tag{8.69}$$

which is the equation governing the dynamic Degrees Of Freedom, and where $\boldsymbol{K}$ is the ***condensed stiffness matrix*** given by:

$$\boldsymbol{K} = \boldsymbol{K}_{\mathrm{mm}} - \boldsymbol{K}_{\mathrm{m0}}\,\boldsymbol{K}_{00}^{-1}\,\boldsymbol{K}_{\mathrm{0m}} \tag{8.70}$$

and

$$\boldsymbol{C}^{\mathrm{P}}_{\mathrm{m0}} = -\,\boldsymbol{K}_{\mathrm{m0}}\,\boldsymbol{K}_{00}^{-1} \tag{8.71}$$

is the ***load-transfer matrix*** transferring loads at zero-mass and nonzero-mass Degrees Of Freedom.

It is noticed that the stiffness matrix $\boldsymbol{K}$ and $\boldsymbol{K}_{\mathrm{mm}}$ are different although both describe the stiffness between dynamic Degrees Of Freedom. For example, the stiffness influence coefficients, K_{11}, K_{21} and K_{31}, of a vertical cantilever beam with three lumped masses are identified in Figure 8.25a, while the stiffness influence coefficients, K_{11}^{mm}, K_{21}^{mm} and K_{31}^{mm} are illustrated in Figure 8.25b, in which all the nodal rotations are restricted.

The MATLAB codes to compute the condensed stiffness matrix and the load-transformation matrix are listed below. The readers who are not interested in MATLAB codes can skip it and jump directly to Example 8.5.

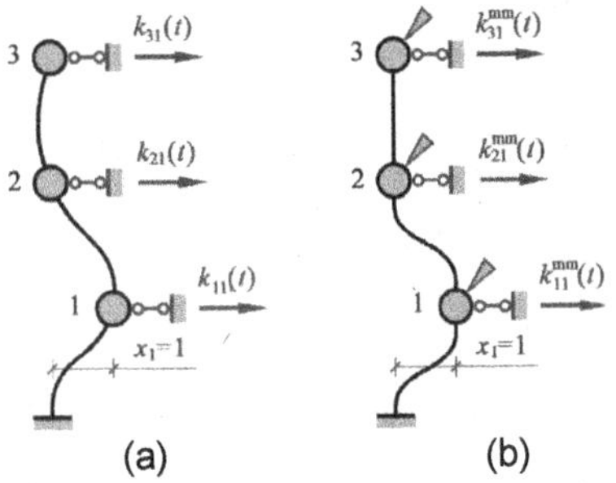

Figure 8.25. (a) The definition of the stiffness influence coefficients, K_{11}, K_{21} and K_{31}, (b) The definition of the stiffness influence coefficients, K_{11}^{mm}, K_{21}^{mm} and K_{31}^{mm}.

```
% Static-Condensation Function
function [K_con,C_pm0]=Kcondensation(K_s,DOF_m)
% K_s: Structural stiffness matrix from the static-matrix-displacement analysis.
% DOF_m: The id set of the DOFs with nonzero masses.
% K_con: The condensed stiffness matrix in the equation of motion.
% C_pm0: The load-transfer matrix.

dimKs=size(K_s);     % dimKs -- The dimension of the stiffness matrix K_ee
N_dof=dimKs(1);      % N_dof -- The number of the elastic DOFs.
DOF_0=setdiff([1:N_dof],DOF_m);   % DOF_0 -- The id set of the DOFs
                                  % with zero mass.


K_mm=K_s(DOF_m,DOF_m);
K_m0=K_s(DOF_m,DOF_0);
K_0m=K_s(DOF_0,DOF_m);
K_00=K_s(DOF_0,DOF_0);

K_con=K_mm-K_m0*K_00^-1*K_0m;
C_pm0=-K_m0*K_00^-1;
```

Listing 8.1. MATLAB codes.

In this function, *Kcondensation.m*, we used two MATLAB library functions, *size* and *setdiff*, whose usage can be found in MATLAB function references.

After putting the file *Kcondensation.m* to the work directory, in the MATLAB command window, the following commands:

```
>> K_ee = [11 12 13; 21 22 23; 31 32 33]
>> DOF_m = [1, 3]
>> [K_con, C_pm0, K_mm, K_m0 , K_00] = Kcondensation (K_ee, DOF_m)
```

we check what happens. Here, we don't input a symmetric K_{ee} in order to identify the output of the function.

EXAMPLE 8.5

Apply the matrix displacement method to formulate the equations describing the motion of the cantilever beam in Example 8.3. The Young's modulus, the moment of inertia and the height of each story are $E = 2.06 \times 10^{11}$ N/m², $I = 1.152 \times 10^{-5}$ m⁴ and $h = 8$m, respectively. Neglect the axial deformation of each member.

Solution:

The deformation of the cantilever beam excited by the base motion $u_g(t)$ of Figure 8.26a is equal to that of the cantilever subjected to the inertial forces, $m\ddot{x}_1$ and $m\ddot{x}_2$, shown in Figure 8.26b. To solve the problem shown in Figure 8.26b by the matrix displacement approach, one can define global coordinates and label all the nodes and elements as shown in Figure 8.26c.

Because three nodes for the matrix displacement analysis are set, the elastic Degrees Of Freedom s are U_1, ..., U_9, in which U_1 and U_4 are the dynamic Degrees Of Freedom and U_i, $i = 2, 3, 5, 6, 7, 8, 9$, are the Degrees Of Freedom with zero masses. The applied boundary condition and neglecting the axial deformation give the following prescribed conditions:

$$U_2 = U_5 = U_7 = U_8 = U_9 = 0 \tag{a}$$

To calculate the element stiffness matrices, let's arbitrarily assign the cross-section area with a numerical value of $A = 9.6 \times 10^{-3}$ m². The element stiffness matrices in element local coordinates are:

$$\bar{k}^{(e)} = 10^5 \times \begin{bmatrix} 2472 & 0 & 0 & -2472 & 0 & 0 \\ 0 & 0.55619 & 2.2248 & & -0.55619 & 2.2248 \\ 0 & 2.2248 & 11.868 & & -2.2248 & 5.9328 \\ -2472 & 0 & 0 & 2472 & 0 & 0 \\ 0 & -0.55619 & -2.2248 & 0 & 0.55619 & -2.2248 \\ 0 & 2.2248 & 5.9328 & 0 & -2.2248 & 11.868 \end{bmatrix}$$

For $e = 1, 2$.

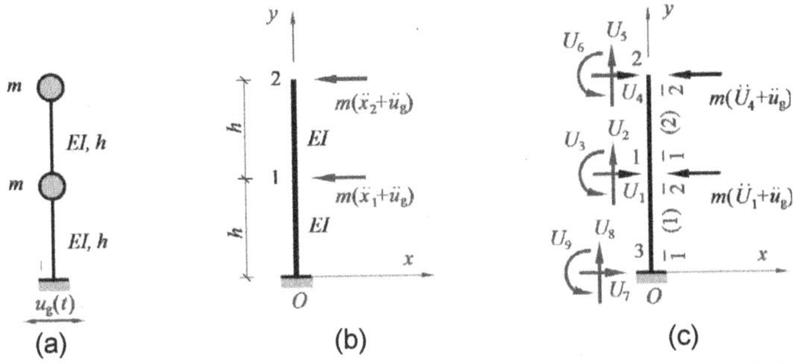

Figure 8.26. (a) A vertical cantilever beam with two lumped masses. (b) The equivalent dynamic equilibrium problem. (c) The label system for the analysis by the matrix displacement method.

The element transformation matrices of coordinates are:

$$\underset{\bar{e}\leftarrow g}{\boldsymbol{T}^{(e)}} = \begin{bmatrix} 0 & 1 & 0 & 0 & 0 & 0 \\ -1 & 0 & 0 & 0 & 0 & 0 \\ 0 & 0 & 1 & 0 & 0 & 0 \\ 0 & 0 & 0 & 0 & 1 & 0 \\ 0 & 0 & 0 & -1 & 0 & 0 \\ 0 & 0 & 0 & 0 & 0 & 1 \end{bmatrix}, \qquad \alpha^{(e)} = \frac{\pi}{2}, \qquad e = 1, 2$$

The element stiffness matrices in global coordinates are:

$$\boldsymbol{k}^{(e)} = \left(\underset{\bar{e}\leftarrow g}{\boldsymbol{T}^{(e)}}\right)^{\mathrm{T}} \bar{\boldsymbol{k}}^{(e)} \underset{\bar{e}\leftarrow g}{\boldsymbol{T}^{(e)}}$$

$$= 10^5 \times \begin{bmatrix} 0.5562 & 0 & -2.2248 & -0.5562 & 0 & -2.2248 \\ 0 & 2472 & 0 & 0 & -2472 & 0 \\ -2.2248 & 0 & 11.866 & 2.2248 & 0 & 5.9328 \\ -0.5562 & 0 & 2.2248 & 0.5562 & 0 & 2.2248 \\ 0 & -2472 & 0 & 0 & 2472 & 0 \\ -2.2248 & 0 & 5.9328 & 2.2248 & 0 & 11.866 \end{bmatrix} \qquad \text{(b)}$$

where $e = 1, 2$.

The element connectivity is:

$$(1)\begin{cases} \bar{1} \rightarrow 3 \\ \bar{2} \rightarrow 1 \end{cases}, \qquad (2)\begin{cases} \bar{1} \rightarrow 1 \\ \bar{2} \rightarrow 2 \end{cases} \qquad \text{(c)}$$

or

$$(1)\begin{cases} U_1^{(1)} = U_7 \\ U_2^{(1)} = U_8 \\ U_3^{(1)} = U_9 \\ U_4^{(1)} = U_1 \\ U_5^{(1)} = U_2 \\ U_6^{(1)} = U_3 \end{cases}, \quad (2)\begin{cases} U_1^{(2)} = U_1 \\ U_2^{(2)} = U_2 \\ U_3^{(2)} = U_3 \\ U_4^{(2)} = U_4 \\ U_5^{(2)} = U_5 \\ U_6^{(2)} = U_6 \end{cases} \qquad \text{(d)}$$

The governing equations of the unconstrained cantilever can be obtained by assembling the element stiffness matrices (b) according to the element connectivity (c) as follows:

$$\begin{bmatrix} m & 0 & 0 & 0 & 0 & 0 & 0 & 0 & 0 \\ 0 & m & 0 & 0 & 0 & 0 & 0 & 0 & 0 \\ 0 & 0 & 0 & 0 & 0 & 0 & 0 & 0 & 0 \\ 0 & 0 & 0 & m & 0 & 0 & 0 & 0 & 0 \\ 0 & 0 & 0 & 0 & m & 0 & 0 & 0 & 0 \\ 0 & 0 & 0 & 0 & 0 & 0 & 0 & 0 & 0 \\ 0 & 0 & 0 & 0 & 0 & 0 & 0 & 0 & 0 \\ 0 & 0 & 0 & 0 & 0 & 0 & 0 & 0 & 0 \\ 0 & 0 & 0 & 0 & 0 & 0 & 0 & 0 & 0 \end{bmatrix} \ddot{\boldsymbol{U}} + \boldsymbol{K}_{9\times 9}^{\text{fr}} \boldsymbol{U} = \begin{Bmatrix} -m\ddot{u}_{\text{g}} \\ -m\ddot{u}_{\text{g}} \\ 0 \\ -m\ddot{u}_{\text{g}} \\ -m\ddot{u}_{\text{g}} \\ 0 \\ 0 \\ 0 \\ 0 \end{Bmatrix} \qquad \text{(e)}$$

Imposing the prescribed conditions (a) on Equation (e) yields:

$$\begin{bmatrix} m & 0 & 0 & 0 \\ 0 & 0 & 0 & 0 \\ 0 & 0 & m & 0 \\ 0 & 0 & 0 & 0 \end{bmatrix} \begin{Bmatrix} \ddot{U}_1 \\ \ddot{U}_3 \\ \ddot{U}_4 \\ \ddot{U}_6 \end{Bmatrix} + \boldsymbol{K}_{\text{ee}} \begin{Bmatrix} U_1 \\ U_3 \\ U_4 \\ U_6 \end{Bmatrix} = \begin{Bmatrix} -m\ddot{u}_{\text{g}} \\ 0 \\ -m\ddot{u}_{\text{g}} \\ 0 \end{Bmatrix} \qquad \text{(f)}$$

where,

$$\boldsymbol{K}_{\text{ee}} = \begin{bmatrix} K_{11}^{\text{fr}} & K_{13}^{\text{fr}} & K_{14}^{\text{fr}} & K_{16}^{\text{fr}} \\ K_{31}^{\text{fr}} & K_{33}^{\text{fr}} & K_{34}^{\text{fr}} & K_{36}^{\text{fr}} \\ K_{41}^{\text{fr}} & K_{43}^{\text{fr}} & K_{44}^{\text{fr}} & K_{46}^{\text{fr}} \\ K_{61}^{\text{fr}} & K_{63}^{\text{fr}} & K_{64}^{\text{fr}} & K_{66}^{\text{fr}} \end{bmatrix} = 10^5 \times \begin{bmatrix} 1.1124 & 0 & -0.5562 & -2.2248 \\ 0 & 23.7312 & 2.2248 & 5.9328 \\ -0.5562 & 2.2248 & 0.5562 & 2.2248 \\ -2.2248 & 5.9328 & 2.2248 & 11.8656 \end{bmatrix}$$

Here, the entries, K_{ij}^{fr}, of the matrix $\boldsymbol{K}_{\text{ee}}$ are calculated by assembling the element stiffness matrix (b) according to the element connectivity (d). For example,

$$K_{13}^{\text{fr}} = k_{46}^{(1)} + k_{13}^{(2)} = 2.2248 \times 10^5 - 2.2248 \times 10^5 = 0$$

To eliminate the unwanted Degrees Of Freedom from Equation (f), U_3 and U_6, one can rewrite it as two partitioned equations:

$$\boldsymbol{M}\begin{Bmatrix}\ddot{U}_1\\ \ddot{U}_4\end{Bmatrix}+\boldsymbol{K}_{\text{mm}}\begin{Bmatrix}U_1\\ U_4\end{Bmatrix}+\boldsymbol{K}_{\text{m0}}\begin{Bmatrix}U_3\\ U_6\end{Bmatrix}=\begin{Bmatrix}-m\ddot{u}_{\text{g}}\\ -m\ddot{u}_{\text{g}}\end{Bmatrix} \tag{g}$$

and

$$\boldsymbol{K}_{\text{0m}}\begin{Bmatrix}U_1\\ U_4\end{Bmatrix}+\boldsymbol{K}_{00}\begin{Bmatrix}U_3\\ U_6\end{Bmatrix}=0 \tag{h}$$

where,

$$\boldsymbol{K}_{\text{mm}}=10^5\times\begin{bmatrix}1.1124 & -0.5562\\ -0.5562 & 0.5562\end{bmatrix} \tag{i}$$

$$\boldsymbol{K}_{00}=10^5\times\begin{bmatrix}23.7312 & 5.9328\\ 5.9328 & 11.8656\end{bmatrix} \tag{j}$$

$$\boldsymbol{K}_{\text{m0}}=10^5\times\begin{bmatrix}0 & -2.2248\\ 2.2248 & 2.2248\end{bmatrix} \tag{k}$$

$$\boldsymbol{K}_{\text{0m}}=\boldsymbol{K}_{\text{0m}}^{\text{T}} \tag{l}$$

Solving Equation (h) for U_3, U_6 and inserting the solution into Equation (g) yields:

$$\boldsymbol{M}\begin{Bmatrix}\ddot{U}_1\\ \ddot{U}_4\end{Bmatrix}+\boldsymbol{K}_{\text{ee}}\begin{Bmatrix}\ddot{U}_1\\ \ddot{U}_4\end{Bmatrix}=\begin{Bmatrix}-m\ddot{u}_{\text{g}}\\ -m\ddot{u}_{\text{g}}\end{Bmatrix}$$

where the condensed stiffness matrix $\boldsymbol{K}_{\text{ee}}$ can be expressed as:

$$\boldsymbol{K}=\boldsymbol{K}_{\text{mm}}-\boldsymbol{K}_{\text{m0}}\boldsymbol{K}_{00}^{-1}\boldsymbol{K}_{\text{0m}}=10^3\times\begin{bmatrix}63.566 & -19.864\\ -19.864 & 7.9457\end{bmatrix}$$

Thus, one has:

$$\begin{bmatrix}m & 0\\ 0 & m\end{bmatrix}\begin{Bmatrix}\ddot{x}_1\\ \ddot{x}_2\end{Bmatrix}+10^3\times\begin{bmatrix}63.566 & -19.864\\ -19.864 & 7.9457\end{bmatrix}\begin{Bmatrix}x_1\\ x_2\end{Bmatrix}=-\begin{Bmatrix}m\ddot{u}_{\text{g}}\\ m\ddot{u}_{\text{g}}\end{Bmatrix}$$

where $x_1 = U_1$ and $x_2 = U_4$. This is the equation of motion of the cantilever beam, which is the same as Equation (8.59) for $E = 2.06 \times 10^{11}$ N/m^2, $I = 1.152 \times 10^{-5}$ m^4 and $h = 8$m.

The following example may develop a deeper understanding of transforming zero-mass loads to nonzero-mass loads.

EXAMPLE 8.6

A simple-supported beam with a lumped mass m at the center, shown in Figure 8.27a, is excited by a dynamic torque at the center and a base motion, $v_g(t)$, at the right roller. The Young's modulus and the moment of inertia of each storey are, $E = 2.06 \times 10^{11}$ N/m^2, $I = 1.152 \times 10^{-5}$ m^4, respectively. Neglecting the axial deformation of

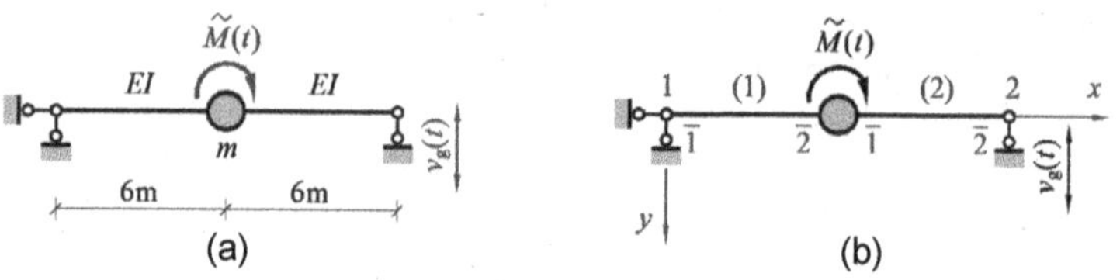

Figure 8.27. (a) A simply-supported beam with a lumped mass at the center, subjected to a dynamic torque and the base motion at the right end. (b) The coordinates and the labelled system used in the analysis.

each member, derive the equation governing the deflection at the center by using the direct stiffness method.

Solution:

Define the coordinates and label all the nodes and members as shown in Figure 8.27b. Since we set three nodes, the elastic Degrees Of Freedom are $U_1, \ldots, U_9$ in which only U_5 is the dynamic Degree Of Freedom. The applied boundary condition and neglecting the axial deformation give the following prescribed conditions:

$$U_1 = U_2 = U_4 = U_7 = 0 \tag{a}$$

and

$$U_8 = v_g(t) \tag{b}$$

To calculate the element stiffness matrices, one can assign each beam element with a specific cross-sectional area, $A = 9.6 \times 10^{-3}$ m². Since the directions of the element local and global coordinates are coincident, the element stiffness matrices are:

$$\boldsymbol{k}^{(e)} = \bar{\boldsymbol{k}}^{(e)} = 10^5 \times \begin{bmatrix} 3296 & 0 & 0 & -3296 & 0 & 0 \\ 0 & 1.3184 & 3.9552 & 0 & -1.3184 & 3.9552 \\ 0 & 3.9552 & 15.821 & 0 & -3.9552 & 7.9104 \\ -3296 & 0 & 0 & 3296 & 0 & 0 \\ 0 & -1.3184 & -3.9552 & 0 & 1.3184 & -3.9552 \\ 0 & 3.9552 & 7.9104 & 0 & -3.9552 & 15.821 \end{bmatrix} \tag{c}$$

for $e = 1, 2$.

The element connectivity can be expressed as:

$$(1)\begin{cases} \bar{1} \to 1 \\ \bar{2} \to 2 \end{cases}, \quad (2)\begin{cases} \bar{1} \to 2 \\ \bar{2} \to 3 \end{cases} \tag{d}$$

or

$$(1)\begin{cases} U_1^{(1)} = U_1 \\ U_2^{(1)} = U_2 \\ U_3^{(1)} = U_3 \\ U_4^{(1)} = U_4 \\ U_5^{(1)} = U_5 \\ U_6^{(1)} = U_6 \end{cases}, \quad (2)\begin{cases} U_1^{(2)} = U_4 \\ U_2^{(2)} = U_5 \\ U_3^{(2)} = U_6 \\ U_4^{(2)} = U_7 \\ U_5^{(2)} = U_8 \\ U_6^{(2)} = U_9 \end{cases} \tag{e}$$

The governing equations of the unconstrained beam can be obtained by assembling the element stiffness matrices (c) according to the element connectivity (d) as follows:

$$\begin{bmatrix} 0 & 0 & 0 & 0 & 0 & 0 & 0 & 0 & 0 \\ 0 & 0 & 0 & 0 & 0 & 0 & 0 & 0 & 0 \\ 0 & 0 & 0 & 0 & 0 & 0 & 0 & 0 & 0 \\ 0 & 0 & 0 & m & 0 & 0 & 0 & 0 & 0 \\ 0 & 0 & 0 & 0 & 0 & 0 & 0 & 0 & 0 \\ 0 & 0 & 0 & 0 & 0 & 0 & 0 & 0 & 0 \\ 0 & 0 & 0 & 0 & 0 & 0 & 0 & 0 & 0 \\ 0 & 0 & 0 & 0 & 0 & 0 & 0 & 0 & 0 \\ 0 & 0 & 0 & 0 & 0 & 0 & 0 & 0 & 0 \end{bmatrix} \ddot{\boldsymbol{U}} + \boldsymbol{K}^{\text{fr}}_{9\times 9}\boldsymbol{U} = \begin{Bmatrix} 0 \\ 0 \\ 0 \\ 0 \\ 0 \\ 0 \\ 0 \\ 0 \\ 0 \end{Bmatrix} \tag{f}$$

After introducing the boundary conditions (a) and (b), Equation (f) can be rewritten as:

$$\begin{bmatrix} 0 & 0 & 0 & 0 \\ 0 & m & 0 & 0 \\ 0 & 0 & 0 & 0 \\ 0 & 0 & 0 & 0 \end{bmatrix} \begin{Bmatrix} \ddot{U}_3 \\ \ddot{U}_5 \\ \ddot{U}_6 \\ \ddot{U}_9 \end{Bmatrix} + \boldsymbol{K}_{\text{ee}} \begin{Bmatrix} U_3 \\ U_5 \\ U_6 \\ U_9 \end{Bmatrix} = \boldsymbol{P}^{\text{bc}} \tag{g}$$

where the stiffness matrix, $\boldsymbol{K}_{\text{ee}}$, of the constrained beam can be written as:

$$\boldsymbol{K}_{\text{ee}} = 10^5 \times \begin{bmatrix} 15.8208 & -3.9552 & 7.9104 & 0 \\ -3.9552 & 2.6368 & 0 & 3.9552 \\ 7.9104 & 0 & 31.6416 & 7.9104 \\ 0 & 3.9552 & 7.9104 & 15.8208 \end{bmatrix} \tag{h}$$

and the nodal loading vector, $\boldsymbol{P}^{\text{bc}}$, induced by the prescribed Degrees Of Freedom is:

$$\boldsymbol{P}^{\text{bc}} = -\begin{Bmatrix} \sum\limits_{j=1,2,8} K^{\text{fr}}_{3j}U_j \\ \sum\limits_{j=1,2,8} K^{\text{fr}}_{5j}U_j \\ \sum\limits_{j=1,2,8} K^{\text{fr}}_{6j}U_j \\ \sum\limits_{j=1,2,8} K^{\text{fr}}_{9j}U_j \end{Bmatrix} = -\begin{Bmatrix} K^{\text{fr}}_{38}U_8 \\ K^{\text{fr}}_{58}U_8 \\ K^{\text{fr}}_{68}U_8 \\ K^{\text{fr}}_{98}U_8 \end{Bmatrix} = -\begin{Bmatrix} 0\cdot U_8 \\ k^{(2)}_{25}\cdot U_8 \\ k^{(2)}_{35}\cdot U_8 \\ k^{(2)}_{65}\cdot U_8 \end{Bmatrix} = \begin{Bmatrix} 0 \\ 1.3184 \\ 3.9552 \\ 3.9552 \end{Bmatrix} \times 10^5 v_{\text{g}}(t) \tag{i}$$

according to Equation (8.47).

To eliminate the unwanted Degrees Of Freedom from Equation (g), U_3, U_6, U_9, one can rewrite it as two partitioned equations:

$$\boldsymbol{M}\{\ddot{U}_5\} + \boldsymbol{K}_{\text{mm}}\{U_5\} + \boldsymbol{K}_{\text{m0}} \begin{Bmatrix} U_3 \\ U_6 \\ U_9 \end{Bmatrix} = \boldsymbol{P}_{\text{m}} \tag{j}$$

and

$$\boldsymbol{K}_{0\mathrm{m}}\{U_5\} + \boldsymbol{K}_{00}\begin{Bmatrix} U_3 \\ U_6 \\ U_9 \end{Bmatrix} = \boldsymbol{P}_0 \tag{k}$$

where,

$$\boldsymbol{P}_{\mathrm{m}} = \{1.3184 \times 10^4\, v_{\mathrm{g}}\} \tag{l}$$

$$\boldsymbol{P}_0 = \begin{Bmatrix} 0 \\ 3.9552\times10^4 v_{\mathrm{g}} \\ 3.9552\times10^4 v_{\mathrm{g}} \end{Bmatrix} \tag{m}$$

and

$$\boldsymbol{K}_{\mathrm{mm}} = 10^5 \times [2.6368]$$

$$\boldsymbol{K}_{\mathrm{m}0} = 10^5 \times [-3.9552 \;\; 0 \;\; 3.9552] \tag{n}$$

$$\boldsymbol{K}_{00} = 10^5 \times \begin{bmatrix} 15.8208 & 7.9104 & 0 \\ 7.9104 & 31.6416 & 7.9104 \\ 0 & 7.9104 & 15.8208 \end{bmatrix} \tag{o}$$

$$\boldsymbol{K}_{0\mathrm{m}} = \boldsymbol{K}_{0\mathrm{m}}^{\mathrm{T}}$$

Solving Equation (k) for U_3, U_6, U_9 and inserting the solution into Equation (j) yields:

$$\boldsymbol{M}\{\ddot{U}_5\} + \boldsymbol{K}_{\mathrm{con}}\{U_5\} = \boldsymbol{P}_{\mathrm{m}} + \boldsymbol{C}_{\mathrm{m}0}^{\mathrm{P}}\boldsymbol{P}_0 \tag{p}$$

where the condensed stiffness matrix $\boldsymbol{K}_{\mathrm{ee}}$ is:

$$\boldsymbol{K} = \boldsymbol{K}_{\mathrm{mm}} - \boldsymbol{K}_{\mathrm{m}0}\,\boldsymbol{K}_{00}^{-1}\,\boldsymbol{K}_{0\mathrm{m}} = [6.592 \times 10^4]$$

and the load-transfer matrix, $\boldsymbol{C}_{\mathrm{m}0}^{\mathrm{P}}$, is:

$$\boldsymbol{C}_{\mathrm{m}0}^{\mathrm{P}} = -\boldsymbol{K}_{\mathrm{m}0}\,\boldsymbol{K}_{00}^{-1} = [0.25 \;\; 0 \;\; -0.25] \tag{q}$$

Thus, the equation of motion of the beam excited by the torque and the base motion is:

$$m\ddot{y} + 6.592 \times 10^4\, y = 3.296 \times 10^3 v_{\mathrm{g}}(t) \tag{r}$$

where $U_5 = y$.

8.2.5 *Damping in Structures*

In the previous sections, we discussed various methods to formulate equations of motions of structures. The equation of motion of a structure is the counterpart (or representation) of a real structure in dynamic analysis. The solution of the equation governing the motion is a theoretical solution of the dynamic response of a structure. If the structural idealization is decent, then the theoretical solution should be a good approximation of the real response of the actual structure. To check the reliability of the idealized system, we may compare a theoretical solution with measured records in experiments. Figure 8.28 illustrates an experiment to observe free vibrations of a steel vertical cantilever beam with a mass density $\rho = 333$ kg/m and a flexural rigidity $EI = 333\ \text{N} \cdot \text{m}^2$.

This cantilever beam may be modeled as a single degree of freedom (SDOF) system, shown in Figure 8.29, whose equation of motion can be expressed as:

$$m\ddot{x} + kx = 0 \tag{8.72}$$

subject to the following condition:

$$\dot{x}(0) = 0 \quad x(0) = 333 \tag{8.73}$$

The particular solution to Equation (8.72) is presented in Figure 8.30. Unlike the measured record shown in Figure 8.28, the theoretical solution continues forever and would never come to rest. This is unrealistic. Intuition suggests that the free vibrations of the cantilever should decay with time just as in the experiment. This indicates that Equation (8.72) is not a good theoretical representation of the actual cantilever beam. The reason for this is because Equation (8.72) does not incorporate the decay feature either.

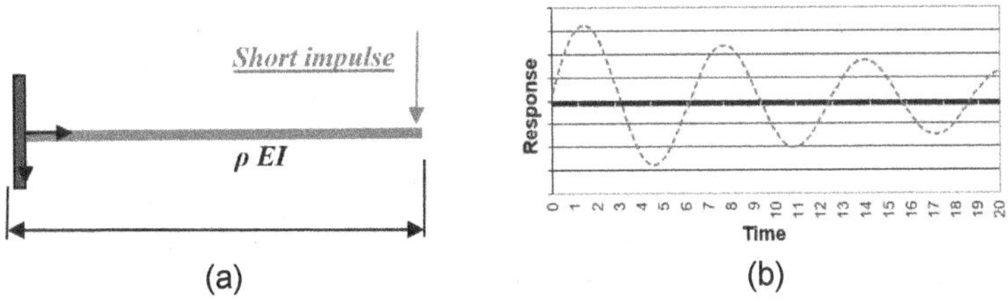

Figure 8.28. (a) A steel vertical cantilever beam, a short impulse applied to the free end of the cantilever gives it an initial velocity to the right. (b) A free vibration record (schematic illustration) at the free end of the steel cantilever beam due to the applied short impulse.

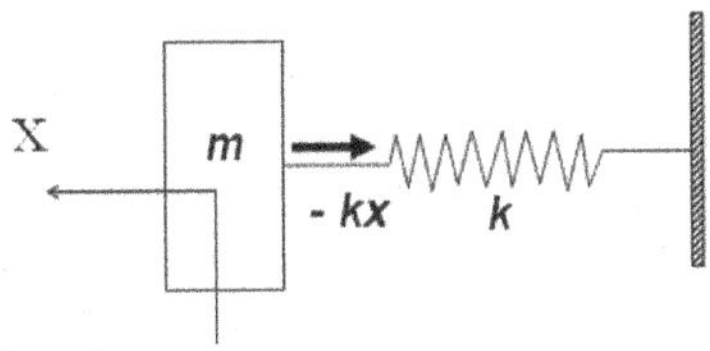

Figure 8.29. The cantilever beam in Figure 8.28 may be modeled as a single degree of freedom (SDOF) system.

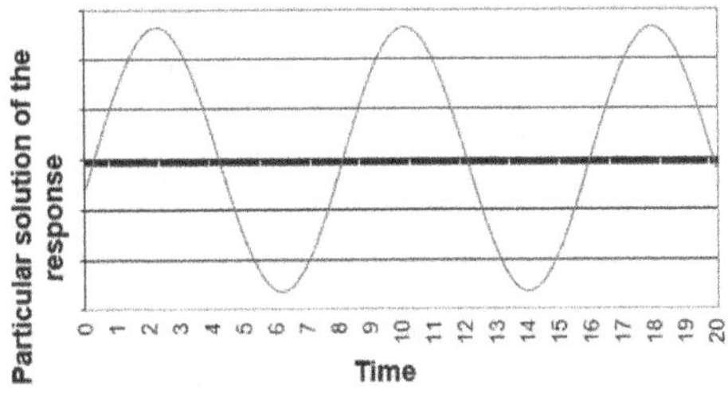

Figure 8.30. The particular solution response (schematic illustration) of the cantilever beam in Figure 8.28.

By investigating the previous process to formulate equations of motion, one would find that it is implicitly assumed that:

1) The skin friction arising from the interaction between the air and the building can be neglected;
2) The dry friction between members of the structure;
3) The building material is purely elastic.

Purely elastic materials do not dissipate energy during the motion of the object. Thus, the assumptions above exclude all the energy-dissipating mechanisms from a theoretical model. This is the theoretical solution that does not incorporate the feature of decaying motions observed during free vibration tests of the cantilever beam.

The dissipation of the kinetic energy and strain energy from an oscillating system is called ***damping***. Generally, in structural engineering, damping may be produced by various energy-dissipating mechanisms, such as the skin-friction between the air and the building, the dry-friction between the members of a structure, and the viscoelasticity of the building materials. However, for most problems in structural dynamics, the energy loss in dry-friction or skin friction can be neglected. Thus, only the energy dissipation due to the material viscoelasticity, called ***viscous damping***, will be considered in this chapter.

There are two ways to introduce viscous damping to a structural idealization, rational or phenomenological. Structural engineering usually adopts the latter approach to introduce viscous damping by adding a viscous damping term, $c\dot{x}$, into an equation of motions, where c is called (viscous) ***damping coefficient***. For example, the equation governing the free vibrations of the SDOF with damping can be written as:

$$m\ddot{x} + c\dot{x} + kx = 0 \tag{8.74}$$

where the damping coefficient c may be determined by experiments. For example, if one assigns the value 333 to the damping coefficient, then, the theoretical solution will have the same decay feature as the measured record.

The next question is how to introduce viscous damping to the equation governing the motion of a MDOF system. Similar to a damped SDOF system, one may include the energy dissipation mechanism in the dynamic model by inserting the damping term, $C\dot{x}$, to the equations of motion, where $\boldsymbol{C}$ is called a ***damping matrix***. For example, the equation describing the damped vibrations of the system with 5 lumped masses can be expressed as:

$$\boldsymbol{M}\ddot{\boldsymbol{x}} + \boldsymbol{C}\dot{\boldsymbol{x}} + \boldsymbol{K}\boldsymbol{x} = 0 \tag{8.75}$$

This immediately raises another question: How can one determine the entries of a damping matrix for a large DOF system? This will be discussed in Example 8.17 in Section 8.3.

In summary, the equation governing the motion of a damped oscillator is obtained by adding a "*phenomenological*" term into the "*rational*" equation of the undamped motions of the oscillator. $c\dot{x}$ is typically called a damping force, but it may be more appropriate to be referred to as a "damping term" because there is no specific *force* corresponding to this term.

8.3 Dynamic Properties of Structures

The dynamic properties of structures may be investigated by experiments, such as full-size building tests and laboratory model tests. As the equation of motion of a structure is the mathematical counterpart of an actual structure, alternatively, we may investigate the dynamic properties of a structure by studying its mathematical representation. Equations of motion may be studied numerically or analytically. Analytical approaches consume few resources, except for intelligence, and usually help us to achieve a deeper understanding of dynamic systems. This section shall explore the essential dynamic properties of structures by analyzing the following differential equation:

$$\boldsymbol{M}\ddot{\boldsymbol{u}} + \boldsymbol{C}\dot{\boldsymbol{u}} + \boldsymbol{K}\boldsymbol{u} = \boldsymbol{P} \tag{8.76}$$

which is the MDOF representation of a structure subjected to dynamic loading.

8.3.1 Vibrations of SDOF Systems: Natural Frequency and Damping Ratio

A simple theoretical model may filter out unwanted information and enable us to focus on the essential properties of a structure. This section shall explore the most important features of a dynamic system, such as the resonance, exponential decay behavior, vibration period and phase lag, by investigating the theoretical solutions to its SDOF model. All these peculiar properties of dynamic systems can be characterized by two important physical quantities, *natural frequency* and *damping ratio*.

The viscously damped vibration of a SDOF system is governed by the following differential equation,

$$m\ddot{u} + c\dot{u} + ku = p(t) \tag{8.77}$$

where, u is the dynamic additional displacement of the mass.

By dividing the equation above with m, one has:

$$\ddot{u} + 2\zeta\omega\dot{u} + \omega^2 u = \frac{p(t)}{m} \tag{8.78}$$

where $\omega \equiv \sqrt{k/m}$ is defined in Equation (8.15) and the damping ratio (a dimensionless measure of damping) is defined as:

$$\zeta \equiv \frac{c}{2m\omega} \tag{8.79}$$

The damped free vibration of this SDOF system is governed by the following homogeneous differential equation:

$$\ddot{u} + 2\zeta\omega\dot{u} + \omega^2 u = 0 \tag{8.80}$$

The solution of Equation (8.78) will be obtained by considering first the homogeneous Equation (8.80). To determine the free-vibration response which is the solution of Equation (8.80), one can calculate the trial solution of the form:

$$u = \mathbf{e}^{st} \tag{8.81}$$

in order to find the value of s such that the trial solution satisfies the homogeneous Equation (8.80). One can implement the trial solution into Equation (8.80), which yields:

$$(s^2 + 2\zeta\omega s + \omega^2)e^{st} = 0 \tag{8.82}$$

which is satisfied for all values of t if:

$$s^2 + 2\zeta\omega s + \omega^2 = 0 \tag{8.83}$$

The solution to the equation above is

$$s_{1,2} = \omega\left(-\zeta \pm \mathrm{i}\sqrt{1-\zeta^2}\right) \tag{8.84}$$

For systems with ***critical damping***, $\zeta = 1$, the radical term in Equation (8.84) vanishes, and both values of s are the same, i.e.,

$$s_1 = s_2 = -\omega \tag{8.85}$$

which leads to the first particular solution of Equation (8.80):

$$u_1 = \mathrm{e}^{-\omega t} \tag{8.86}$$

for $\zeta = 1$. Since,

$$u_2(t) = t\mathrm{e}^{-\omega t} \tag{8.87}$$

satisfies Equation (8.83) for $\zeta = 1$. This is the second particular solution linearly independent of the first one. Thus, the general solution of Equation (8.80) in this special case can be expressed as:

$$u_c(t) = (C_1 + C_2 t)\,\mathrm{e}^{-\omega t} \tag{8.88}$$

The particular solution to the initial conditions $u(0) = u_0$ and $\dot{u}(0) = \dot{u}_0$ is:

$$u(t) = [u_0\,(1 - \omega t) + \dot{u}_0 t]\mathrm{e}^{-\omega t} \tag{8.89}$$

which is portrayed graphically as the blue solid line in Figure 8.31. In this case, the system returns to its equilibrium position within the shortest time without oscillating.

For systems with ***overdamping***, $\zeta > 1$, Equation (8.84) yields two linearly independent solutions to Equation (8.83):

$$u_1(t) = \mathrm{e}^{-\left(\zeta-\sqrt{\zeta^2-1}\right)\omega t} \quad \text{and} \quad u_2(t) = \mathrm{e}^{-\left(\zeta+\sqrt{\zeta^2-1}\right)\omega t} \tag{8.90}$$

which leads to the general solution of Equation (8.80):

$$u_c(t) = C_1\mathrm{e}^{-\left(\zeta-\sqrt{\zeta^2-1}\right)\omega t} + C_2\mathrm{e}^{-\left(\zeta+\sqrt{\zeta^2-1}\right)\omega t} \tag{8.91}$$

The rest of this presentation will treat the systems with ***underdamping*** ($\zeta < 1$), because the damping ratio of a structure is usually less than 0.2. In this case,

$$u_1(t) = \mathrm{e}^{s_1 t} \quad \text{and} \quad u_2(t) = \mathrm{e}^{s_2 t} \tag{8.92}$$

there are two linearly independent solutions to Equation (8.80). Then, its general solution can be expressed as:

$$u_c(t) = C_1\mathrm{e}^{s_1 t} + C_2\mathrm{e}^{s_2 t} \tag{8.93}$$

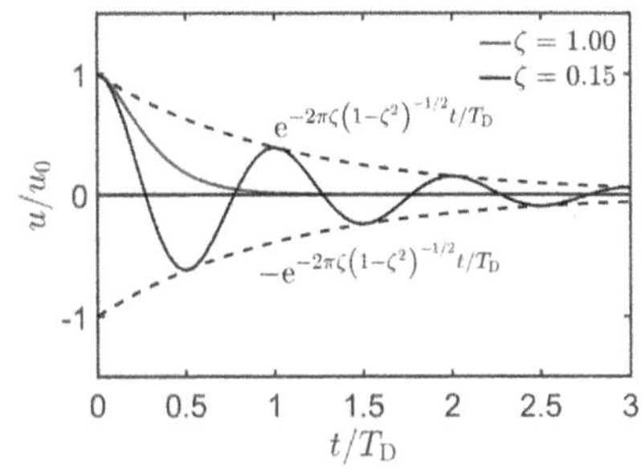

Figure 8.31. Free-vibration responses of systems with critical damping, $\zeta = 1$ and underdamping, $\zeta = 0.15$. $T_D \equiv 2\pi/\omega_D$ is the damped period and close to the natural period, T, of the system. The blue line denotes the critically-damped response, while the black solid line denotes the underdamped response.

which after inserting Equation (8.84) becomes

$$u_c(t) = e^{-\zeta\omega t}(C_1 e^{i\omega_D t} + C_2 e^{-i\omega_D t}) \tag{8.94}$$

where C_1 and C_2 are complex constants as yet undetermined and:

$$\omega_D \equiv \omega\sqrt{1-\zeta^2} \tag{8.95}$$

is called a damped frequency.

Notice that $u_{1,2}$ are only purely mathematical solutions of Equation (8.80) but not a free-vibration response of the SDOF system because they are complex. Although a vibration response must be a solution of the governing equation of motion, not all the solutions of the equation of motion are the feasible responses of a vibrating system. Equation (8.94) shows that a free-vibration response can be expressed as a linear combination of two purely mathematical solutions. One can express it as a linear combination of two linearly-independent vibration responses. To this end, the term in the parentheses in Equation (8.94) can be rewritten in terms of trigonometric functions. Substitution of Euler's formula

$$e^{i\theta} = \cos\theta + i \cdot \sin\theta \tag{8.96}$$

into Equation (8.94) yields:

$$u_c(t) = e^{\zeta\omega t}[(C_1 + C_2)\cos\omega_D t + i\,(C_1 - C_2)\sin\omega_D t] \tag{8.97}$$

which indicates that C_1 and C_2 must be a complex conjugate pair because $u(t)$ is real. Introducing

$$A_1 \equiv C_1 + C_2 \quad \text{and} \quad A_2 \equiv C_1 - C_2 \tag{8.98}$$

into Equation (8.97) gives

$$u_c(t) = e^{\zeta\omega t}[A_1\cos\omega_D t + i\,A_2\sin\omega_D t] \tag{8.99}$$

where A_1 and A_2 are real constants yet undetermined. After considering initial conditions $u_c(0) = u_0$ and $\dot{u}_c(0) = \dot{u}_0$, A_1 and A_2 can be expressed as:

$$A_1 = u_0 \quad \text{and} \quad A_2 = \frac{\dot{u}_0 + \zeta\omega u_0}{\omega_D} \tag{8.100}$$

The free-vibration response of an underdamped system is plotted as the black solid line in Figure 8.31, in which the dashed lines denote the envelope curves, $\pm e^{-\zeta\omega t}$, of the underdamped response.

As typical engineering structures such as steel structures, reinforced-concrete structures, and so on – all have a damping ratio much less than 1 (e.g., $\zeta = 0.02$ for steel structures and $\zeta = 0.05$ for reinforced-concrete structures the damped frequency), ω_D, is very close to the natural frequency, ω:

$$\omega_D \approx \omega \tag{8.101}$$

The following example illustrates how a damping ratio can be determined by experiments.

EXAMPLE 8.7

Figure 8.32a shows the measured record of the top-displacement response observed during free vibration tests of the vertical cantilever beam shown in Figure 8.32b. If this cantilever beam is modeled as a single-degree-of-freedom system with a lumped mass at the top as shown in Figure 8.32c, then the equation governing its decaying motion is:

$$\ddot{u} + 2\zeta\omega\dot{u} + \omega^2 u = 0$$

where the damping ratio ζ is the physical quantity representing the energy-dissipation mechanism of the cantilever beam. Determine the value of ζ according to the experiment data given in Figure 8.32.

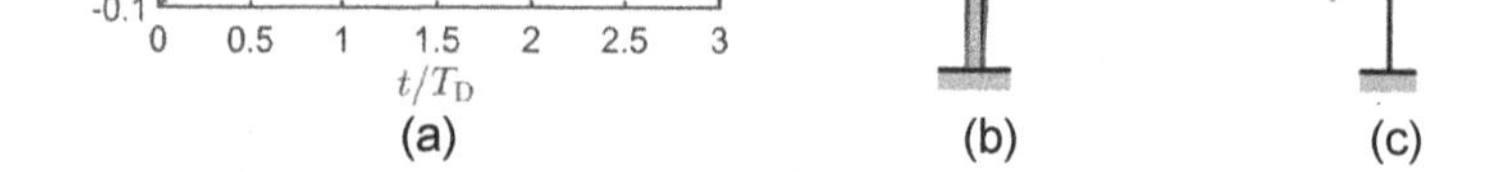

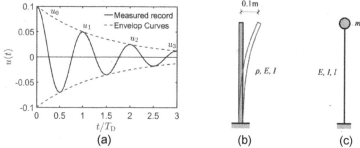

Figure 8.32. (a) The measured record of a free-vibration response. (b) The laboratory model. (c) A theoretical model with a lumped mass.

***Solution*:**

As shown in Figure 8.32a, let u_i denote the intersections of the damped-vibration response and its envelope. u_{i+1} is the peak within a full vibration period T_D after u_i. Since the envelope curve is $e^{-\zeta\omega t}$, the ratio of successive peaks can be expressed as:

$$\frac{u_i}{u_{i+1}} = e^{\zeta\omega T_D} = e^{2\pi\zeta\omega/T_D} = e^{2\pi\zeta/\sqrt{1-\zeta^2}} \tag{8.102}$$

where the last equality is obtained by using Equation (8.95). If ζ is small, $\sqrt{1-\zeta^2} \approx 1$ and the ratio of successive peaks can be approximated by

$$\frac{u_i}{u_{i+1}} \approx e^{2\pi\zeta} \tag{8.103}$$

Taking the natural logarithm of this equation gives:

$$\zeta = \frac{1}{2\pi} \ln\left(\frac{u_i}{u_{i+1}}\right) \tag{8.104}$$

The data given in Figure 8.32a shows that:

$$\frac{u_0}{u_1} = \frac{0.1}{0.05} = 2 \tag{8.105}$$

Substituting this ratio value into Equation (8.104) yields:

$$\zeta = \frac{\ln 2}{2\pi} = 0.11 \tag{8.106}$$

which ensures the SDOF model to have the same decaying motions as the original laboratory model.

Next, we shall discuss the resonance phenomenon of a vibrating system by investigating the harmonic vibration. Including damping the equation governing the forced vibrations of a SDOF system subject to harmonic loading is:

$$m\ddot{u} + c\dot{u} + ku = p_0 \sin(\tilde{\omega}t + \tilde{\varphi}) \tag{8.107}$$

or

$$\ddot{u} + 2\zeta\omega\dot{u} + \omega^2 u = \gamma \sin(\tilde{\omega}t + \tilde{\varphi}) \tag{8.108}$$

where,

$$\gamma \equiv \frac{p_0}{m} \tag{8.109}$$

The general solution can be expressed as:

$$u(t) = u_c(t) + u_P(t) \tag{8.110}$$

where u_P is a particular solution of Equation (8.108) and u_c is the free vibration response given by Equation (8.98).

To seek a particular solution of Equation (8.108), consider a trial solution of the form as follows:

$$u_P = B_1 \sin(\tilde{\omega}t + \tilde{\varphi}) + B_1 \cos(\tilde{\omega}t + \tilde{\varphi}) \tag{8.111}$$

Substitution of the trial solution (8.110) into Equation (8.108) gives:

$$\left[\left(\omega^2 + \tilde{\omega}^2\right)B_1 - 2\zeta\omega\tilde{\omega}B_2 - \frac{p_0}{m}\right]\sin\left(\tilde{\omega}t + \tilde{\varphi}\right) + \left[2\zeta\omega\tilde{\omega}B_1 + \left(\omega^2 + \tilde{\omega}^2\right)B_2\right]\cos\left(\tilde{\omega}t + \tilde{\varphi}\right) = 0 \tag{8.112}$$

As $\sin(\tilde{\omega}t + \tilde{\varphi})$ and $\cos(\tilde{\omega}t + \tilde{\varphi})$ are linearly independent, the coefficients of the sine and cosine terms must vanish for the existence of Equation (8.112). This requirement provides two equations in B_1 and B_2, which, after dividing by ω^2 and introducing the relation $k = \omega^2 m$, become:

$$\begin{cases} \left[1 - \left(\dfrac{\tilde{\omega}}{\omega}\right)^2\right] B_1 - \left(2\zeta\dfrac{\tilde{\omega}}{\omega}\right) B_2 = \dfrac{p_0}{k} \\ \left(2\zeta\dfrac{\tilde{\omega}}{\omega}\right) B_1 + \left[1 - \left(\dfrac{\tilde{\omega}}{\omega}\right)^2\right] B_2 = 0 \end{cases} \tag{8.113}$$

Solving Equation (8.113) for B_1 and B_2 gives:

$$\begin{cases} B_1 = \dfrac{p_0}{k}\dfrac{1-(\tilde{\omega}/\omega)^2}{\left[1-(\tilde{\omega}/\omega)^2\right]^2+\left[2\zeta(\tilde{\omega}/\omega)\right]^2} = \gamma\dfrac{1-(\tilde{\omega}/\omega)^2}{\omega^2\left[1-(\tilde{\omega}/\omega)^2\right]^2+(2\zeta\tilde{\omega})^2} \\ B_2 = \dfrac{p_0}{k}\dfrac{-2\zeta\tilde{\omega}/\omega}{\left[1-(\tilde{\omega}/\omega)^2\right]^2+\left[2\zeta(\tilde{\omega}/\omega)\right]^2} = \gamma\dfrac{-2\zeta\tilde{\omega}/\omega}{\omega^2\left[1-(\tilde{\omega}/\omega)^2\right]^2+(2\zeta\tilde{\omega})^2} \end{cases} \tag{8.114}$$

All this shows that u_P described by Equation (8.111) is a particular solution of Equation (8.108) for B_1, B_2 given by Equation (8.114).

Finally, the general vibration response of a viscously damped SDOF system to harmonic excitation can be expressed as:

$$\begin{aligned} u(t) &= \underbrace{e^{-\zeta\omega t}\left(G_1\cos\omega_D t + G_2\sin\omega_D t\right)}_{\text{transient term}} + \underbrace{B_1\sin\left(\tilde{\omega}t+\tilde{\varphi}\right)+B_2\cos\left(\tilde{\omega}t+\tilde{\varphi}\right)}_{\text{steady-state term}} \\ &\equiv u_{\text{tra}}(t) + u_{\text{ste}} \end{aligned} \tag{8.115}$$

where B_1 and B_2 are given by Equation (8.114), and G_1 and G_2 are constants to be determined by the following initial conditions:

$$u(0) = u_0 \quad \text{and} \quad \dot{u}(0) = \dot{u}_0 \tag{8.116}$$

Equation (8.115) shows that the total response of a viscously damped system to external excitation includes two parts, the ***transient term*** and the ***steady-state term***. Since the transient term decays quickly, after a short period, the steady-state term independent initial conditions are a decent approximation of the response of the system starting from any initial condition. In general, a steady-state response of a dynamic system may be defined as follows.

Definition 9.1 (Steady-State Response) *A time-function, $u_s(t)$, independent from initial conditions is called a* ***steady-state response*** *of a dynamic system, if, for a given tolerance ε, there exists a time t_s such that*

$$\left|u(t, u_0, \dot{u}_0) - u_s(t)\right| \le \varepsilon \tag{8.117}$$

for $t \ge t_s$ and any u_0, $\dot{u}_0 \in \boldsymbol{R}$, where $u(t, u_0, \dot{u}_0)$ is a response of the system starting from any initial condition u_0, $\dot{u}_0$.

Definition 9.1 shows that a steady-state response is independent from initial conditions and is a good approximation of the response of a system after some period. It should be noticed that not all dynamic systems exhibit steady-state responses. For example, any response of an undamped system depends on its initial conditions forever. In many cases, we have to consider the effect of initial conditions on a response. The response of a system starting from some given initial condition is usually termed a ***transient response***.

Specifically, for a response starting from rest, $u_0 = 0$ and $\dot{u}_0 = 0$, G_1 and G_2 can be expressed as:

$$\begin{cases} G_1 = -B_1\sin\tilde{\omega} - B_2\cos\tilde{\omega} \\ G_2 = \dfrac{1}{\omega_D}\left(G_1\zeta\omega - B_1\tilde{\omega}\cos\tilde{\omega} + B_2\tilde{\omega}\sin\tilde{\omega}\right) \end{cases} \tag{8.118}$$

Since the transient part decays quickly, one usually focuses on the ***steady-state response*** in the form as follows:

$$u_{\text{ste}}(t) = u_{\max}\sin(\tilde{\omega}t + \tilde{\varphi} - \varphi) \tag{8.119}$$

The response amplitude associated with the equation above is:

$$u_{\max} = \frac{p_0}{k}\frac{1}{\sqrt{\left[1-(\tilde{\omega}/\omega)^2\right]^2+\left[2\zeta(\tilde{\omega}/\omega)\right]^2}} = \frac{\gamma}{\omega^2\sqrt{\left[1-(\tilde{\omega}/\omega)^2\right]^2+\left[2\zeta(\tilde{\omega}/\omega)\right]^2}} \tag{8.120}$$

and the associated phase is:

$$\varphi = \begin{cases} \tan^{-1}\left[\dfrac{2\zeta(\tilde{\omega}/\omega)}{1-(\tilde{\omega}/\omega)^2}\right], & \textit{when } \tilde{\omega}/\omega \le 1 \\ \pi + \tan^{-1}\left[\dfrac{2\zeta(\tilde{\omega}/\omega)}{1-(\tilde{\omega}/\omega)^2}\right], & \textit{when } \tilde{\omega}/\omega > 1 \end{cases} \tag{8.121}$$

by which the steady-state response lags behind the excitation. Plots of φ versus ζ and $\tilde{\omega}/\omega$ are shown in Figure 8.33a. If a response and the excitation have the same algebraic sign, then the response is said to be ***in phase*** with the excitation. Otherwise, it is said to be ***out of phase*** relative to the excitation. Figure 8.33a shows that the displacement response is in phase with the dynamic loading when $\tilde{\omega}/\omega \le 1$, and out of phase with the excitation when $\tilde{\omega}/\omega > 1$.

To measure the difference between the results from static analysis and dynamic analysis, one can introduce the following physical quantity:

$$\beta_{\text{d}} \equiv \frac{u_{\max}}{u_{\text{st}}} \tag{8.122}$$

where $u_{\max}$ is the maximum dynamic displacement and u_{st} called the ***static displacement*** is the displacement of the structure to the equivalent static load. The parameter β_{d} measures the difference between the solutions to the same problem from the dynamic analysis and the static analysis, and is called the ***deformation magnification factor***, or the ***deformation response factor***. Similarly, one may define

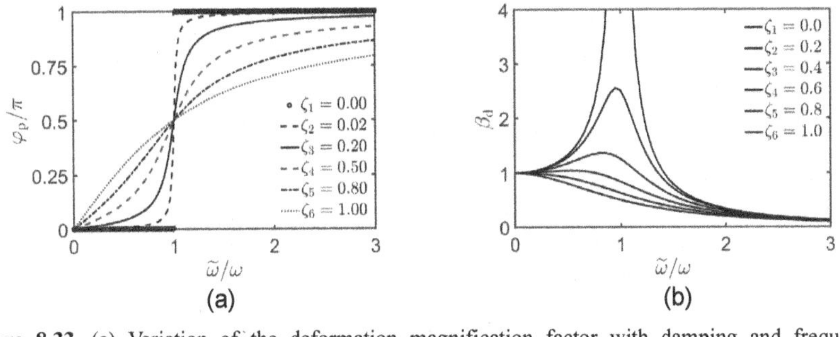

Figure 8.33. (a) Variation of the deformation magnification factor with damping and frequency. (b) Variation of the phase with damping and frequency.

the ***velocity magnification factor***, β_v, and the ***acceleration magnification factor***, β_a, because a dynamic system involves various response measures. They are called ***dynamic magnification factors***. The term dynamic magnification factor is used to often denote a deformation magnification factor.

For the SDOF system subjected to the harmonic load, one may take the amplitude, p_0, of the excitation as the equivalent static load. In this case, β_d can be expressed as

$$\beta_d = \frac{u_{max}}{p_0/k} = \frac{1}{\sqrt{\left[1-(\tilde{\omega}/\omega)^2\right]^2 + \left[2\zeta(\tilde{\omega}/\omega)\right]^2}} \tag{8.123}$$

which is plotted in Figure 8.33(b).

From Equation (8.123), it is evident that the steady-state response amplitude of an undamped system tends toward infinity as the excitation frequency, $\tilde{\omega}$, approaches the natural frequency, ω; this feature can be seen in Figure 8.33b for the undamped case $\zeta = 0$. For small dampings, it is seen in this same figure that the maximum steady-state response amplitude occurs at an excitation frequency slightly less than the natural frequency. Nevertheless, the condition resulting when the excitation frequency equals the natural frequency is called ***resonance***.

If the dynamic magnification factor is close to 1, the problem can then be taken as a static one. Otherwise, it should be viewed as a dynamic problem. It is interesting to note that a dynamic magnification factor may be less than 1 when the ratio $\tilde{\omega}/\omega$ is large enough because of the response lag excitation.

In addition, the decay rate and the period of free vibrations of the dynamic system are independent of the initial conditions, but completely characterized by the damping ratio and the natural frequency according to Equation (8.99).

8.3.2 Undamped Free Vibrations of MDOF Systems: Normal Modes

To explore complete properties of dynamic systems, one has to establish a more complicated model. This section will study another peculiar feature, *normal modes*, of a dynamic system by investigating the undamped free vibrations of a MDOF model.

The undamped free vibration of a n-DOF system is governed by the following homogeneous differential equation:

$$\boldsymbol{M}\ddot{\boldsymbol{u}} + \boldsymbol{K}\boldsymbol{u} = 0 \tag{8.124}$$

To find the general solution of Equation (8.124), one has to find $2n$ linearly independent solutions. One can assume that the equation has the following particular solution of the form:

$$\boldsymbol{u}(t) = \boldsymbol{\phi} \sin \omega t \tag{8.125}$$

where ω is a constant and $\boldsymbol{\phi} = [\phi_1,..., \phi_n]^T$ is a time-independent vector as yet undetermined.

Before solving for ω and $\boldsymbol{\phi}$, it will be useful to consider their physical interpretation. The trial solution (8.128) describes a pattern of undamped free vibrations in which all the mass points move sinusoidally with the same frequency, ω, and with the phase. The ratio between the displacements of two arbitrary mass points is independent of time and described by $\boldsymbol{\phi}$. For example, the ratio between the displacements of the *i*th and *j*th mass points is ϕ_i/ϕ_j, which is a constant. ω is called a ***modal frequency*** or ***natural circular frequency***, and $\boldsymbol{\phi}$ is referred to as the corresponding ***modal vector*** or ***modal shape vector***. (ω, $\boldsymbol{\phi}$) is called a ***modal pair*** or ***eigen pair*** of the system.

To find ω and $\boldsymbol{\phi}$ such that the trial solution (8.125) satisfies Equation (8.124), substitute it into Equation (8.124) and one then obtains:

$$-\omega^2\boldsymbol{M}\boldsymbol{\phi}\sin\omega t + \boldsymbol{K}\boldsymbol{\phi}\sin\omega t = 0 \tag{8.126}$$

which gives the following ***equation of modes***

$$(\boldsymbol{K} - \omega^2\boldsymbol{M})\boldsymbol{\phi} = 0 \tag{8.127}$$

or

$$\boldsymbol{K}\boldsymbol{\phi} = \omega^2\boldsymbol{M}\boldsymbol{\phi} \tag{8.128}$$

This implies that finding modes involves solving a ***generalized eigenvalue equation*** in mathematics, whose eigenvalue and eigenvector are ω^2 and $\boldsymbol{\phi}$, respectively.

A nontrivial solution Equation (8.127) is possible when the modal frequency ω satisfies the following ***frequency equation***:

$$|\boldsymbol{K} - \omega^2\boldsymbol{M}| = 0 \tag{8.129}$$

where $|\cdot|$ denotes a determinant of a matrix. In mathematics, the frequency Equation (8.129) is also called the characteristic equation.

If $\boldsymbol{M}$ is a lumped-mass matrix, Equation (8.129) can then be rewritten as:

$$\begin{vmatrix} k_{11} - m_1\omega^2 & k_{12} & \cdots & k_{1n} \\ k_{21} & k_{22} - m_2\omega^2 & \cdots & k_{2n} \\ \vdots & \vdots & \ddots & \vdots \\ k_{n1} & k_{n2} & \cdots & k_{nn} - m_n\omega^2 \end{vmatrix} = 0 \tag{8.130}$$

The frequency equation of a n DOF system is an algebraic equation of the nth degree in the frequency parameter ω^2. All the roots of the eigenvalue, ω^2, are positive, because both mass matrix, $\boldsymbol{K}$, and the stiffness matrix, $\boldsymbol{M}$, are positive definite. Although multiple roots are possible, it is beyond the scope of this book. Here, we consider just the case that the frequency equation has n distinct roots, $\omega_1^2 < \omega_2^2 < \cdots < \omega_n^2$, which determine n distinct positive ω_i.

When a modal frequency ω_i is determined, Equation (8.127) can be solved for the corresponding modal vector (or the eigenvector) $\boldsymbol{\phi}_i$ within a multiplicative constant. To this end substitute the modal frequency, ω_i, into Equation (8.127) and one obtains:

$$(\boldsymbol{K} - \omega_i^2\boldsymbol{M})\boldsymbol{\phi}_i = 0 \tag{8.131}$$

The equation above cannot determine the absolute amplitude value of the modal vector, $\boldsymbol{\phi}_i$, because the modal frequencies (or the eigenvalues) are evaluated from the condition that the associated equation governing eigen-modes has numerous solutions. However, the shape of the modal vector can be determined by assigning a value to a DOF and solving for all the DOFs in terms of the prescribed DOF. For this purpose, it may be assumed that the first coordinate of $\boldsymbol{\phi}_i$ has a unit amplitude; that is

$$\phi_{1j} = 1 \tag{8.132}$$

in which ϕ_{1j} denotes the first coordinate of the modal vector $\boldsymbol{\phi}_j$. Combining Equations (8.131) and (8.132) gives a specific solution of $\boldsymbol{\phi}_i$, whose first coordinate is 1. The additional equation, like Equation (8.132), which is used to specify a modal vector, is called a ***normalizing condition***. The process to specify the modal vector by an additional normalizing condition is called the ***normalization*** of modal vectors.

Other normalizations are also frequently used, e.g., sometimes it is convenient to normalize each modal vector so that its largest coordinate is unity. The most widely used normalization, however, involves adjusting each modal amplitude to satisfy the following condition

$$\boldsymbol{\phi}_i^{\mathrm{T}} \boldsymbol{M} \boldsymbol{\phi}_i = 1 \tag{8.133}$$

In the next section, we will prove that modal vectors are linearly independent. This implies that one can obtain n linearly independent solutions to Equation (8.124). It is easy to verify that $\boldsymbol{\phi}_i \cos \omega_i t$ is also a solution of Equation (8.124) when $(\omega_i, \boldsymbol{\phi}_i)$ is a modal pair of the system. Since $\sin \omega_i t$ and $\cos \omega_i t$ are linearly independent, one can immediately obtain $2n$ linearly independent solutions of Equation (8.124). Thus, its general solution can be expressed as:

$$u(t) = \sum_{i=1}^{n} \left(A_i \boldsymbol{\phi}_i \sin \omega_i t + B_i \boldsymbol{\phi}_i \cos \omega_i t \right) \tag{8.134}$$

where the real constants A_i and B_i can be evaluated using the initial conditions $\boldsymbol{u}(0) = \boldsymbol{u}_0$ and $\dot{\boldsymbol{u}}(0) = \dot{\boldsymbol{u}}_0$.

In summary, a modal pair $(\omega_i, \boldsymbol{\phi}_i)$ is a set of parameters describing a pattern of undamped free vibrations which can be mathematically expressed as:

$$u_i(t) = A_i\, \boldsymbol{\phi}_i \sin \omega_i t + B_i\, \boldsymbol{\phi}_i \cos \omega_i t \tag{8.135}$$

Such a pattern of motion is called a ***mode*** of the system. For any given A_i and B_i, Equation (8.135) governs a specific vibration of the ith mode of the system. Any free-vibration response of an undamped system can be expressed as a linear combination of its modes. The rest of this chapter shows that modes are particularly important in dynamic analysis, and are often expressed as the modal vectors in the form of a ***modal matrix*** or ***mode shape matrix***, $\boldsymbol{\Phi}$,

$$\boldsymbol{\Phi} = \begin{bmatrix} \boldsymbol{\phi}_1 & \boldsymbol{\phi}_2 & \cdots & \boldsymbol{\phi}_n \end{bmatrix} = \begin{bmatrix} \phi_{11} & \phi_{12} & \cdots & \phi_{1n} \\ \phi_{21} & \phi_{22} & \cdots & \phi_{2n} \\ \vdots & \vdots & \ddots & \vdots \\ \phi_{n1} & \phi_{n2} & \cdots & \phi_{nn} \end{bmatrix} \tag{8.136}$$

whose jth column is the jth modal vector of the system, and whose entry ϕ_{ij} denotes the ith coordinate of the jth modal vector. The n eigenvalues ω_i^2 can be assembled into a diagonal matrix $\boldsymbol{\Omega}^2$, which is called the ***spectral matrix*** of the generalized eigenvalue problem:

$$\boldsymbol{\Omega}^2 = \begin{bmatrix} \omega_1^2 & & & \\ & \omega_2^2 & & \\ & & \ddots & \\ & & & \omega_n^2 \end{bmatrix} \tag{8.137}$$

The natural circular frequency, ω_i, may be convenient for theoretical discussions. However, engineers prefer to use the ***natural cyclic frequency***, f_i, defined as follows:

$$f_i = \frac{2\pi}{\omega_i} \tag{8.138}$$

whose unit is hertz (Hz) (cycles per second). f_i is obviously related to the natural period T_i through:

$$f_i = \frac{1}{T_i} \tag{8.139}$$

The term *natural frequency* (or natural frequency) applies to both ω_i and f_i.

The following example illustrates the procedure to analytically determine all the modal pairs of a dynamic system.

EXAMPLE 8.8

Consider a vertical cantilever beam illustrated in Figure 8.34, for which the lumped masses, the Young's modulus, the moment of inertia and the height of each storey are $m_1 = 500$ kg, $m_2 = 300$ kg, $E = 2.06 \times 10^{11}$ N/m², $I = 1.152 \times 10^{-5}$ m⁴ and $h = 8$m, respectively. Neglecting the axial deformation of each member, determine all the modal pairs, and calculate the natural cyclic frequencies and natural periods of the system.

Solution:

Example 8.5 shows that the undamped free vibrations of this system are governed by the following homogeneous equation:

$$\boldsymbol{M}\ddot{\boldsymbol{x}} + \boldsymbol{K}\boldsymbol{x} = \boldsymbol{0}$$

where:

$$\boldsymbol{M} = \begin{bmatrix} 500 & 0 \\ 0 & 300 \end{bmatrix} \quad \text{and} \quad \boldsymbol{K} = 10^3 \times \begin{bmatrix} 63.566 & -19.864 \\ -19.864 & 7.9457 \end{bmatrix} \tag{a}$$

The equation describing the modes of this system is:

$$(\boldsymbol{K} - \omega^2\boldsymbol{M})\boldsymbol{\phi} = \boldsymbol{0} \tag{b}$$

which yields the following frequency equation:

$$|\boldsymbol{K} - \omega^2\boldsymbol{M}| = \boldsymbol{0} \tag{c}$$

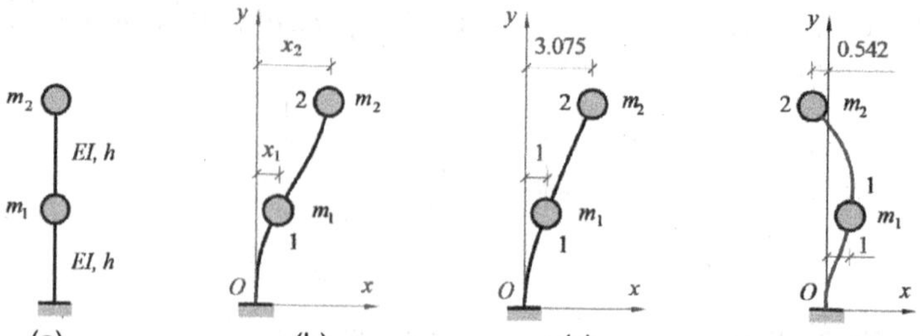

Figure 8.34. (a) A vertical cantilever beam, (b) Coordinate system, (c) 1st eigenmode, (d) 2nd eigenmode.

Plugging Equation (a) into the frequency equation (c) and expanding the determinant leads to:

$$150000\left(\omega^2\right)^2 - 23042650\left(\omega^2\right) + \frac{552489351}{5} = 0 \tag{d}$$

which provides two natural circular frequencies:

$$\begin{cases} \omega_1 = 2.226028185667449 \approx 20226 & (\text{rad/sec}) \\ \omega_2 = 12.192721811936854 \approx 12.193 & (\text{rad/sec}) \end{cases} \tag{e}$$

Substituting the roots (e) into Equation (b) and combining them with the following normalizing condition:

$$\phi_{1j} = 0, j = 1, 2$$

leads to the following modal matrix:

$$\boldsymbol{\Phi} = \begin{bmatrix} 1.000 & 1.000 \\ 3.075 & -0.542 \end{bmatrix}$$

The modal vectors ϕ_1 and ϕ_2 are illustrated graphically in Figure 8.34(c) and Figure 8.34(d), respectively.
The natural cyclic frequencies and the natural periods are respectively

$$f_1 = 0.35\ (\text{Hz}) \quad \text{and} \quad f_2 = 1.94\ (\text{Hz})$$

and

$$T_1 = 2.82\ (\text{sec.}) \quad \text{and} \quad T_2 = 0.52\ (\text{sec.})$$

Since the frequency equation is nonlinear, it is strongly recommended to solve it by numerical methods for practical applications. This computational task can be easily accomplished by computational code functions such as the MATLAB function *eig*, which utilizes the normalizing condition (8.133) to fix the generalized eigenvectors.

8.3.3 Properties of Modes

Four essential properties of eignen-modes are as follows:

Theorem 8.1 *Both the mass and stiffness matrices of oscillating systems in structural engineering are symmetric and positive definite.*

Theorem 8.2 *All the natural frequency parameters ω^2 of a dynamic system have positive definite mass and stiffness matrices.*

Theorem 8.3 (Orthogonality of modes) *If $\boldsymbol{\phi}_i$ and $\boldsymbol{\phi}_j$ are two modal vectors with distinct modal frequencies, ω_i and ω_j, then they satisfy the following **M**– and **K–Orthogonality** conditions*:

$$\boldsymbol{\phi}_i^{\mathrm{T}} \boldsymbol{M} \boldsymbol{\phi}_j = 0 \tag{8.140}$$

and

$$\boldsymbol{\phi}_i^{\mathrm{T}} \boldsymbol{K} \boldsymbol{\phi}_j = 0 \tag{8.141}$$

where $\boldsymbol{M}$ and $\boldsymbol{K}$ are the mass and stiffness matrices.

Theorem 8.4 (Linear independence of modes) *The modal vectors corresponding to different natural frequencies are linearly independent.*

The first step in dynamic analysis is to define space-time coordinates to describe the object. Such a coordinate system is called a ***physical coordinate system*** because each of the spatial coordinates is a *real* displacement of a mass point. The ***physical coordinates x*** are relative to the following standard basis of $\boldsymbol{R}^n$:

$$\hat{x}_i = \begin{Bmatrix} 0 \\ \vdots \\ 0 \\ 1 \\ 0 \\ \vdots \\ 0 \end{Bmatrix} \begin{matrix} 1 \\ \vdots \\ i-1 \\ i \\ i+1 \\ \vdots \\ 0 \end{matrix} \qquad \text{for } i = 1, 2, \cdots, n \tag{8.142}$$

i.e., a displacement response is represented by:

$$\sum_{i=1}^{n} x_i \hat{x}_i = \begin{bmatrix} \hat{x}_1 & \hat{x}_2 & \cdots & \hat{x}_n \end{bmatrix} x \tag{8.143}$$

Theorem 8.4 shows that a set of modal vectors, $\boldsymbol{\phi}_1, \cdots, \boldsymbol{\phi}_n$, is also a basis of $\boldsymbol{R}$. The space spanned by the modal vectors is called a ***modal space***, while the space spanned by the standard basis $\hat{x}_i$ is called a ***physical space***. $\boldsymbol{x}$ is the representation of a displacement response in the physical space. In the modal space, the same displacement response can be represented as:

$$\sum_{i=1}^{n} \eta_i \phi_i = \begin{bmatrix} \phi_1 & \phi_2 & \cdots & \phi_n \end{bmatrix} \eta \tag{8.144}$$

where,

$$\boldsymbol{\eta} \equiv [\eta_1 \ \eta_2 \ \cdots \ \eta_n]^{\mathrm{T}} \tag{8.145}$$

are the coordinates of the displacement response relative to the modal basis $\boldsymbol{\phi}_i$, $i = 1, 2, \ldots, n$, and are called the ***modal coordinates*** of the displacement response. The modal coordinates are a type of ***generalized coordinates***.

The displacement-response representations (8.143)–(8.145) and Equation (8.142) yield:

$$x = \Phi\eta \tag{8.146}$$

which shows that the modal matrix is the ***change-of-coordinates matrix*** from the modal space to the physical space. Left-multiplication by ϕ^{-1} converts $\boldsymbol{x}$ into its modal coordinate vector $\boldsymbol{\eta}$:

$$\boldsymbol{\eta} = \Phi^{-1}\boldsymbol{x} \tag{8.147}$$

By substituting the change-of-coordinate relation, Equation (8.147), into the dynamic equilibrium equation governing the motions of a undamped system we get,

$$\boldsymbol{M}\ddot{\boldsymbol{x}} + \boldsymbol{K}\boldsymbol{x} = \boldsymbol{P} \tag{8.148}$$

which leads to its representation in the modal space as follows:

$$M\Phi\ddot{\eta} + K\Phi\eta = P \tag{8.149}$$

By left-multiplying both sides of Equation (8.149) with $\boldsymbol{\Phi}^{\mathrm{T}}$ provides:

$$\bar{\boldsymbol{M}}\ddot{\boldsymbol{\eta}} + \boldsymbol{K}\boldsymbol{\eta} = \bar{\boldsymbol{P}} \tag{8.150}$$

or

$$\bar{M}_i\ddot{\eta}_i + \bar{K}_i\eta_i = \bar{P}_i, \quad i = 1, 2, \cdots, n \tag{8.151}$$

where,

$$\bar{\boldsymbol{M}} \equiv \begin{bmatrix} \bar{m}_1 & & & \\ & \bar{m}_2 & & \\ & & \ddots & \\ & & & \bar{m}_n \end{bmatrix} \equiv \boldsymbol{\Phi}^{\mathrm{T}}\boldsymbol{M}\boldsymbol{\Phi} = \begin{bmatrix} \boldsymbol{\phi}_1^{\mathrm{T}}\boldsymbol{M}\boldsymbol{\phi}_1 & & & \\ & \boldsymbol{\phi}_2^{\mathrm{T}}\boldsymbol{M}\boldsymbol{\phi}_2 & & \\ & & \ddots & \\ & & & \boldsymbol{\phi}_n^{\mathrm{T}}\boldsymbol{M}\boldsymbol{\phi}_n \end{bmatrix} \tag{8.152}$$

is the ***generalized mass matrix***;

$$\bar{\boldsymbol{K}} \equiv \begin{bmatrix} \bar{k}_1 & & & \\ & \bar{k}_2 & & \\ & & \ddots & \\ & & & \bar{k}_n \end{bmatrix} \equiv \boldsymbol{\Phi}^{\mathrm{T}}\boldsymbol{K}\boldsymbol{\Phi} = \begin{bmatrix} \boldsymbol{\phi}_1^{\mathrm{T}}\boldsymbol{K}\boldsymbol{\phi}_1 & & & \\ & \boldsymbol{\phi}_2^{\mathrm{T}}\boldsymbol{K}\boldsymbol{\phi}_2 & & \\ & & \ddots & \\ & & & \boldsymbol{\phi}_n^{\mathrm{T}}\boldsymbol{K}\boldsymbol{\phi}_n \end{bmatrix} \tag{8.153}$$

is the ***generalized stiffness matrix***; and

$$\bar{\boldsymbol{P}} \equiv \begin{bmatrix} \bar{P}_1 & \bar{P}_2 & \cdots & \bar{P}_n \end{bmatrix}^{\mathrm{T}} \equiv \boldsymbol{\Phi}^{\mathrm{T}}\boldsymbol{P} = \begin{bmatrix} \boldsymbol{\phi}_1^{\mathrm{T}}\boldsymbol{P} & \boldsymbol{\phi}_2^{\mathrm{T}}\boldsymbol{P} & \cdots & \boldsymbol{\phi}_n^{\mathrm{T}}\boldsymbol{P} \end{bmatrix}^{\mathrm{T}} \tag{8.154}$$

All this shows that the vibration of an undamped system can be represented by a set of uncoupled differential equations in a modal space. This is why eigen-modes are particularly important in dynamic analysis.

If the modal vector is normalized by the condition (8.133), then the modal matrix satisfies

$$\boldsymbol{\Phi}^{\mathrm{T}}\boldsymbol{M}\,\boldsymbol{\Phi} = \boldsymbol{I} \tag{8.155}$$

which is called the ***mass-orthonormal condition***.

The equation of modes, Equation (8.128), gives:

$$\boldsymbol{K}\boldsymbol{\phi}_k = \omega_k^2 \boldsymbol{M}\boldsymbol{\phi}_k, \quad k = 1, 2, \cdots, n \tag{8.156}$$

which can be rewritten as:

$$\boldsymbol{K}\,\boldsymbol{\Phi} = \boldsymbol{\Omega}^2 \boldsymbol{M}\,\boldsymbol{\Phi} \tag{8.157}$$

By left-multiplying both sides of Equation (8.157) with $\boldsymbol{\Phi}^{\mathrm{T}}$ and adopting the mass-orthonormal condition, one obtains:

$$\boldsymbol{\Phi}^{\mathrm{T}} \boldsymbol{K}\,\boldsymbol{\Phi} = \boldsymbol{\Omega}^2 \tag{8.158}$$

where we used the mass-orthonormality (8.155).

In this case, the equation of motion in modal space given in Equation (8.149) can be expressed as:

$$\ddot{\boldsymbol{\eta}} + \boldsymbol{\Omega}^2 \ddot{\boldsymbol{\eta}} = \boldsymbol{\Phi}^{\mathrm{T}} \boldsymbol{P} \tag{8.159}$$

or

$$\ddot{\eta} + \omega_i^2 \eta_i = \boldsymbol{\phi}_i^{\mathrm{T}} \boldsymbol{P} \tag{8.160}$$

8.3.4 Rayleigh Damping Matrix

Lord Rayleigh suggested that the damping may be assumed to be a linear combination of the mass and the stiffness matrices, i.e.,

$$C \equiv a_0 \boldsymbol{M} + a_1 \boldsymbol{K} \tag{8.161}$$

where the constants a_0 and a_1 can be determined by experiments.

A damping having the form of (8.161) is called a ***Rayleigh damping***. Although it may cause some theoretical trouble, it is still widely used in structural engineering. A major advantage of Rayleigh damping is that it preserves the orthogonality of modes; that is to say, modal vectors satisfy

$$\boldsymbol{\phi}_i^{\mathrm{T}} \boldsymbol{C} \boldsymbol{\phi}_j = 0, \quad \text{when } i \neq j \tag{8.162}$$

which can be obtained immediately from Equation (8.161), the $\boldsymbol{M}$- and $\boldsymbol{K}$-orthogonality of modes.

Analogous to the generalized mass and generalized stiffness, we can define the ***generalized damping matrix*** as

$$\bar{\boldsymbol{C}} \equiv \begin{bmatrix} \bar{c}_1 & & & \\ & \bar{c}_2 & & \\ & & \ddots & \\ & & & \bar{c}_n \end{bmatrix} \equiv \boldsymbol{\Phi}^{\mathrm{T}} \boldsymbol{C} \boldsymbol{\Phi} = \begin{bmatrix} \boldsymbol{\phi}_1^{\mathrm{T}} \boldsymbol{C} \boldsymbol{\phi}_1 & & & \\ & \boldsymbol{\phi}_2^{\mathrm{T}} \boldsymbol{C} \boldsymbol{\phi}_2 & & \\ & & \ddots & \\ & & & \boldsymbol{\phi}_n^{\mathrm{T}} \boldsymbol{C} \boldsymbol{\phi}_n \end{bmatrix} \tag{8.163}$$

In the modal space, the vibration of a damped system is also governing by n uncoupled differential equations:

$$\bar{\boldsymbol{M}} \ddot{\boldsymbol{\eta}} + \bar{\boldsymbol{C}} \dot{\boldsymbol{\eta}} + \bar{\boldsymbol{K}} \boldsymbol{\eta} = \bar{\boldsymbol{P}} \tag{8.164}$$

or

$$\bar{m}_i \ddot{\eta}_i + \bar{c}_i \dot{\eta}_i + \bar{k}_i \eta_i = \bar{p}_i, \quad i = 1, 2, \cdots, n \tag{8.165}$$

in which $\bar{m}_i$, $\bar{k}_i$ and $\bar{c}_i$ are called the generalized mass, stiffness and damping coefficients of mode i.

Equation (8.165) may be rewritten as:

$$\ddot{\eta}_i + 2\zeta_i\omega_i\dot{\eta}_i + \omega_i^2\eta_i = \frac{\bar{p}_i}{\bar{m}_i}, \qquad i = 1, 2, \cdots, n \tag{8.166}$$

where,

$$\zeta_i \equiv \frac{\bar{c}_i}{2\bar{m}_i\omega_i} = \frac{1}{2\omega_i}\frac{\boldsymbol{\phi}_i^{\mathrm{T}}\boldsymbol{C}\boldsymbol{\phi}_i}{\boldsymbol{\phi}_i^{\mathrm{T}}\boldsymbol{M}\boldsymbol{\phi}_i} \tag{8.167}$$

is the damping ratio of mode i.

Substituting Equations (8.161) and (8.163) into Equation (8.167) gives

$$\zeta_i = \frac{a_0}{2}\frac{1}{\omega_i} + \frac{a_1}{2}\omega_i \tag{8.168}$$

For modes i and j, Equation (8.168) provides the following two algebraic equations:

$$\begin{cases} \dfrac{1}{2\omega_i}a_0 + \dfrac{\omega_i}{2}a_1 = \zeta_i \\ \dfrac{1}{2\omega_j}a_0 + \dfrac{\omega_j}{2}a_1 = \zeta_j \end{cases} \tag{8.169}$$

which can be used to determine the coefficients a_0 and a_1 from specified damping ratios ζ_1 and ζ_2 for the ith and jth modes, respectively.

EXAMPLE 8.10

Consider a dynamic system with mass and stiffness matrices

$$\boldsymbol{M} = 300 \times \begin{bmatrix} 1.0 & 0.0 & 0.0 \\ 0.0 & 0.8 & 0.0 \\ 0.0 & 0.0 & 0.5 \end{bmatrix} \ (\mathrm{kg}); \qquad \boldsymbol{K} = 6\times10^3 \times \begin{bmatrix} 1.0 & -0.5 & 0 \\ -0.5 & 1.0 & -0.5 \\ 0 & -0.5 & 0.5 \end{bmatrix} \ (\mathrm{N/m})$$

In structural engineering, the first and second modes are often assumed to have the same damping ratio ζ_1. Derive a Rayleigh damping matrix such that the damping ratio is 2% for the first and second modes. Calculate the damping ratio for the third mode.

***Solution*:**

The natural frequencies and the modal matrix of the system are respectively

$$\omega_1 = 1.7246; \quad \omega_2 = 4.4721; \quad \omega_3 = 6.4827; \quad (\mathrm{rad/sec})$$

and

$$\boldsymbol{\Phi} = \begin{bmatrix} -0.02503 & 0.04714 & 0.02201 \\ -0.04262 & -0.0 & -0.04848 \\ -0.05007 & -0.04714 & 0.04402 \end{bmatrix}$$

Substituting the natural frequencies and the damping ratios into Equation (8.169) gives:

$$\begin{bmatrix} 0.2899 & 0.8623 \\ 0.1118 & 2.2361 \end{bmatrix} \begin{Bmatrix} a_0 \\ a_1 \end{Bmatrix} = \begin{Bmatrix} 0.02 \\ 0.02 \end{Bmatrix}$$

whose solutions are:

$$a_0 = 0.04979, \quad a_1 = 0.00645$$

Substituting a_0 and a_1 Equation (8.161) gives:

$$\boldsymbol{C} = \begin{bmatrix} 53.6656 & -19.3649 & 0 \\ -19.3649 & 50.6785 & -19.3649 \\ 0 & -19.3649 & 26.8328 \end{bmatrix}, \qquad (\text{rad/sec})$$

Substituting $\boldsymbol{C}$, $\boldsymbol{M}$, $\boldsymbol{\phi}_3$ and ω_3 into Equation (8.167) gives $\zeta_3 = 0.025$.

8.4 Analysis of Dynamic Responses Using the Mode Superposition Method

In Section 9.2, we derived the equations governing the motion of a structure

$$\boldsymbol{M}\ddot{\boldsymbol{u}}(t) + \boldsymbol{C}\dot{\boldsymbol{u}}(t) + \boldsymbol{K}\boldsymbol{u}(t) = \boldsymbol{P}(t) \tag{8.170}$$

with the initial conditions,

$$\boldsymbol{u}(0) = \boldsymbol{u}_0, \quad \dot{\boldsymbol{u}}(0) = \dot{\boldsymbol{u}}_0 \tag{8.171}$$

where $\boldsymbol{M}$, $\boldsymbol{C}$, and $\boldsymbol{K}$ are the mass, damping, and stiffness matrices; $\boldsymbol{P}(t)$ is the vector of excitations; and $\boldsymbol{u}(t)$ is the displacement vector.

Mathematically, Equations (8.170) and (8.171) represent an ordinary differential equation (***ODE***) ***initial value problem*** (also called a ***Cauchy problem***). The solution to an ODE initial value problem can be obtained by many procedures, which are, in principle, divided into two categories: analytical and numerical approaches. The mostly effective and frequently used analytical method is the *mode superposition approach*, which shall be discussed in this section.

8.4.1 Transient Responses of Uncoupled MDOF Systems to Combined Excitations

In general, the stiffness matrix of a dynamic system is not diagonal, which means that Equation (8.170) represents a *coupled* MDOF system. However, the transient solution of a coupled MDOF system is based on the transient response of an uncoupled MDOF system subjected to combined excitations. This section shall study the transient response $\boldsymbol{\eta}(t)$ of the following uncoupled system:

$$\ddot{\boldsymbol{\eta}}(t) + 2\boldsymbol{\zeta}\boldsymbol{\Omega}\dot{\boldsymbol{\eta}}(t) + \boldsymbol{\Omega}^2\boldsymbol{\eta}(t) = \boldsymbol{p}(t) = \boldsymbol{p}_0\boldsymbol{f}(t) \tag{8.172}$$

Starting from the initial conditions

$$\boldsymbol{\eta}(0) = \boldsymbol{\eta}_0, \quad \dot{\boldsymbol{\eta}}(0) = \dot{\boldsymbol{\eta}}_0 \tag{8.173}$$

where $\boldsymbol{\eta}(t) \equiv [\eta_1(t), \cdots, \eta_n(t)]^{\mathrm{T}}, \boldsymbol{f}(t) \equiv [f_1(t), \cdots, f_n(t)]^{\mathrm{T}}$, and

$$\boldsymbol{\zeta} \equiv \begin{bmatrix} \zeta_1 & & \\ & \ddots & \\ & & \zeta_n \end{bmatrix}, \quad \boldsymbol{\Omega} \equiv \begin{bmatrix} \omega_1 & & \\ & \ddots & \\ & & \omega_n \end{bmatrix}, \quad \boldsymbol{p}_0 \equiv \begin{bmatrix} p_{011} & \cdots & p_{01n} \\ \vdots & \ddots & \vdots \\ p_{0n1} & \cdots & p_{0nn} \end{bmatrix} \tag{8.174}$$

To determine the transient response of a system (8.172), it is necessary to introduce the following theorems.

Theorem 8.5 *Let L be a second-order linear differential operator. The transient solution u(t) of the second-order differential equation:*

$$L(u) = \sum_{i=1}^{n} p_{0i} f_i(t) \tag{8.175}$$

starting from the condition of standing still (rest):

$$u(0) = u_0, \quad \dot{u}(0) = \dot{u}_0 \tag{8.176}$$

which can be expressed as:

$$u(t) = \sum_{i=1}^{n} p_{0i} u_i(t) \tag{8.177}$$

where $u_i(t)$, $i = 1,\ldots, n$, are the transient solutions of the following equations:

$$L(u) = f_i(t), \quad i = 1, 2 \tag{8.178}$$

starting from the initial conditions (8.176).

Proof: Because L is a linear operator,

$$L\left(\sum_{i=1}^{n} p_{0i} u_i(t)\right) = \sum_{i=1}^{n} p_{0i} L(u_i) = \sum_{i=1}^{n} p_{0i} f_i(t) \tag{8.179}$$

where the second equality is obtained by utilizing Equation (8.178).

Equation (8.179) shows that the linear combination (8.177) is a solution of Equation (8.175). In addition, if $u_i(0) = 0$, then,

$$u(0) = \sum_{i=1}^{n} p_{0i} u_i(0) = 0 \tag{8.180}$$

which shows that the solution (8.177) satisfies the initial conditions (8.176). Thus, this theorem is true.

Theorem 8.6 *The transient solution u(t) of the second-order differential equation*

$$L(u) = f(t) \tag{8.181}$$

starting from the initial conditions is,

$$u(0) = u_0, \quad \dot{u}(0) = \dot{u}_0 \tag{8.182}$$

which can be expressed as:

$$u(t) = u_r(t) + u_h(t) \tag{8.183}$$

where $u_r(t)$ is the transient solution of the forced equation (8.181) starting from the standing still (rest)condition,

$$u(0) = 0, \quad \dot{u}(0) = 0 \tag{8.184}$$

and $u_{\text{h}}(t)$ *is the transient solution of the homogeneous equation:*

$$L(u) = 0 \tag{8.185}$$

from the initial condition (8.182).

***Proof*:** Because L is linear,

$$L(u_{\text{r}} + u_{\text{h}}) = L(u_{\text{r}}) + L(u_{\text{h}}) = f(t) \tag{8.186}$$

where the second equality is obtained by using the following equations:

$$L(u_{\text{r}}) = f(t), \quad L(u_{\text{h}}) = 0 \tag{8.187}$$

Equation (8.186) shows that $u_{\text{r}} + u_{\text{h}}$ is a solution of Equation (8.181). In addition, because

$$\begin{cases} u_{\text{r}}(0) = 0 \\ \dot{u}_{\text{r}}(0) = 0 \end{cases} \quad \text{and} \quad \begin{cases} u_{\text{h}}(0) = u_0 \\ \dot{u}_{\text{h}}(0) = \dot{u}_0 \end{cases} \tag{8.188}$$

one has,

$$u(0) = u_{\text{r}}(0) + u_{\text{h}}(0) = u_0 \tag{8.189}$$

$$\dot{u}(0) = \dot{u}_{\text{r}}(0) + \dot{u}_{\text{h}}(0) = \dot{u}_0 \tag{8.190}$$

which shows that $u = u_{\text{r}} + u_{\text{h}}$ is a solution starting from the initial conditions given by (8.182). Thus, this theorem is true.

Theorems 8.5 and 8.6 lead to the following theorem which can be used to determine the transient response of the system (8.173).

Theorem 8.7 *The transient response,* $\boldsymbol{\eta}(t)$*, of the system* (8.172) *starting from the initial conditions* (8.173) *can be expressed as:*

$$\boldsymbol{\eta}(t) = \boldsymbol{\xi}_{\text{h}} + \sum_{j=1}^{n} \text{diag}\left(\boldsymbol{p}_{0j}\right)\boldsymbol{\xi}_j(t) \tag{8.191}$$

where $\boldsymbol{P}_{0i} = [p_{01i}, p_{02i}, \cdots, p_{ni}]^{\text{T}}$ *is the ith column of the matrix* $\boldsymbol{P}_0$*, and*

$$\text{diag}\left(\boldsymbol{p}_{0i}\right) = \begin{bmatrix} p_{01i} & & & \\ & p_{02i} & & \\ & & \ddots & \\ & & & p_{0ni} \end{bmatrix} \tag{8.192}$$

if $\boldsymbol{\xi}_j(t) \equiv [\xi_{1j}(t) \;\; \xi_{2j}(t) \;\; \cdots \;\; \xi_{nj}(t)]^{\text{T}}$ *are the transient responses of the n forced systems*

$$\ddot{\boldsymbol{\xi}}(t) + 2\boldsymbol{\zeta\Omega}\dot{\boldsymbol{\xi}}(t) + \boldsymbol{\Omega}^2\dot{\boldsymbol{\xi}}(t) = \mathbf{1}f_j(t), \quad j = 1, 2, \cdots, n \tag{8.193}$$

starting from the standing still (rest) condition,

$$\boldsymbol{\xi}(0) = \mathbf{0}, \quad \dot{\boldsymbol{\xi}}(0) = \mathbf{0} \tag{8.194}$$

where $\mathbf{1}$ *is a vector of order n with each entry equal to unity, and if* $\boldsymbol{\xi}_{\text{h}}(t) \equiv [\xi_{1\text{h}}(t) \;\; \xi_{2\text{h}}(t) \;\; \cdots \;\; \xi_{n\text{h}}(t)]$*, the free vibrations dynamic equilibrium of the system can be expressed as:*

$$\ddot{\boldsymbol{\xi}}(t) + 2\boldsymbol{\zeta\Omega}\dot{\boldsymbol{\xi}}(t) + \boldsymbol{\Omega}^2\dot{\boldsymbol{\xi}}(t) = \mathbf{0} \tag{8.195}$$

starting from the initial conditions,

$$\boldsymbol{\xi}(0) = \boldsymbol{\eta}_0, \quad \dot{\boldsymbol{\xi}}(0) = \dot{\boldsymbol{\eta}}_0 \tag{8.196}$$

***Proof*:** The load vector $\boldsymbol{P}(t)$ in Equation (8.172) can be expressed as:

$$\begin{aligned} \boldsymbol{p}(t) = \boldsymbol{p}_0 \boldsymbol{f}(t) &= \begin{Bmatrix} p_{011} f_1(t) \\ p_{021} f_1(t) \\ \vdots \\ p_{0n1} f_1(t) \end{Bmatrix} + \begin{Bmatrix} p_{012} f_2(t) \\ p_{022} f_2(t) \\ \vdots \\ p_{0n2} f_2(t) \end{Bmatrix} + \cdots \begin{Bmatrix} p_{01n} f_n(t) \\ p_{02n} f_n(t) \\ \vdots \\ p_{0nn} f_n(t) \end{Bmatrix} \\ &= \sum_{i=1}^{n} \mathrm{diag}(\boldsymbol{p}_{0i}) \mathbf{1} f_i(t) \end{aligned} \tag{8.197}$$

Theorems 8.6–8.7 and Equation (8.197) lead to Equation (8.191)

Specifically, Equation (8.99) gives:

$$\xi_{hi}(t) = e^{-\zeta_i \omega_i t} (A_i \cos \omega_{Di} t + B_i \sin \omega_{Di} t) \tag{8.198}$$

where,

$$A_i = u_{0i}; \qquad B_i = \frac{\dot{u}_{0i} + \zeta_i \omega_i u_{0i}}{\omega_{Di}}; \qquad \omega_{Di} = \omega_i \sqrt{1 - \zeta_i^2} \tag{8.199}$$

If the components of $\boldsymbol{f}(t)$ are harmonic, i.e.,

$$\boldsymbol{f}(t) = [\sin(\tilde{\omega}_1 t + \tilde{\varphi}_1) \ \sin(\tilde{\omega}_2 t + \tilde{\varphi}_2) \ \cdots \ \sin(\tilde{\omega}_n t + \tilde{\varphi}_n)]^{\mathrm{T}} \tag{8.200}$$

then,

$$\xi_{ij}(t) = e^{-\zeta_i \omega_i t} (G_{ij} \cos \omega_{Di} t + H_{ij} \sin \omega_{Di} t) + C_{ij} \sin(\tilde{\omega}_j t + \tilde{\varphi}_j) + D_{ij} \cos(\tilde{\omega}_j t + \tilde{\varphi}_j) \tag{8.201}$$

where we utilized Equation (8.115), and

$$C_{ij} = \alpha_{ij} \frac{\left[1 - (\tilde{\omega}_j / \omega_i)^2\right]}{\omega_i^2 \left[1 - (\tilde{\omega}_j / \omega_i)^2\right]^2 + (2\zeta_i \tilde{\omega}_j)^2} \tag{8.202}$$

$$D_{ij} = \alpha_{ij} \frac{-2\zeta_i \tilde{\omega}_j / \omega_i}{\omega_i^2 \left[1 - (\tilde{\omega}_j / \omega_i)^2\right]^2 + (2\zeta_i \tilde{\omega}_j)^2} \tag{8.203}$$

$$G_{ij} = -C_{ij} \sin \tilde{\varphi}_j - D_{ij} \cos \tilde{\varphi}_j \tag{8.204}$$

$$H_{ij} = \frac{1}{\omega_{D_i}} \left(G_{ij} \zeta_i \omega_i - C_{ij} \tilde{\omega}_j \cos \tilde{\varphi}_j + D_{ij} \tilde{\omega}_j \sin \tilde{\varphi}_j \right) \tag{8.205}$$

$$\alpha_{ij} = \begin{cases} 1 & \text{when } \tilde{\omega}_j \neq 0 \\ 0 & \text{when } \tilde{\omega}_j = 0 \end{cases} \tag{8.206}$$

EXAMPLE 8.11

Determine the vibration of an undamped dynamic system

$$\boldsymbol{M\ddot{u}} + \boldsymbol{Ku} = \boldsymbol{P}_0\boldsymbol{f}(t) \tag{a}$$

starting from the initial conditions

$$\boldsymbol{u}_0 = [0.05\ \ -0.08]^{\mathrm{T}}\ (\mathrm{m}),\quad \boldsymbol{\dot{u}}_0 = [-1.5\ \ 0.5]^{\mathrm{T}}\ (\mathrm{m/sec}) \tag{b}$$

where

$$\boldsymbol{M} = \begin{bmatrix} 300 & \\ & 200 \end{bmatrix}\ (\mathrm{kg})\quad \boldsymbol{K} = 5\times10^4 \times \begin{bmatrix} 1 & \\ & 0.8 \end{bmatrix}\ (\mathrm{N/m})$$

$$\boldsymbol{P}_0 = \begin{bmatrix} 750 & 450 \\ -840 & 640 \end{bmatrix}\ (\mathrm{N})\quad \boldsymbol{f}(t) = \begin{Bmatrix} \sin(3t+0.3) \\ \sin(4t+0.6) \end{Bmatrix}$$

Solution: The natural frequencies of the system are

$$\omega_1 = \sqrt{\frac{k_{11}}{m_{11}}} = 12.910,\qquad \omega_2 = \sqrt{\frac{k_{22}}{m_{22}}} = 14.142$$

Equation (a) can be rewritten in the following frequency form:

$$\boldsymbol{\ddot{u}}(t) + \boldsymbol{\Omega}^2\boldsymbol{u} = \boldsymbol{p}_0\boldsymbol{f}(t) \tag{c}$$

where,

$$p_c = M^{-1}P_0 = [2.5\ 1.5 - 4.2\ 3.2],\quad \Omega = [12.910\ \ 14.142]$$

Equations (8.198) and (8.198) show that the *free* vibration, $\xi_{\mathrm{h}}(t)$, of the system (c) starting from the initial conditions (b) can be expressed as

$$\boldsymbol{\xi}_{\mathrm{h}} = \begin{Bmatrix} A_1 \cos\omega_1 t + B_1 \sin\omega_1 t \\ A_2 \cos\omega_2 t + B_2 \sin\omega_2 t \end{Bmatrix}$$

where we used $\zeta_i = 0$, $\omega_{\mathrm{D}i} = \omega_i$, $i = 1, 2$, and

$$\begin{cases} A_1 = 0.05 \\ A_2 = -0.08' \end{cases}\qquad \begin{cases} B_1 = -0.11619 \\ B_2 = 0.035355 \end{cases}$$

Equations (8.201)–(8.205) show that the *forced* vibration, ξ_j, of the system

$$\ddot{\boldsymbol{\xi}}(t) + \boldsymbol{\Omega}^2\boldsymbol{\xi}(t) = \mathbf{1}f_j(t),\quad j = 1, 2$$

starting from *rest*, $\boldsymbol{\xi}_0 = \dot{\boldsymbol{\xi}}_0 = 0$, can be expressed as:

$$\xi_{ij}(t) = G_{ij}\cos\omega_i t + H_{ij}\sin\omega_i t + C_{ij}\sin(\omega_i t + \tilde{\varphi}_j),\quad i, j = 1, 2$$

where we used $\zeta_1 = \zeta_2 = 0$, and

$$\boldsymbol{C} = 10^{-3}\begin{bmatrix} 6.3425 & 6.6372 \\ 5.2356 & 5.4348 \end{bmatrix},\ \boldsymbol{G} = 10^{-3}\begin{bmatrix} -1.8743 & -3.7476 \\ -1.5472 & -3.0687 \end{bmatrix},\ \boldsymbol{H} = 10^{-3}\begin{bmatrix} -1.4080 & -1.6973 \\ -1.0610 & -1.2687 \end{bmatrix}$$

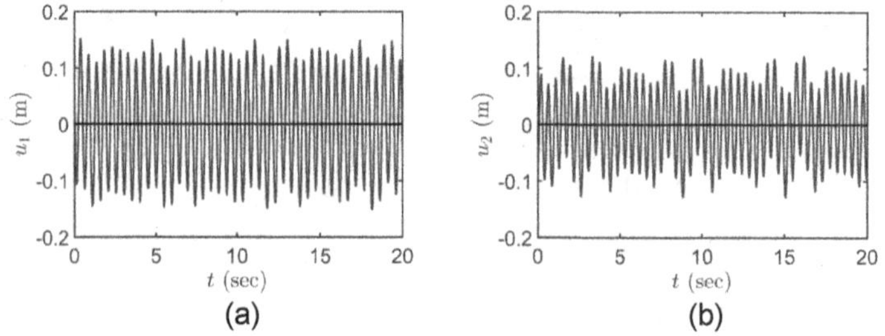

Figure 8.35. The results of Example 8.11.

Finally, Theorem 8.8 shows that the forced vibration of a system (c) starting from the initial conditions (b) can be expressed as:

$$\boldsymbol{u}(t) = \xi_{\mathrm{h}} + \sum_{j=1}^{n} \operatorname{diag}(\boldsymbol{p}_{0j})\xi_j(t) \tag{8.207}$$

which is also the transient response of the system (a) starting from the initial conditions (b), and is plotted in Figure 8.35.

8.4.2 Steady-State Responses of Uncoupled MDOF Systems to Combined Excitations

This section investigates the steady-state response $\boldsymbol{u}_{\mathrm{s}}(t)$ of the following uncoupled MDOF system subjected to the combined harmonic excitations:

$$\ddot{\boldsymbol{u}}(t) + 2\boldsymbol{\zeta\Omega}\dot{\boldsymbol{u}}(t) + \boldsymbol{\Omega}^2\boldsymbol{u}(t) = \boldsymbol{p}_0\boldsymbol{f}(t) \tag{8.208}$$

where,

$$\boldsymbol{p}_0\boldsymbol{f} = \begin{bmatrix} p_{\mathrm{c}11} & \cdots & p_{\mathrm{c}1n} \\ \vdots & \ddots & \vdots \\ p_{\mathrm{c}n1} & \cdots & p_{\mathrm{c}nn} \end{bmatrix} \begin{Bmatrix} f_1(t) \\ \vdots \\ f_n(t) \end{Bmatrix} \tag{8.209}$$

To formulate the expression of the steady-state response $\boldsymbol{u}_{\mathrm{s}}(t)$, one has to propose the following theorem.

Theorem 8.8 (Steady-State Response) *Let L be a linear differential operator. If $u_{\mathrm{s}i}(t)$ is the steady-state solution of the differential equation:*

$$L(u) = f_i(t) \tag{8.210}$$

then the steady-state solution $u_{\mathrm{s}}(t)$ of the differential equation can be expressed as:

$$L(u) = \sum_{i=1}^{n} p_{0i} f_i(t) \tag{8.211}$$

$u_{\mathrm{s}i}(t)$ can be expressed as a linear combination of $u_{\mathrm{s}i}$, i.e.,

$$u_{\mathrm{s}}(t) = \sum_{i=1}^{n} p_{0i} u_{\mathrm{s}i}(t) \tag{8.212}$$

Proof: The linearity of L gives,

$$L\left(\sum_{i=1}^{n} p_{0i}u_{si}\right) = \sum_{i=1}^{n} p_{0i}L(u_{si}) \tag{8.213}$$

Since u_{si} is a steady-state solution of Equation (8.210), Equation (8.213) and the steady-state-response definition show that, for any tolerance $\varepsilon > 0$, there exists a time t_s such that

$$\left|L\left(\sum_{i=1}^{n} p_{0i}u_{si}\right) - \sum_{i=1}^{n} p_{0i}f_i(t)\right| \le \varepsilon, \quad \text{for } t \ge t_s \tag{8.214}$$

This shows that $u_s(t)$ described by Equation (8.212) is a steady-state solution of Equation (8.214).

Theorem 8.9 immediately leads to the following theorem, which can be used to determine the steady-state response of a system (8.208).

Theorem 8.9 (steady-state response) *If $\xi_{sj}(t)$ is the steady-state solution of the system*

$$\ddot{\xi}(t) + 2\zeta\boldsymbol{\Omega}\dot{\xi}(t) + \boldsymbol{\Omega}^2\xi(t) = \mathbf{1}f_j(t) \tag{8.215}$$

where **1** *is the vector of order n with each entry equal to unity, then the steady-state response of system can be expressed as:*

$$\eta(t) = \sum_{j=1}^{n} \operatorname{diag}(\boldsymbol{p}_{0j})\xi_{sj}(t) \tag{8.216}$$

where,

$$\xi_{sj}(t) \equiv [\xi_{s1j}(t) \ \ \xi_{s2j}(t) \ \ \cdots \ \ \xi_{snj}(t)]^{\mathrm{T}} \tag{8.217}$$

Specifically, if $\boldsymbol{f}(t)$ is harmonic, i.e.,

$$\boldsymbol{f}(t) = [\sin(\tilde{\omega}_1 t + \tilde{\varphi}_1) \ \ \sin(\tilde{\omega}_2 t + \tilde{\varphi}_2) \ \ \cdots \ \ \sin(\tilde{\omega}_n t + \tilde{\varphi}_n)]^{\mathrm{T}} \tag{8.218}$$

then, Equations (8.213) and (8.119) show that $\xi_{sij}(t)$ can be expressed as:

$$\begin{aligned}\xi_{sij}(t) &= C_{ij}\sin(\tilde{\omega}_j t + \tilde{\varphi}_j) + D_{ij}\cos(\tilde{\omega}_j t + \tilde{\varphi}_j) \\ &= A_{ij}\sin(\tilde{\omega}_j t + \varphi_{ij})\end{aligned} \tag{8.219}$$

where C_{ij} and D_{ij} are described by Equations (8.202) and (8.203),

$$A_{ij} = \frac{\alpha_{ij}}{\omega_i^2\sqrt{\left[1-(\tilde{\omega}_j/\omega_i)^2\right]^2 + \left[2\zeta_i(\tilde{\omega}_j/\omega_i)\right]^2}} \tag{8.220}$$

$$\varphi_{ij} = \begin{cases} \tilde{\varphi}_j - \tan^{-1}\left[\dfrac{2\zeta_i(\tilde{\omega}_j/\omega_i)}{1-(\tilde{\omega}_j/\omega_i)^2}\right], & \text{when } \tilde{\omega}_j/\omega_i \le 1 \\ \tilde{\varphi}_j - \pi - \tan^{-1}\left[\dfrac{2\zeta_i(\tilde{\omega}_j/\omega_i)}{1-(\tilde{\omega}_j/\omega_i)^2}\right], & \text{when } \tilde{\omega}_j/\omega_i > 1 \end{cases} \tag{8.221}$$

$$\alpha_{ij} = \begin{cases} 1 & \text{when } \tilde{\omega}_j \neq 0 \\ 0 & \text{when } \tilde{\omega}_j = 0 \end{cases} \tag{8.222}$$

EXAMPLE 8.12

Determine the steady-state response of the following system with damping ratio $\zeta_1 = \zeta_2 = 2\%$. The dynamic equilibrium is:

$$\boldsymbol{M}\ddot{\boldsymbol{u}} + \boldsymbol{C}\dot{\boldsymbol{u}} + \boldsymbol{K}\boldsymbol{u} = \boldsymbol{P}_0\boldsymbol{f}(t) \tag{a}$$

where $\boldsymbol{C}$ is the Rayleigh damping matrix, and

$$\boldsymbol{M} = \begin{bmatrix} 300 & \\ & 200 \end{bmatrix} \text{(kg)} \quad \boldsymbol{K} = 5\times10^4 \times \begin{bmatrix} 1 & \\ & 0.8 \end{bmatrix} \text{(N/m)}$$

$$\boldsymbol{P}_0 = \begin{bmatrix} 750 & 450 \\ -840 & 640 \end{bmatrix} \text{(N)} \quad \boldsymbol{f}(t) = \begin{Bmatrix} \sin 8.2t \\ \sin 2.5t \end{Bmatrix}$$

Solution: The original system can be rewritten as

$$\ddot{\boldsymbol{u}}(t) + 2\boldsymbol{\zeta}\boldsymbol{\Omega}\dot{\boldsymbol{u}}(t) + \boldsymbol{\Omega}^2\boldsymbol{u}(t) = \boldsymbol{p}_0\boldsymbol{f}(t) \tag{b}$$

where,

$$\boldsymbol{\zeta} = \begin{bmatrix} 0.02 & \\ & 0.02 \end{bmatrix}; \ \boldsymbol{\Omega} = \begin{bmatrix} 12.910 & 1 \\ & 14.142 \end{bmatrix}; \ \boldsymbol{p}_0 = \begin{bmatrix} 2.5 & 1.5 \\ -4.2 & 3.2 \end{bmatrix}; \ \boldsymbol{f}(t) = \begin{Bmatrix} \sin 8.2t \\ \sin 2.5t \end{Bmatrix}$$

Equations (8.219)–(8.221) show that the steady-state response, $\boldsymbol{\xi}_{sj}(t)$, of the system is:

$$\ddot{\boldsymbol{\xi}}(t) + 2\boldsymbol{\zeta}\boldsymbol{\Omega}\dot{\boldsymbol{\xi}}(t) + \boldsymbol{\Omega}^2\boldsymbol{\xi}(t) = \mathbf{1}f_j(t), \quad j = 1, 2$$

$\xi_{sij}(t)$ can be expressed as:

$$\xi_{sij}(t) = A_{ij}\, sin(\tilde{\omega}_j t + \varphi_{ij}), \quad i, j = 1, 2$$

where,

$$\boldsymbol{A} = 10^{-3} \times \begin{bmatrix} 10.049 & 6.2336 \\ 7.5278 & 5.1612 \end{bmatrix}, \quad \tilde{\boldsymbol{\omega}} = \begin{Bmatrix} 8.2 \\ 2.5 \end{Bmatrix}, \quad \boldsymbol{\varphi} = 10^{-2} \times \begin{bmatrix} -4.2563 & -0.8048 \\ -3.4926 & -0.7299 \end{bmatrix}$$

Theorem 8.10 shows that the steady-state response of system (b) can be expressed as:

$$\begin{aligned} \boldsymbol{u}_s(t) &= \sum_{j=1}^{2} \text{diag}(\boldsymbol{p}_{0j})\boldsymbol{\xi}_{sj}(t) \\ &= \sum_{j=1}^{2} \text{diag}(\boldsymbol{p}_{0j})\text{diag}(A_{1j}, A_{2j}) \begin{Bmatrix} \sin(\tilde{\omega}_j + \varphi_{1j}) \\ \sin(\tilde{\omega}_j + \varphi_{2j}) \end{Bmatrix} \\ &= \begin{Bmatrix} 0.025121\sin(\tilde{\omega}_1 + \varphi_{11}) + 0.0093503\sin(\tilde{\omega}_2 + \varphi_{12}) \\ -0.031617\sin(\tilde{\omega}_1 + \varphi_{21}) + 0.0165160\sin(\tilde{\omega}_2 + \varphi_{22}) \end{Bmatrix} \end{aligned}$$

which is also the steady-state response of system (a), and is illustrated in Figure 8.36.

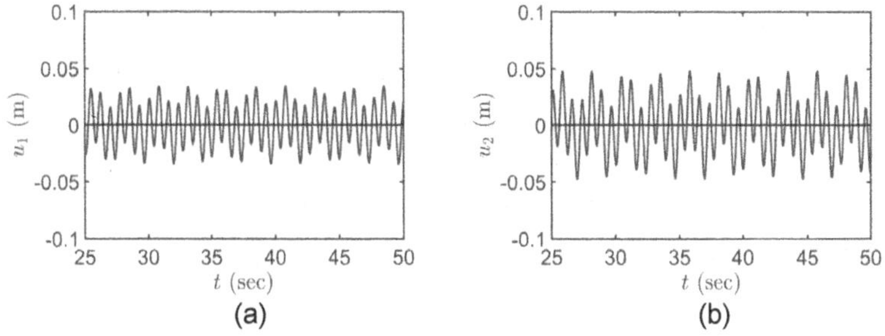

Figure 8.36. The results of Example 8.12.

8.4.3 Responses of Coupled MDOF Systems: Mode Superposition Method

The displacement responses of a coupled n-DOF system with damping are governed by the following equations:

$$\boldsymbol{M}\ddot{\boldsymbol{u}} + \boldsymbol{C}\dot{\boldsymbol{u}} + \boldsymbol{K}\boldsymbol{u} = \boldsymbol{P}_0 f(t) \tag{8.223}$$

with the initial conditions,

$$\boldsymbol{u}(0) = \boldsymbol{u}_0 \quad \text{and} \quad \dot{\boldsymbol{u}}(0) = \dot{\boldsymbol{u}}_0 \tag{8.224}$$

where $\boldsymbol{C}$ and $\boldsymbol{K}$ are not diagonal matrices.

Substituting Equation (8.146), the change-of-coordinates equation, into Equation (8.223) gives:

$$\boldsymbol{M\Phi}\ddot{\boldsymbol{\eta}} + \boldsymbol{C\Phi}\dot{\boldsymbol{\eta}} + \boldsymbol{K\Phi}\boldsymbol{\eta} = \boldsymbol{P}_0 f(t) \tag{8.225}$$

where $\boldsymbol{\eta}$ is the modal coordinate vector, i.e., the representation of the displacement in the modal space, and,

$$\boldsymbol{u} = \boldsymbol{\Phi\eta} \tag{8.226}$$

By multiplying each term of Equation (8.225) with $\boldsymbol{\Phi}^{\mathrm{T}}$, one has:

$$\boldsymbol{\Phi}^{\mathrm{T}}\boldsymbol{M\Phi}\ddot{\boldsymbol{\eta}} + \boldsymbol{\Phi}^{\mathrm{T}}\boldsymbol{C\Phi}\dot{\boldsymbol{\eta}} + \boldsymbol{\Phi}^{\mathrm{T}}\boldsymbol{K\Phi}\boldsymbol{\eta} = \boldsymbol{\Phi}^{\mathrm{T}}\boldsymbol{P}_0 f(t) \tag{8.227}$$

Due to the orthogonality of mode shapes, Equations (8.227) are uncoupled and can be rewritten as:

$$\bar{\boldsymbol{M}}\ddot{\boldsymbol{\eta}} + \bar{\boldsymbol{C}}\boldsymbol{\Phi}\dot{\boldsymbol{\eta}} + \bar{\boldsymbol{K}}\boldsymbol{\eta} = \bar{\boldsymbol{P}}_0 f(t) \tag{8.228}$$

where the generalized mass, damping and stiffness matrices are:

$$\bar{\boldsymbol{M}} = \boldsymbol{\Phi}^{\mathrm{T}}\boldsymbol{M\Phi}, \quad \bar{\boldsymbol{C}} = \boldsymbol{\Phi}^{\mathrm{T}}\boldsymbol{C\Phi}, \quad \bar{\boldsymbol{K}} = \boldsymbol{\Phi}^{\mathrm{T}}\boldsymbol{K\Phi} \tag{8.229}$$

It is obvious that they are all diagonal, and the ***generalized load-coefficient matrix, expressed as,***

$$\bar{\boldsymbol{P}}_0 = \boldsymbol{\Phi}^{\mathrm{T}}\boldsymbol{P}_0 \tag{8.230}$$

is not necessarily diagonal.

Equation (8.228) can be rewritten as:

$$\ddot{\boldsymbol{\eta}}(t) + 2\boldsymbol{\zeta\Omega}\dot{\boldsymbol{\eta}}(t) + \boldsymbol{\Omega}^2\boldsymbol{\eta}(t) = \boldsymbol{p}_0 f(t) \tag{8.231}$$

or

$$\ddot{\eta} + 2\zeta_i\omega_i\dot{\eta}_i + \omega_i^2\eta_i = \sum_{j=1}^{n} p_{0ij} f_j(t) \qquad i = 1, 2, \cdots, n \tag{8.232}$$

where,

$$\boldsymbol{p}_0 = \bar{\boldsymbol{M}}^{-1}\bar{\boldsymbol{P}}_0 = \bar{\boldsymbol{M}}^{-1}\boldsymbol{\Phi}^{\mathrm{T}}\boldsymbol{P}_0 \tag{8.233}$$

In the modal space, the initial conditions (8.224) can be represented as:

$$\boldsymbol{\eta}(0) = \boldsymbol{\eta}_0 = \boldsymbol{\Phi}^{-1}\boldsymbol{u}_0 \quad \text{and} \quad \dot{\boldsymbol{\eta}}(0) = \dot{\boldsymbol{\eta}}_0 = \boldsymbol{\Phi}^{-1}\dot{\boldsymbol{u}}_0 \tag{8.234}$$

Equation (8.231) and the initial conditions (8.234) are the complete mathematical description, in the modal space, of the original dynamic system (8.223). All these show that a coupled MDOF system can be represented as an uncoupled MDOF system subjected to combined excitations in the modal. The transient and the steady-state responses of the uncoupled MDOF system (8.231) can be obtained by the methods presented in Sections (8.4.1) and (8.4.2). Substituting the transient or the steady-state response into the change-of-coordinates equation (8.226) gives the transient or the steady-state response of the original dynamic system (8.223).

Table 8.1 summarizes the procedure for applying the mode superposition method to determine the transient response of a coupled MDOF system of the form (8.223) with damping-ratio matrix ζ. The process for determining the steady-state response of a coupled MDOF system is summarized in Table 8.2.

EXAMPLE 8.13

Determine the transient response of the following coupled system

$$\boldsymbol{M}\ddot{\boldsymbol{u}} + \boldsymbol{C}\dot{\boldsymbol{u}} + \boldsymbol{K}\boldsymbol{u} = \boldsymbol{P}_0\boldsymbol{f}(t) \tag{a}$$

starting from the initial conditions,

$$\boldsymbol{u}_0 = [0.05 \;\; -0.08]^{\mathrm{T}} \text{ (m)}, \quad \dot{\boldsymbol{u}}_0 = [-0.8 \;\; 0.5]^{\mathrm{T}} \text{ (m/sec)} \tag{b}$$

where $\boldsymbol{C}$ is the Rayleigh damping matrix, and,

$$\boldsymbol{M} = 500 \times \begin{bmatrix} 1.0 & \\ & 0.6 \end{bmatrix} \text{(kg)} \quad \boldsymbol{K} = 6.36 \times 10^4 \times \begin{bmatrix} 1 & -0.5 \\ -0.5 & 0.6 \end{bmatrix} \text{(N/m)}$$

$$\boldsymbol{P}_0 = 10^3 \times \begin{bmatrix} 1.0 & \\ & 1.5 \end{bmatrix} \text{(N)} \quad \boldsymbol{f}(t) = \begin{Bmatrix} \sin 4.0t \\ \sin 2.5t \end{Bmatrix}$$

The first and second modes are assumed to have the same damping ratio 2%.

Solution: (1) Modal analysis.
Solving Equation

$$(\boldsymbol{K} - \omega^2\boldsymbol{M})\boldsymbol{\phi} = \boldsymbol{0}$$

gives,

$$\boldsymbol{\Omega} = \begin{bmatrix} \omega_1 & \\ & \omega_2 \end{bmatrix} = \begin{bmatrix} 6.7151 & \\ & 14.467 \end{bmatrix}, \quad \boldsymbol{\Phi} = \begin{bmatrix} 1 & 1 \\ 1.291 & -1.291 \end{bmatrix}$$

where we used the normalization condition, $\phi_{1j} = 1, j = 1, 2$, to determine the modal shape matrix $\boldsymbol{\Phi}$.

Table 8.1. Determine transient responses by mode superposition method.

The following data are given: $\boldsymbol{M}$, $\boldsymbol{K}$, $\boldsymbol{\zeta}$, $\boldsymbol{P}_0$, $\boldsymbol{f}(t)$, $\boldsymbol{u}_0$, $\dot{\boldsymbol{u}}_0$.

1. Modal analysis

The modal-shape and spectral matrices. $\boldsymbol{\Phi}$ and $\boldsymbol{\Omega}$, are determined by solving

$$(\boldsymbol{K} - \omega^2 \boldsymbol{M})\boldsymbol{\phi} = 0$$

2. Formulate the modal equation and initial conditions

(a) Calculate the load-coefficient matrix: $\boldsymbol{p}_0 = \bar{\boldsymbol{M}}^{-1}\boldsymbol{\Phi}^{\mathrm{T}}\boldsymbol{P}_0$

(b) The modal equation in the frequency form is

$$\ddot{\boldsymbol{\eta}}(t) + 2\boldsymbol{\zeta}\boldsymbol{\Omega}\dot{\boldsymbol{\eta}}(t) + \boldsymbol{\Omega}^2\boldsymbol{\eta}(t) = \boldsymbol{p}_0\boldsymbol{f}(t)$$

(c) The initial conditions in terms of modal coordinates are

$$\boldsymbol{\eta}(0) = \boldsymbol{\eta}_0 = \boldsymbol{\Phi}^{-1}\boldsymbol{u}_0, \quad \dot{\boldsymbol{\eta}}(0) = \dot{\boldsymbol{\eta}}_0 = \boldsymbol{\Phi}^{-1}\dot{\boldsymbol{u}}_0$$

3. Calculate the standard transient solutions $\boldsymbol{\xi}_{\mathrm{h}}(t)$ and $\boldsymbol{\xi}_{\mathrm{h}}(t)$, $j = 1, 2, \cdots, n$,

(a) The ith component of $\boldsymbol{\xi}_{\mathrm{h}}$ can be expressed as

$$\xi_{\mathrm{h}}(t) = \mathrm{e}^{-\zeta_i\omega_i t}(A_i \cos\omega_{\mathrm{D}i}t + B_i \sin\omega_{\mathrm{D}i}t)$$

where,

$$A_i = u_{0i}; \quad B_i = \frac{\dot{u}_{0i} + \zeta_i\omega_i u_{0i}}{\omega_{\mathrm{D}i}}; \quad \omega_{\mathrm{D}i} \equiv \omega_i\sqrt{1-\zeta_i^2}$$

(b) $\boldsymbol{\xi}_j(t)$ is determined by solving

$$\ddot{\boldsymbol{\xi}}(t) + 2\boldsymbol{\zeta}\boldsymbol{\Omega}\dot{\boldsymbol{\xi}}(t) + \boldsymbol{\Omega}^2\boldsymbol{\xi}(t) = \boldsymbol{1}f_j(t), \quad j = 1, 2, \cdots, n$$

With the initial conditions,

$$\boldsymbol{\xi}(0) = \boldsymbol{0}, \quad \dot{\boldsymbol{\xi}}(0) = \boldsymbol{0}$$

If $\boldsymbol{f}(t)$ can be expressed as

$$\boldsymbol{f}(t) = [\sin(\tilde{\omega}_1 t + \tilde{\varphi}_1) \ \sin(\tilde{\omega}_2 t + \tilde{\varphi}_2) \ \cdots \ \sin(\tilde{\omega}_n t + \tilde{\varphi}_n)]^{\mathrm{T}}$$

Then the ith component of $\boldsymbol{\xi}_j$ can be expressed as,

$$\xi_{ij}(t) = \mathrm{e}^{-\zeta_i\omega_i t}(G_{ij}\cos\omega_{\mathrm{D}i}t + H_{ij}\sin\omega_{\mathrm{D}i}t) + C_{ij}\sin(\omega_j t + \tilde{\varphi}_j) + D_{ij}\cos(\omega_j t + \tilde{\varphi}_j)$$

where,

$$C_{ij} = \frac{\alpha_{ij}\left[1-\left(\tilde{\omega}_j/\omega_i\right)^2\right]}{\omega_i^2\left[1-\left(\tilde{\omega}_j/\omega_i\right)^2\right]^2 + \left(2\zeta_i\tilde{\omega}_j\right)^2} \qquad D_{ij} = \frac{-2\alpha_{ij}\zeta_i\tilde{\omega}_j/\omega_i}{\omega_i^2\left[1-\left(\tilde{\omega}_j/\omega_i\right)^2\right]^2 + \left(2\zeta_i\tilde{\omega}_j\right)^2}$$

$$G_{ij} = -C_{ij}\sin\tilde{\varphi}_j - D_{ij}\cos\tilde{\varphi}_j \qquad H_{ij} = \frac{G_{ij}\zeta_i\omega_i - C_{ij}\tilde{\omega}_j\cos\tilde{\varphi}_j + D_{ij}\tilde{\omega}_j\sin\tilde{\varphi}_j}{\omega_{\mathrm{D}i}}$$

and $\alpha_{ij} = 1$ for $\tilde{\omega}_j \neq 0$; $\alpha_{ij} = 0$ for $\tilde{\omega}_j = 0$

4. The transient solution $\boldsymbol{\eta}(t)$ of the modal equation is given by

$$\boldsymbol{\eta}(t) = \boldsymbol{\xi}_{\mathrm{h}} + \sum_{j=1}^{n}\mathrm{diag}\left(\boldsymbol{p}_{0j}\right)\boldsymbol{\xi}_j(t)$$

5. Calculate the transient solution in the naive space by using the change-of-coordinates equation,

$$\boldsymbol{u}(t) = \boldsymbol{\Phi}\boldsymbol{\eta}(t)$$

(2) Formulate the modal equation and initial conditions.

Calculate the load-coefficient matrix in modal space:

$$\boldsymbol{p}_0 = \bar{\boldsymbol{M}}^{-1}\boldsymbol{\Phi}^{\mathrm{T}}\boldsymbol{P}_0 = \begin{bmatrix} 1.0000 & 1.9365 \\ 1.0000 & -1.9365 \end{bmatrix}$$

Table 8.2. Determine transient responses by mode superposition method.

The following data are given: $\boldsymbol{M}, \boldsymbol{K}, \zeta, \boldsymbol{P}_0, \boldsymbol{f}(t)$.

1. Modal analysis

The modal-shape and spectral matrices. $\boldsymbol{\Phi}$ and $\boldsymbol{\Omega}$, are determined by solving

$$(\boldsymbol{K}-\omega^2\boldsymbol{M})\boldsymbol{\phi}=\boldsymbol{0}$$

2. Formulate the modal equation and initial conditions

(a) Calculate the load-coefficient matrix: $\boldsymbol{p}_0 = \bar{\boldsymbol{M}}^{-1}\boldsymbol{\Phi}^{\mathrm{T}}\boldsymbol{P}_0$

(b) The modal equation in the frequency form is

$$\ddot{\boldsymbol{\eta}}(t)+2\zeta\boldsymbol{\Omega}\dot{\boldsymbol{\eta}}(t)+\boldsymbol{\Omega}^2\boldsymbol{\eta}(t)=\boldsymbol{p}_0\boldsymbol{f}(t)$$

3. Calculate the standard transient solutions $\xi_{sj}(t), j=1,2,\cdots,n$,

$\xi_{sj}(t)$ is the steady-state solution of the equation:

$$\ddot{\xi}(t)+2\zeta\boldsymbol{\Omega}\dot{\xi}(t)+\boldsymbol{\Omega}^2\xi(t)=\boldsymbol{1}f_j(t)$$

If $\boldsymbol{f}(t)$ has the form:

$$\boldsymbol{f}(t)=[\sin(\tilde{\omega}_1 t+\tilde{\varphi}_1)\ \ \sin(\tilde{\omega}_2 t+\tilde{\varphi}_2)\ \ \cdots\ \ \sin(\tilde{\omega}_n t+\tilde{\varphi}_n)]^{\mathrm{T}}$$

Then the ith component of ξ_{sj} can be expressed as

$$\xi_{sj}(t)=A_{ij}\sin(\tilde{\omega}_j t+\varphi_{ij})$$

where,

$$A_{ij}=\frac{\alpha_{ij}}{\omega_i^2\left[1-\left(\tilde{\omega}_j/\omega_i\right)^2\right]^2+\left[2\zeta_i\left(\tilde{\omega}_j/\omega_i\right)\right]^2}$$

$$\varphi_{ij}=\begin{cases}\tilde{\varphi}_j-\tan^{-1}\left[\dfrac{2\zeta_i\left(\tilde{\omega}_j/\omega_i\right)}{1-\left(\tilde{\omega}_j/\omega_i\right)^2}\right], & \text{when } \tilde{\omega}_j/\omega_i\le 1\\[2ex] \tilde{\varphi}_j-\pi-\tan^{-1}\left[\dfrac{2\zeta_i\left(\tilde{\omega}_j/\omega_i\right)}{1-\left(\tilde{\omega}_j/\omega_i\right)^2}\right], & \text{when } \tilde{\omega}_j/\omega_i> 1\end{cases}$$

and $\alpha_{ij}=1$ for $\tilde{\omega}_j\neq 0$; $\alpha_{ij}=0$ for $\tilde{\omega}_j=0$

4. The transient solution $\boldsymbol{\eta}_s(t)$ of the modal equation is given by,

$$\boldsymbol{\eta}_s(t)=\boldsymbol{\xi}_h+\sum_{j=1}^{n}\mathrm{diag}\left(\boldsymbol{p}_{0j}\right)\boldsymbol{\xi}_{sj}(t)$$

5. Calculate the transient solution in the naive space by using the change-of-coordinates equation,

$$\boldsymbol{u}_s(t)=\boldsymbol{\Phi}\boldsymbol{\eta}_s(t)$$

The initial-value problem (a)–(b) may be represented by the following modal equation:

$$\ddot{\boldsymbol{\eta}}(t)+2\zeta\boldsymbol{\Omega}\dot{\boldsymbol{\eta}}(t)+\boldsymbol{\Omega}^2\boldsymbol{\eta}(t)=\boldsymbol{p}_0\boldsymbol{f}(t) \tag{c}$$

together with the modal initial conditions,

$$\begin{cases}\boldsymbol{\eta}(0)=\boldsymbol{\Phi}^{-1}\boldsymbol{u}_0=\left[-5.9839\times10^{-3}\quad 5.5984\times10^{-2}\right]^{\mathrm{T}}\\ \dot{\boldsymbol{\eta}}(0)=\boldsymbol{\Phi}^{-1}\dot{\boldsymbol{u}}_0=\left[-0.20635\quad -0.59365\right]^{\mathrm{T}}\end{cases} \tag{d}$$

where $\boldsymbol{\eta}(t)$ is the displacement vector of the system (a) in terms of the modal coordinates, and

$$\boldsymbol{\zeta} \equiv \begin{bmatrix} \zeta_1 & \\ & \zeta_2 \end{bmatrix} = \begin{bmatrix} 0.02 & \\ & 0.02 \end{bmatrix}$$

(3) Determine the standard free and forced vibrations, $\xi_h(t)$ and $\xi_j(t)$ of the uncoupled system (c).

Since, $\boldsymbol{f}(t) = [\sin 4.0t \;\; \sin 2.5t]^T$, the free vibration, $\xi_h(t)$, of the uncoupled system (c) starting from the initial conditions (d) can be expressed as:

$$\boldsymbol{\xi}_h = \begin{Bmatrix} e^{-\zeta_1\omega_1 t}\left(A_1 \cos\omega_{D1}t + B_1 \sin\omega_{D1}t\right) \\ e^{-\zeta_2\omega_2 t}\left(A_2 \cos\omega_{D2}t + B_2 \sin\omega_{D2}t\right) \end{Bmatrix}$$

where,

$$\begin{cases} A_1 = -0.0059839 \\ A_2 = 0.055984 \end{cases} \quad \begin{cases} B_1 = -0.030855 \\ B_2 = -0.039922 \end{cases} \quad \begin{cases} \omega_{D1} = 6.7138 \\ \omega_{D2} = 14.465 \end{cases}$$

Similarly, the *i*th component of the forced vibration, $\xi_j(t)$, of the uncoupled systems

$$\ddot{\xi}(t) + 2\boldsymbol{\zeta\Omega}\dot{\xi}(t) + \boldsymbol{\Omega}^2\xi(t) = \mathbf{1}f_j(t), \quad j = 1, 2$$

starting from rest, $\xi(0) = \mathbf{0}$ and $\dot{\xi}(0) = \mathbf{0}$, can be expressed as:

$$\xi_{ij}(t) = e^{-\zeta_i\omega_i t}\,(G_{ij} \cos\omega_{Di}t + H_{ij} \sin\omega_{Di}t) + C_{ij}\sin(\omega_j t + \tilde{\varphi}_j) + D_{ij}\cos(\omega_j t + \tilde{\varphi}_j)$$

where $i, j = 1, 2$, and

$$\boldsymbol{C} = 10^{-3} \times \begin{bmatrix} 34.326 & 25.737 \\ 5.1724 & 4.9245 \end{bmatrix}, \qquad \boldsymbol{D} = 10^{-4} \times \begin{bmatrix} -12.677 & -4.4494 \\ -0.61937 & -0.35086 \end{bmatrix}$$

$$\boldsymbol{G} = 10^{-4} \times \begin{bmatrix} 12.677 & 4.4494 \\ 0.61937 & 0.35086 \end{bmatrix}, \qquad \boldsymbol{H} = 10^{-3} \times \begin{bmatrix} -20.426 & -9.5748 \\ -1.4291 & -0.85042 \end{bmatrix}$$

(4) Calculate the transient solution of the modal equation (c) by

$$\eta(t) = \xi_h + \sum_{j=1}^{n} \operatorname{diag}\left(\boldsymbol{p}_{0j}\right)\xi_j(t)$$

(5) Calculate the transient displacement response of the system by

$$\boldsymbol{u}(t) = \boldsymbol{\Phi\eta}(t) \tag{8.235}$$

which is plotted in Figure 8.37.

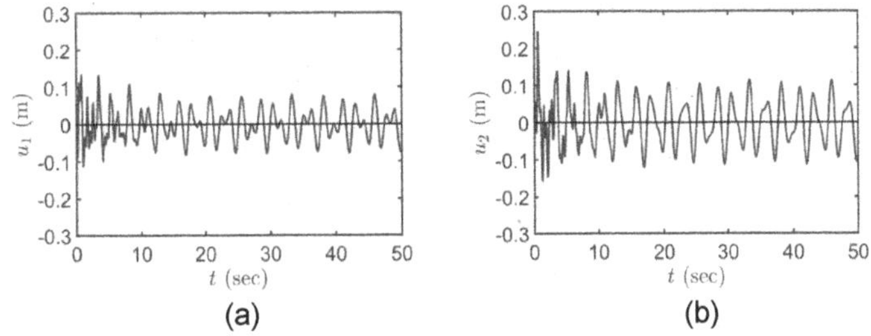

Figure 8.37. The transient response of the system (a) in Example 8.13 starting from the initial conditions (b).

EXAMPLE 8.14

Determine the steady-state response of the system in Example 8.13.

Solution: (1) Repeating the steps (1)–(2) in Example 8.13 leads to the following modal equation:

$$\ddot{\boldsymbol{\eta}}(t) + 2\boldsymbol{\zeta\Omega}\dot{\boldsymbol{\eta}}(t) + \boldsymbol{\Omega}^2\boldsymbol{\eta}(t) = \boldsymbol{p}_0\boldsymbol{f}(t)$$

where,

$$\boldsymbol{\zeta} = \begin{bmatrix} 0.02 & \\ & 0.02 \end{bmatrix}, \quad \boldsymbol{\Omega} = \begin{bmatrix} 6.7151 & \\ & 14.467 \end{bmatrix}, \quad \boldsymbol{p}_0 = \begin{bmatrix} 1.0000 & 1.9365 \\ 1.0000 & -1.9365 \end{bmatrix}$$

(2) Calculate the standard steady-state response $\xi_{sj}(t)$.

Since, $\boldsymbol{f}(t) = [\sin 4.0t \;\; \sin 2.5t]^{\mathrm{T}}$, the ith component of the steady-state solution of the following equation

$$\ddot{\boldsymbol{\xi}}(t) + 2\boldsymbol{\zeta\Omega}\dot{\boldsymbol{\xi}}(t) + \boldsymbol{\Omega}^2\boldsymbol{\xi}(t) = \mathbf{1}f_j(t), \quad j = 1, 2$$

can be expressed as,

$$\xi_{sij}(t) = A_{ij}\sin(\tilde{\omega}_j t + \varphi_{ij})$$

where,

$$\boldsymbol{A} = 10^{-2} \times \begin{bmatrix} 3.4349 & 2.5741 \\ 0.51727 & 0.49246 \end{bmatrix}, \quad \tilde{\boldsymbol{\omega}} = \begin{Bmatrix} 4.0 \\ 2.5 \end{Bmatrix}, \quad \boldsymbol{\varphi} = 10^{-2} \times \begin{bmatrix} -3.6914 & -1.7286 \\ -1.1974 & -0.71247 \end{bmatrix}$$

(3) Calculate the steady-state solution $\boldsymbol{\eta}_s(t)$ of the modal equation (a).

$$\begin{aligned} \boldsymbol{\eta}_s(t) &= \boldsymbol{\xi}_h + \sum_{j=1}^{n} \operatorname{diag}(\boldsymbol{p}_{0j})\boldsymbol{\xi}_{sj}(t) \\ &= \begin{bmatrix} 0.034349 & \\ & 0.0051727 \end{bmatrix} \begin{Bmatrix} \sin(4t - 0.036914) \\ \sin(4t - 0.011974) \end{Bmatrix} \\ &+ \begin{bmatrix} 0.049847 & \\ & -0.0095364 \end{bmatrix} \begin{Bmatrix} \sin(2.5t - 0.017286) \\ \sin(2.5t - 0.0071247) \end{Bmatrix} \end{aligned}$$

(4) Calculate the steady-state displacement, $\boldsymbol{u}_s(t)$, response of the original system.

$$\begin{aligned} \boldsymbol{u}_s(t) &= \boldsymbol{\Phi}\boldsymbol{\eta}_s(t) \\ &= 10^{-2} \times \begin{bmatrix} 3.4349 & 0.51727 \\ 4.4345 & -0.66780 \end{bmatrix} \begin{Bmatrix} \sin(4t - 0.036914) \\ \sin(4t - 0.011974) \end{Bmatrix} \\ &+ 10^{-2} \times \begin{bmatrix} 4.9847 & -0.95364 \\ 6.4352 & 1.2311 \end{bmatrix} \begin{Bmatrix} \sin(2.5t - 0.017286) \\ \sin(2.5t - 0.0071247) \end{Bmatrix} \end{aligned}$$

which is plotted in Figure 8.38.

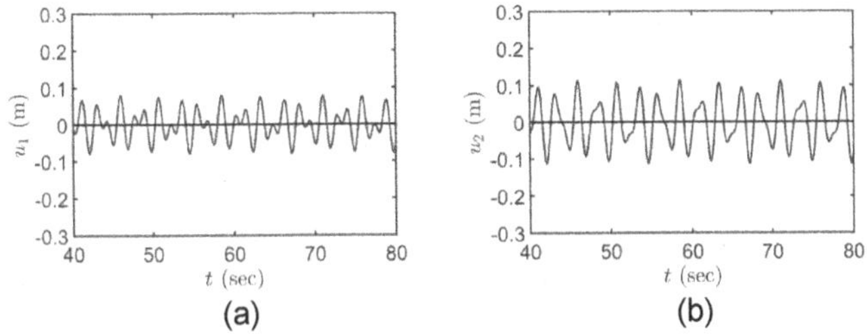

Figure 8.38. The steady-state response of system (a).

8.5 Analysis of the Dynamic Response Using the MATLAB ODE Solver: "ode45"

As we mentioned in Section 8.4, the initial value problem described by (8.170) and (8.171) can also be solved by numerical approaches, such as the central difference method and Newmark method, which are employed extensively in the dynamic analysis of structures and can be found in references [1], [2] and [3]. On the other hand, MATLAB has an extensive library of functions for solving ordinary differential equations, including a number of numerical ODE initial value problem solvers. In this section, we will focus on the build-in function *ode45*, which is an implementation of Runge-Kutta 4th/5th-order.

Runge-Kutta 4th-order method is a numerical technique used to approximate solutions of first-order ODE initial problems of the form,

$$\dot{y} = f(t, y), \quad y(0) = y_0 \tag{8.236}$$

where $y(t) \equiv [y_1(t), y_2(t), \cdots, y_n(t)]^{\mathrm{T}}$ is a vector of unknown functions, $y_i(t)$, of time t, and f is a given vector-valued function of y. As an example, let's consider a system of first-order ordinary differential equations of dimension 3:

$$\begin{cases} \dot{y}_1 = y_2 y_3 + \sin t \\ \dot{y}_2 = -y_1 y_3 + 1.5\cos(2t + 0.5) \\ \dot{y}_3 = -0.77 y_1 y_2 \end{cases} \tag{8.237}$$

with the initial conditions:

$$y_1(0) = 0, \quad y_2(0) = 1, \quad y_3(0) = 1 \tag{8.238}$$

When calling the MATLAB built-in function *ode45*, a computer implementation of Runge-Kutta 4th-order method, to simulate the system (8.237), one has to create a MATLAB user-defined function (UDF) and *ode45functions*, containing the three vector-valued functions in system (8.237). Then, by running the code in the List, one can obtain the solution to the initial value problem (8.237)–(8.238).

In structure dynamics, the initial value problem is described by a system of the *second*-order ordinary differential equation:

$$\boldsymbol{M}\ddot{\boldsymbol{u}} + \boldsymbol{C}\dot{\boldsymbol{u}} + \boldsymbol{K}\boldsymbol{u} = \boldsymbol{P}(t) \tag{8.239}$$

with the initial conditions,

$$\boldsymbol{u}(0) = \boldsymbol{u}_0, \quad \dot{\boldsymbol{u}}(0) = \dot{\boldsymbol{u}}_0 \tag{8.240}$$

However, the Runge-Kutta 4th-order method was originally developed to solve *first*-order differential equations. To apply the Runge-Kutta method to simulate the dynamic system (8.239), one has to rewrite it in the form of (8.236). For this purpose, one can define a new 2-dimensional vector, $\boldsymbol{y}$, of unknown functions as follows

$$y_{2i-1} \equiv u_i \quad \text{and} \quad y_{2i} \equiv \dot{u}_i, \quad i = 1, 2, \cdots, n \tag{8.241}$$

and let,

$$\boldsymbol{y} \equiv \begin{Bmatrix} y_1 \\ y_2 \\ \vdots \\ y_{2n} \end{Bmatrix}, \quad \boldsymbol{y}_{\text{od}} \equiv \begin{Bmatrix} y_1 \\ y_3 \\ \vdots \\ y_{2n-1} \end{Bmatrix}, \quad \boldsymbol{y}_{\text{ev}} \equiv \begin{Bmatrix} y_2 \\ y_4 \\ \vdots \\ y_{2n} \end{Bmatrix} \tag{8.242}$$

The second order differential Equations (8.239) with the initial conditions (8.240) can be rewritten as the first order differential equations:

$$\dot{\boldsymbol{y}} = \boldsymbol{F}(\boldsymbol{y}, t) \equiv \begin{Bmatrix} f_1(\boldsymbol{y}, t) \\ f_2(\boldsymbol{y}, t) \\ \vdots \\ f_n(\boldsymbol{y}, t) \end{Bmatrix} \tag{8.243}$$

with the initial conditions:

$$\boldsymbol{y}_{\text{od}}(0) = \boldsymbol{u}_0 \quad \text{and} \quad \boldsymbol{y}_{\text{ev}}(0) = \dot{\boldsymbol{u}}_0 \tag{8.244}$$

where,

$$\begin{cases} \boldsymbol{f}_{\text{od}} = \boldsymbol{y}_{\text{ev}} \\ \boldsymbol{f}_{\text{ev}} = \boldsymbol{M}^{-1}\left[\boldsymbol{P}(t) - \boldsymbol{C}\boldsymbol{y}_{\text{ev}} + \boldsymbol{K}\boldsymbol{y}_{\text{od}}\right] \end{cases} \tag{8.245}$$

and,

$$\boldsymbol{f}_{\text{od}} \equiv [f_1, f_3, \cdots f_{2n-1}]^{\text{T}}, \qquad \boldsymbol{f}_{\text{ev}} \equiv [f_2, f_4, \cdots f_{2n}]^{\text{T}} \tag{8.246}$$

8.6 Steady-State Responses to Separable Excitations

In many cases, the loads applied on a structure may have the same time variation $p(t)$. This implies that the load vector can be separated into the product of a time-varying function $f(t)$ and their spatial distribution vector $\boldsymbol{P}_0$, independent of time. In this case, the equation of motion can be written as:

$$\boldsymbol{M}\ddot{\boldsymbol{u}} + \boldsymbol{C}\dot{\boldsymbol{u}} + \boldsymbol{K}\boldsymbol{u} = \boldsymbol{P}_0 f(t) \tag{8.247}$$

A typical example of separable loads is earthquake excitation. If all the Degrees Of Freedom of the system are in the direction of the ground motion, and if the earthquake excitation is identical at all the structural supports, then the ***effective earthquake loading vector*** can be expressed as,

$$\boldsymbol{P}_{\text{eff}}(t) = -\boldsymbol{M}\boldsymbol{1}\ddot{u}_{\text{g}}(t) \tag{8.248}$$

where $\boldsymbol{1}$ is a vector of order n with each entry equal to unity.

Equation (8.248) can be expressed in terms of modal coordinates $\eta_i(t)$ as follows:

$$\ddot{\eta}_i + 2\zeta_i\omega_i\dot{\eta}_i + \omega_i^2\eta_i = \gamma_i f(t), \qquad i = 1, 2, \cdots, n \tag{8.249}$$

and

$$\ddot{\boldsymbol{\eta}} + 2\boldsymbol{\zeta}\boldsymbol{\omega}\dot{\boldsymbol{\eta}} + \boldsymbol{\omega}^2\boldsymbol{\eta} = \boldsymbol{\Gamma} f(t) \tag{8.250}$$

where $\boldsymbol{\zeta}$, $\boldsymbol{\omega}$ and $\boldsymbol{\omega}^2$ are diagonal matrices defined as

$$\boldsymbol{\zeta} \equiv \mathrm{diag}[\zeta_1, \zeta_2, \cdots, \zeta_n]; \ \boldsymbol{\omega} \equiv \mathrm{diag}[\omega_1, \omega_2, \cdots, \omega_n]; \ \boldsymbol{\omega}^2 \equiv \mathrm{diag}\left[\omega_1^2, \omega_2^2, \cdots, \omega_n^2\right] \tag{8.254}$$

and the amplitude factor, γ_i, of the generalized loading of mode i is called the ***modal participation factor*** or the ***participation factor of mode*** i, and can be expressed as

$$\gamma_i \equiv \frac{\boldsymbol{\phi}_i^{\mathrm{T}}\boldsymbol{P}_0}{\overline{m}_i} = \frac{\boldsymbol{\phi}_i^{\mathrm{T}}\boldsymbol{P}_0}{\boldsymbol{\phi}_i^{\mathrm{T}}\boldsymbol{M}\boldsymbol{\phi}_i} \tag{8.251}$$

The solution $\boldsymbol{\eta}$ of Equation (8.249) can be written as:

$$\eta_i(t) = \gamma_i\xi_i(t), \qquad i = 1, 2, \cdots, n \tag{8.252}$$

where ξ_i is the standard solution to the following equation:

$$\ddot{\xi}_i + 2\zeta_i\omega_i\dot{\xi}_i + \omega_i^2\xi_i = f(t), \qquad i = 1, 2, \cdots, n \tag{8.253}$$

Finally, the solution of Equation (8.249) can be rewritten as

$$u_j = \sum_{i=1}^{n}\phi_{ji}\eta_i(t) = \sum_{i=1}^{n}\phi_{ji}\gamma_i\xi_i(t), \ \ i = 1, 2, \cdots, n \tag{8.254}$$

where we used Equations (8.226) and (8.253). Equation (9.260) can be rewritten in the matrix form

$$u = \boldsymbol{D}_{\mathrm{mm}}\boldsymbol{\xi}(t) \tag{8.255}$$

where $\boldsymbol{\xi}(t) \equiv [\xi_1(t), \xi_2(t), \cdots, \xi_n(t)]^{\mathrm{T}}$ and

$$\boldsymbol{D}_{\mathrm{mm}} \equiv \begin{bmatrix} \phi_{11}\gamma_1 & \cdots & \phi_{1n}\gamma_n \\ \vdots & \ddots & \vdots \\ \phi_{n1}\gamma_1 & \cdots & \phi_{nn}\gamma_n \end{bmatrix} \tag{8.256}$$

is called a ***mode-decomposition matrix*** of the response of a MDOF system subjected to separable loading.

Equations (8.252) and (8.257) show that, although the modal participation factors depend on the normalization condition, a mode-decomposition matrix is independent of the normalization condition used to determine the modal matrix.

All these show that, if one creates a standard database to store the solutions of SDOF systems to frequently used excitations, the solution of a MDOF system to the corresponding separable excitation can then be easily obtained from Equation (8.255) or (8.256).

EXAMPLE 8.15

Consider a 2-DOF dynamic system

$$\boldsymbol{M}\ddot{\boldsymbol{u}} + \boldsymbol{C}\dot{\boldsymbol{u}} + \boldsymbol{K}\boldsymbol{u} = \boldsymbol{P}_0 \sin 4t \tag{a}$$

where,

$$M = \begin{bmatrix} 300 & 0 \\ 0 & 200 \end{bmatrix} (\text{kg}); \quad K = 5 \times 10^3 \times \begin{bmatrix} 1 & -0.6 \\ -0.6 & 1 \end{bmatrix} (\text{N/m}); \quad P_0 = \begin{Bmatrix} 150 \\ 0 \end{Bmatrix} (\text{N})$$

and C is the Rayleigh damping matrix. The first and second modes are assumed to have the same damping ratio 1%. Find the steady-state response of the dynamic system (a).

***Solution*:**

The spectral and modal matrices of the dynamic system (a) are:

$$\Omega^2 = \begin{bmatrix} \omega_1^2 & \\ & \omega_2^2 \end{bmatrix} = \begin{bmatrix} 2.81^2 & \\ & 5.81^2 \end{bmatrix}; \quad \Phi^2 = \begin{bmatrix} \phi_{11} & \phi_{12} \\ \phi_{21} & \phi_{22} \end{bmatrix} = \begin{bmatrix} 1 & 1 \\ 0.877 & -1.71 \end{bmatrix}$$

Equation (8.252) shows that the participation factors of modes 1 and 2 are respectively

$$\gamma_1 = 0.33052; \quad \gamma_2 = 0.16948$$

The modal equations of the dynamic system (a) can be expressed as:

$$\ddot{\eta}_i + 2\zeta_i \omega_i \dot{\eta}_i + \omega_i^2 \eta_i = \gamma_i \sin 4t$$

where $\zeta_i = 0.01$ and $i = 1, 2$.

Equation (8.256) shows that the solution of Equation (a) can be expressed as

$$u = \begin{bmatrix} \phi_{11}\gamma_1 & \phi_{12}\gamma_2 \\ \phi_{21}\gamma_1 & \phi_{22}\gamma_2 \end{bmatrix} \begin{Bmatrix} \xi_1(t) \\ \xi_2(t) \end{Bmatrix} = \begin{bmatrix} 0.33052 & 0.16948 \\ 0.28987 & -0.28987 \end{bmatrix} \begin{Bmatrix} \xi_1(t) \\ \xi_2(t) \end{Bmatrix}$$

where,

$$\xi_1(t) = 0.97 \sin(4t - 0.3155\pi); \quad \xi_2(t) = 1.90 \sin(4t - 0.0027\pi)$$

are obtained from Equation (8.119), and are the solutions of the following equations:

$$\ddot{\xi}_i + 2\zeta_i \omega_i \dot{\xi}_i + \omega_i^2 \xi_i = \sin 4t, \quad i = 1, 2$$

Finally, the response of the dynamic system (8.252) can be expressed as

$$u = \begin{bmatrix} 0.33052 & 0.16948 \\ 0.28987 & -0.28987 \end{bmatrix} \begin{Bmatrix} 0.97 \sin(4t - 0.3155\pi) \\ 1.90 \sin(4t - 0.0027\pi) \end{Bmatrix} (\text{m})$$

8.6.1 Response Spectra

In general, a ***response spectrum*** is a curve of the peak responses (displacement, velocity, or acceleration) of a family of oscillators versus the indexes of the oscillators. As an example, consider the SDOF oscillators subject to the harmonic excitation $\sin \tilde{\omega} t$, whose responses are governed by the following equation:

$$\ddot{\xi} + 2\zeta\omega\dot{\xi} + \omega^2 \xi = \sin \tilde{\omega} t \tag{8.257}$$

which can be characterized by (or indexed) by the damping ratio ζ, the natural frequency ω, and the excitation frequency $\tilde{\omega}$.

Equations (8.108) and (8.119) shows that the steady-state solution of Equation (8.258) is:

$$\xi_{\text{ste}}(t) = \xi_{\max} \sin(\tilde{\omega} t - \varphi) \tag{8.258}$$

where,

$$\xi_{\max} = \frac{1}{\sqrt{\left[1-(\tilde{\omega}/\omega)^2\right]^2 + \left[2\zeta(\tilde{\omega}/\omega)\right]^2}} \tag{8.259}$$

The curve shown in Figure 8.39 or Equation (8.260) describes a map from the family of SDOF oscillators for their peak responses. Such a map is called the ***response spectrum*** of SDOF oscillators to the harmonic excitation for a given damping ratio. The displacement response spectrum to the harmonic excitation can be analytically expressed as,

$$S_{\text{d},\zeta}(\tilde{\omega}/\omega) = \frac{1}{\sqrt{\left[1-(\tilde{\omega}/\omega)^2\right]^2 + \left[2\zeta(\tilde{\omega}/\omega)\right]^2}} \tag{8.260}$$

For any excitation, one can create such a response spectrum.

8.6.2 Peak Responses to Separable Excitations and Modal Combination Rules

The peak responses of MDOF systems to separable excitations, governed by,

$$\boldsymbol{M}\ddot{\boldsymbol{u}} + \boldsymbol{C}\dot{\boldsymbol{u}} + \boldsymbol{K}\boldsymbol{u} = \boldsymbol{P}_0 f(t) \tag{8.261}$$

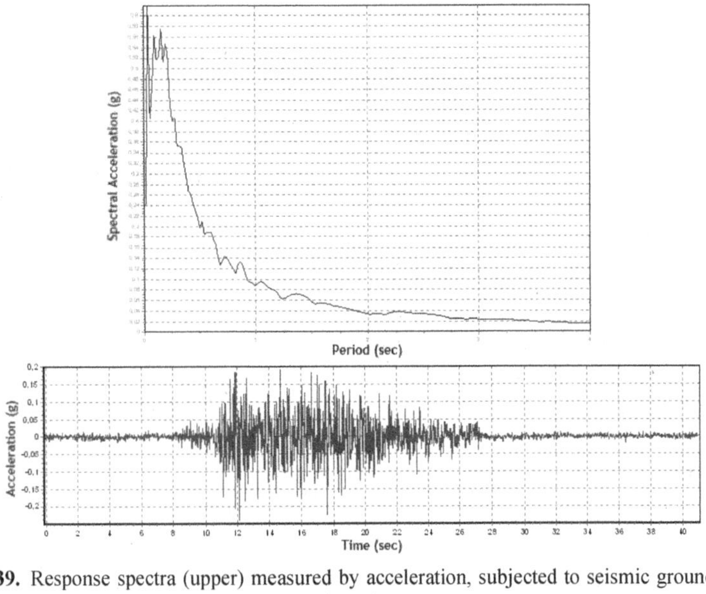

Figure 8.39. Response spectra (upper) measured by acceleration, subjected to seismic ground shaking (lower).

can be approximated by combining the modal superposition method, the response spectrum database, and the modal combination rules to be introduced in this section.

As an example, consider again the harmonic excitation $f(t) = \sin \tilde{\omega} t$, and rewrite Equation (8.261) as

$$\boldsymbol{M}\ddot{\boldsymbol{u}} + \boldsymbol{C}\dot{\boldsymbol{u}} + \boldsymbol{K}\boldsymbol{u} = \boldsymbol{P}_0 \sin \tilde{\omega} t \tag{8.262}$$

which can be represented in terms of modal coordinates as follows:

$$\ddot{\eta} + 2\zeta\omega\dot{\eta} + \omega^2\eta = \Gamma \sin \tilde{\omega} t \tag{8.263}$$

By using the displacement-response spectrum, $S_{d,\zeta}$, of SDOF systems subjected to harmonic loading, the peak value $\eta_{i,\max}$ of the ith component of the steady-state solution of Equation (8.264) can be expressed as:

$$\eta_{i,\max} = \gamma_i S_{d,\zeta_i}\left(\tilde{\omega}/\omega_i\right), \qquad i = 1, 2, \cdots, n \tag{8.264}$$

where $S_{d,\zeta}$ is described by Equation (8.261) or Figure 8.39.

Equation (8.146) shows that the jth component of the steady-state solution of Equation (8.262) can be expressed as:

$$\begin{aligned} x_j(t) &= \sum_{i=1}^{n} \phi_{ji}\eta_i(t) \\ &= \sum_{i=1}^{n} \phi_{ji}\gamma_i S_{d,\zeta}\left(\tilde{\omega}/\omega\right)\sin\left(\tilde{\omega}t - \varphi_i\right) \\ &= \sum_{i=1}^{n} \phi_{ji}\eta_{i,\max} \sin\left(\tilde{\omega}t - \varphi_i\right) \end{aligned} \tag{8.265}$$

where Equation (8.259) was used to determine η, and the phase angle φ_i is governed by Equation (8.121). The summation of all modal peaks provides an upper bound to the peak value of the total response, i.e.,

$$x_{j,\max} \le \sum_{i=1}^{n} \left|\phi_{ji}\eta_{i,\max}\right|, \qquad j = 1, 2, \cdots, n \tag{8.266}$$

This upper-bound value is usually too conservative to be used as an approximation of the jth Degree Of Freedom peak value, $x_{i,\max}$, of the system. Therefore, this *absolute sum* modal combination rule is not popular in structural design applications.

As a better alternative, the ***square-root-of-sum-of-squares*** (SRSS) rule for modal combination, is widely used in structural engineering, which can be expressed as:

$$x_{j,\max} \approx \sqrt{\sum_{i=1}^{n} \left(\phi_{ji}\eta_{i,\max}\right)^2}, \qquad j = 1, 2, \cdots, n \tag{8.267}$$

This rule was originally developed by E. Rosenblueth in 1951.

In general,

$$x_{j,\max} \approx \sqrt{\sum_{i=1}^{s} \left(\phi_{ji}\gamma_i S_f\left(\zeta_i, \omega_i\right)\right)^2}, \qquad j = 1, 2, \cdots, s \tag{8.268}$$

where $s \le n$ is the number of modes used to approximate the peak response.

EXERCISES

8.1 Apply the matrix displacement method to formulate the equations of undamped motion of the structures shown in Figure 8.40. Use static condensation to eliminate the Degrees Of Freedom with zero masses. The Young's modulus, the cross-sectional area, and the moment of inertia of each member are $E = 2.06 \times 10^{11}$ N/m², $A = 0.0096$ m², $I = 1.152 \times 10^{-5}$ m⁴, respectively.

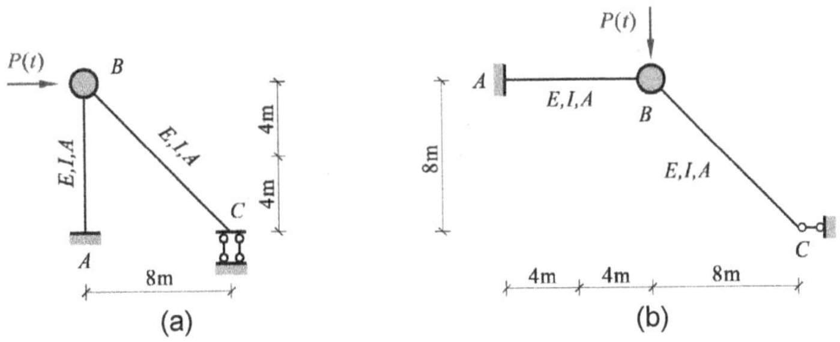

Figure 8.40. Exercise 8.1.

8.2 A structure has the following mass and stiffness matrices:

$$\boldsymbol{M} = \begin{bmatrix} 300 & 0.0 \\ 0.0 & 200 \end{bmatrix} (\text{kg}); \qquad \boldsymbol{K} = 5\times10^3 \times \begin{bmatrix} 1.0 & -0.6 \\ -0.6 & 1.0 \end{bmatrix} (\text{N/m})$$

Determine the modes and the Rayleigh damping matrix of this structure, if the first and second modes are assumed to have the same damping ratio of 1%.

8.7 Appendix for Structural Dynamics

8.7.1 Steady-State Responses to Space-time Coupled Excitations

Since, an arbitrary loading vector $\boldsymbol{P}(t)$ can be expressed as a superposition of a series of separable loads, the equation of motion of a forced MDOF system can be expressed as:

$$\boldsymbol{M}\ddot{\boldsymbol{u}} + \boldsymbol{C}\dot{\boldsymbol{u}} + \boldsymbol{K}\boldsymbol{u} = \sum_{j=1}^{r} \boldsymbol{P}_j(t) \tag{8.269}$$

where $r \le n$ and $\boldsymbol{P}_j$, $j = 1, 2, \cdots, r$, are separable loads and can be written as

$$\boldsymbol{P}_j(t) = \boldsymbol{P}_{0j} f_j(t) \tag{8.270}$$

Equation (8.270) can be expressed in terms of the modal coordinates as follows:

$$\bar{\boldsymbol{M}}\ddot{\boldsymbol{\eta}} + \bar{\boldsymbol{C}}\dot{\boldsymbol{\eta}} + \bar{\boldsymbol{K}}\boldsymbol{\eta} = \sum_{j=1}^{r} \boldsymbol{\Phi}^{\mathrm{T}} \boldsymbol{P}_{0j} f_j(t) = \sum_{j=1}^{r} \begin{Bmatrix} \boldsymbol{\phi}_1^{\mathrm{T}} \boldsymbol{P}_{0j} \\ \vdots \\ \boldsymbol{\phi}_i^{\mathrm{T}} \boldsymbol{P}_{0j} \\ \vdots \\ \boldsymbol{\phi}_n^{\mathrm{T}} \boldsymbol{P}_{0j} \end{Bmatrix} f_j(t) \tag{8.271}$$

If one defines γ_{ij} as,

$$\gamma_{ij} \equiv \frac{\boldsymbol{\phi}_i^{\mathrm{T}} \boldsymbol{P}_{0j}}{\bar{m}_i} = \frac{\boldsymbol{\phi}_i^{\mathrm{T}} \boldsymbol{P}_{0j}}{\boldsymbol{\phi}_i^{\mathrm{T}} \boldsymbol{M} \boldsymbol{\phi}_i} \tag{8.272}$$

Equation (8.272) can then be rewritten as:

$$\ddot{\eta} + 2\zeta\omega\dot{\eta} + \omega^2\eta = \sum_{j=1}^{r} \begin{Bmatrix} \gamma_{i1} \\ \gamma_{i2} \\ \vdots \\ \gamma_{in} \end{Bmatrix} f_j(t) \equiv \sum_{j=1}^{r} \boldsymbol{\Gamma}_j f_j(t) \equiv \sum_{j=1}^{r} \boldsymbol{p}_j(t) \tag{8.273}$$

or

$$\ddot{\eta}_i + 2\zeta_i\omega_i\dot{\eta}_i + \omega_i^2\eta_i = \sum_{j=1}^{r} \gamma_{ij} f_j(t), \qquad i = 1, 2, \cdots, n \tag{8.274}$$

The solution $\boldsymbol{\eta}(t)$ of Equation (8.274) can be expressed as:

$$\boldsymbol{\eta}(t) = \sum_{j=1}^{r} \mathrm{diag}(\boldsymbol{\Gamma}_j)\boldsymbol{\xi}_j(t) \tag{8.275}$$

where,

$$\mathrm{diag}(\boldsymbol{\Gamma}_j) \equiv \begin{bmatrix} \gamma_{1j} & & & \\ & \gamma_{2j} & & \\ & & \ddots & \\ & & & \gamma_{nj} \end{bmatrix} \tag{8.276}$$

and $\xi_j(t)$ is governed by the following equation:

$$\ddot{\boldsymbol{\xi}} + 2\zeta\omega\dot{\boldsymbol{\xi}} + \omega^2\boldsymbol{\xi} = \mathbf{1} f_j(t) \tag{8.277}$$

$\mathbf{1}$ is a vector of order n with each entry equal to unity.

By substituting Equation (8.276) into Equation (8.226), one can obtain:

$$\boldsymbol{x}(t) = \boldsymbol{\Phi}\sum_{j=1}^{r} \mathrm{diag}(\boldsymbol{\Gamma}_j)\boldsymbol{\xi}_j(t) = \sum_{j=1}^{r} \left[\boldsymbol{\Phi}\mathrm{diag}(\boldsymbol{\Gamma}_j)\right]\boldsymbol{\xi}_j(t) \tag{8.278}$$

where $\boldsymbol{x}(t)$ is the response of system (8.270), and $\xi_j(t)$ may be obtained from a standard database storing the solutions of SDOF systems.

EXAMPLE 8.16

As shown in Figure 8.41a, a vertical cantilever beam with two lumped masses is subjected to two harmonic loads applied at points 1 and 2. The Young's modulus, the moment of inertia and the height of each storey are $m_1 = 500$ kg, $m_2 = 300$ kg, $E = 2.06 \times 10^{11}$ N/m^2, respectively. $I = 1.152 \times 10^{-5}$ m^4 and $h = 8$m. The damping ratio is 2% for the first and second modes. Neglecting the axial deformation of each member, determine the steady-state displacement responses of the system subject to the excitation frequencies and the excitation amplitudes:

$$\begin{Bmatrix} \tilde{\omega}_1 \\ \tilde{\omega}_2 \end{Bmatrix} = \begin{Bmatrix} 5.2 \\ 7.5 \end{Bmatrix} \text{ (ran/sec)}; \qquad \begin{Bmatrix} p_{01} \\ p_{02} \end{Bmatrix} = \begin{Bmatrix} 100 \\ 150 \end{Bmatrix} \text{ (N)} \tag{8.279}$$

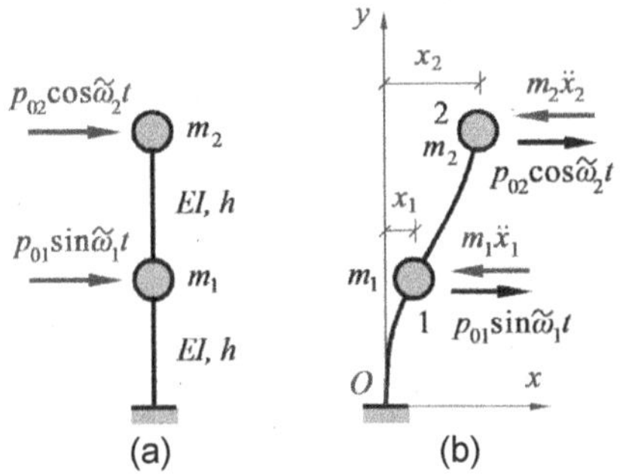

Figure 8.41. (a) An oscillator with Rayleigh damping excited by two harmonic loadings. (b) The deformed configuration of the system at time t.

***Solution*:**

(1) Establish the equation of motion as,

$$\boldsymbol{M}\ddot{\boldsymbol{x}} + \boldsymbol{C}\dot{\boldsymbol{x}} + \boldsymbol{K}\boldsymbol{x} = \boldsymbol{P}(t) = \begin{Bmatrix} 100\sin\tilde{\omega}_1 t \\ 150\cos\tilde{\omega}_2 t \end{Bmatrix} \tag{8.280}$$

where $\boldsymbol{C}$ is the Rayleigh damping matrix;

$$\boldsymbol{M} = 500\times\begin{bmatrix} 1.0 & 0 \\ 0 & 0.6 \end{bmatrix}; \qquad \boldsymbol{K} = 6.36\times10^3\times\begin{bmatrix} 1 & -0.313 \\ -0.313 & 0.125 \end{bmatrix} \tag{8.281}$$

The space-time coupled loading $\boldsymbol{P}(t)$ can be decomposed into the sum of two separable excitations, i.e.,

$$\begin{aligned} \boldsymbol{P}(t) &= \begin{Bmatrix} 100 \\ 0 \end{Bmatrix}\sin\tilde{\omega}_1 t + \begin{Bmatrix} 0 \\ 150 \end{Bmatrix}\sin\left(\tilde{\omega}_1 t + \frac{\pi}{2}\right) \\ &\equiv \boldsymbol{P}_{01}\sin\tilde{\omega}_1 t + \boldsymbol{P}_{02}\sin\left(\tilde{\omega}_2 t + \frac{\pi}{2}\right) \equiv \boldsymbol{P}_1(t) + \boldsymbol{P}_2(t) \end{aligned} \tag{8.282}$$

(2) Modal analysis

$$\left|\boldsymbol{K} - \omega^2\boldsymbol{M}\right| = 0 \tag{8.283}$$

$$\omega_1 = 2.21; \qquad \omega_2 = 12.20, \ (\text{rad/sec}) \tag{8.284}$$

$$\left(\boldsymbol{K} - \omega^2\boldsymbol{M}\right)\boldsymbol{\phi} = 0 \tag{8.285}$$

With normalization condition:

$$\phi_{1j} = 1 \tag{8.286}$$

$$\boldsymbol{\Phi} = \begin{bmatrix} 1 & 1 \\ 3.072 & -0.543 \end{bmatrix} \tag{8.287}$$

$$\bar{\boldsymbol{M}} = \boldsymbol{\Phi}^{\mathrm{T}}\boldsymbol{M}\boldsymbol{\Phi} = 10^3\times\begin{bmatrix} 3.331 & 0 \\ 0 & 0.588 \end{bmatrix} \tag{8.288}$$

(3) Establish the modal equation of motion:

$$\ddot{\eta}+2\zeta\omega\dot{\eta}+\omega^2\eta=\Gamma_1 f_1(t)+\Gamma_2 f_2(t) \tag{8.289}$$

where $\boldsymbol{\eta}\equiv[\eta_1,\eta_2]^{\mathrm{T}}$, and

$$[\Gamma_1\quad\Gamma_2]=\begin{bmatrix}\gamma_{11} & \gamma_{12}\\ \gamma_{21} & \gamma_{22}\end{bmatrix}=\begin{bmatrix}0.030 & 0.138\\ 0170 & -0.138\end{bmatrix};\quad \begin{Bmatrix}f_1(t)\\ f_2(t)\end{Bmatrix}\equiv\begin{Bmatrix}\sin\tilde{\omega}_1 t\\ \sin\left(\tilde{\omega}_1 t+\dfrac{\pi}{2}\right)\end{Bmatrix} \tag{8.290}$$

Here, Equation (8.273) was used to calculate γ_{ij}.

(4) Solutions of the modal equation:

The solution $\boldsymbol{\eta}(t)$ of Equation (8.290) can be written as

$$\eta(t)=\mathrm{diag}(\Gamma_1)\xi_1(t)+\mathrm{diag}(\Gamma_1)\xi_2(t) \tag{8.291}$$

where $\xi_i(t)$, $i=1, 2$, are the solutions of the following equations:

$$\ddot{\xi}+2\zeta\omega\dot{\xi}+\omega^2\dot{\xi}=\begin{Bmatrix}1\\1\end{Bmatrix}f_i(t),\quad i=1,2 \tag{8.292}$$

Equation (8.119) gives:

$$\xi_1(t)=\begin{bmatrix}0.221 & 0\\ 0 & 1.222\end{bmatrix}\begin{Bmatrix}\sin(\tilde{\omega}_1 t-0.316\pi)\\ \sin(\tilde{\omega}_1 t-0.002\pi)\end{Bmatrix} \tag{8.293}$$

$$\xi_2(t)=\begin{bmatrix}0.095\\ 1.606\end{bmatrix}\begin{Bmatrix}\sin(\tilde{\omega}_2 t-0.183\pi)\\ \sin(\tilde{\omega}_2 t-0.496\pi)\end{Bmatrix} \tag{8.294}$$

(5) The real-displacement responses of the cantilever beam:

$$\begin{aligned}x(t)&=\Phi\eta(t)\\ &=\sum_{j=1}^{n}\left[\Phi\mathrm{diag}(\Gamma_j)\right]\xi_j(t)\\ &=\begin{bmatrix}0.030 & 0.170\\ 0.092 & 0.092\end{bmatrix}\xi_1(t)+\begin{bmatrix}0.138 & -0.138\\ 0.425 & 0.075\end{bmatrix}\xi_2(t)\\ &=\begin{bmatrix}0.007 & 0.208\\ 0.020 & -0.113\end{bmatrix}\begin{Bmatrix}\sin(\tilde{\omega}_1 t-0.316\pi)\\ \sin(\tilde{\omega}_1 t-0.002\pi)\end{Bmatrix}+\begin{bmatrix}0.013 & -0.222\\ 0.041 & 0.121\end{bmatrix}\begin{Bmatrix}\sin(\tilde{\omega}_2 t-0.183\pi)\\ \sin(\tilde{\omega}_2 t-0.496\pi)\end{Bmatrix}\end{aligned}$$

where Equation (8.279) was used to calculate the response of system $\boldsymbol{x}(t)$.

Chapter 9
Limit Loads of Structures

9.1 Introduction

The ***limit analysis*** of structures is concerned with determining the ***ultimate load*** of a structure, which is the maximum load that the structure can sustain before collapse. It is also known as the ***limit load***, the ***failure load***. The limit analysis is an integrated part of plastic analysis.

Figure (9.1a) shows a simply supported rectangular beam subjected to two loads, λ and 2λ, where λ is called a common load factor or simply a load factor. As the load factor is gradually increased, the two loads increase quasi-statically and proportionally, the term "quasi-statically" indicates that the loading process is sufficiently slow for all dynamic effects to be ignored. The term "proportionally" implies that the ratio of the two loads at any time is a constant. Such a system of loads is called proportional loading. For prudence and for simplicity, the beam is assumed to be made of an elastic-perfectly-plastic material with a stress-strain relationship shown in Figure 9.1b. The Young's modulus, the yield strength, the width and height of the beam are $E = 2.06 \times 10^{11}$ Pa, $\sigma_y = 235$ Mpa, $b = 0.15$ m and $h = 0.3$ m, respectively.

Next, the behavior of the beam as the load factor increases will be discussed.

Figure 9.1c shows the moment diagram of the beam subjected to increasing proportional loads, which shows that the absolute maximum moment occurs at section 3. This implies that the maximum normal stress occurs at the upper and lower edges of section 3. As the load factor is gradually increased from zero, the beam is stressed first in a purely elastic state, whose normal-stress distribution on section 3 is shown in Figure 9.2a. By increasing the load factor until the maximum stress reaches the yield strength σ_y, the normal-stress distribution reaches a new state shown in Figure 9.2b. The moment on section 3 at this stage is called a ***yield moment***, denoted by M_y. When the load factor λ is increased further, section 3 undergoes a plastification process, in which the yield zone spreads and becomes larger and larger (Figure 9.2e). Figure 9.2c shows the normal-stress distribution at this stage, which is referred to as the partial plasticity state. As one continues increasing the load factor, eventually, the upper and lower plastic regions meet. At this instant, all material above the neutral axis of section 3 has yielded in compression and all material below has

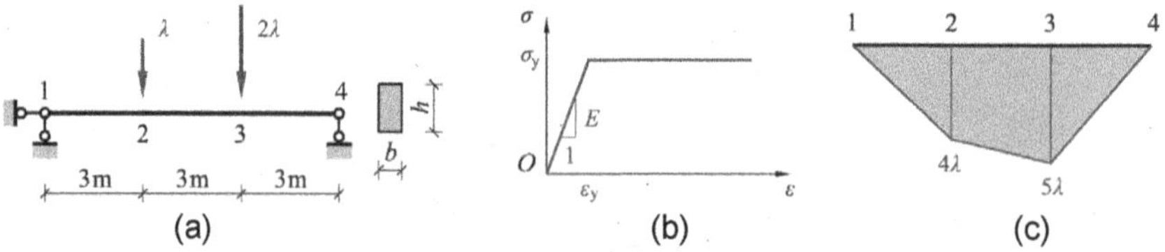

Figure 9.1. (a) A rectangular beam subjected to proportional loads. (b) The stress strain constitutive relationship for the beam material. (c) Moment diagram of the beam subjected to the increasing proportional loads.

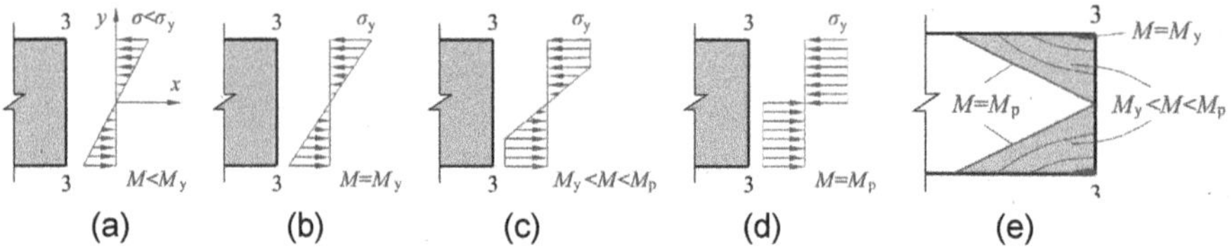

Figure 9.2. Normal stress distribution on section 3 (Figure 9.1c) for a rectangular cross section: (a) elastic. (b) elasticity-initial yielding (yielding at extreme fibers). (c) partial plasticity. (d) full plasticity. (e) yield zone spreading in partial plasticity state given in (c).

yielded in tension, and a ***plastic hinge*** has formed, reaching the full plasticity stage. Then the section cannot offer additional resistance to deformation, and an increase in curvature continues at a constant ***plastic moment*** M_P, or called (less commonly) ***ultimate moment***. Now the simply-supported beam becomes a mechanism and reaches its ***limit state*** or ***collapse state***. The load factor, λ_c, corresponding to this state is called a ***collapse load factor*** or an ***ultimate load factor***. The proportional loads described by a collapse load factor is called ***collapse loading***. ***Collapse loading*** is known variously as the ***failure loading***, the ***ultimate loading***, and the ***limit loading***.

The definition of yield moment shows that it is independent of the loading, and totally depends on the yield strength and the cross-sectional dimensions. For this uniform beam, the yield moment M_y is the same for all the sections, and can be expressed as:

$$M_y = \int_{-h/2}^{h/2} y\left[\sigma(y)b\mathrm{d}y\right] = \int_{-h/2}^{h/2} \frac{2b\sigma_y}{h} y^2 \mathrm{d}y = \sigma_y\left(\frac{1}{6}bh^2\right) = 5.29\times10^5\,\mathrm{N\cdot m} \quad (9.1)$$

where $\sigma(y)$ is the stress at height y; σ_y is the yield strength, and can be determined by:

$$\sigma(y) = \frac{\sigma_y}{h/2} y \quad (9.2)$$

As soon as the moment at section 3 reaches the yield moment, the beam enters the elastoplastic stage. The load factor, λ_y, at this time is called the ***elastic-limit load factor***, and can be expressed as:

$$\lambda_y = \frac{1}{5} M_y = 1.05\times10^5\,\mathrm{N} \quad (9.3)$$

The plastic moment is the moment capacity limit of the cross section of a beam. For this uniform beam, the plastic moment M_P is the same for all the sections, and can be expressed as,

$$M_p = \sigma_y \left(\frac{1}{4} bh^2 \right) = 7.93 \times 10^5 \, \text{N} \cdot \text{m} \tag{9.4}$$

The collapse load factor λ_c can be expressed as,

$$\lambda_c = \frac{1}{5} M_p = 1.58 \times 10^5 \, \text{N} \tag{9.5}$$

The $\lambda - \Delta_3$ curve is shown in Figure 9.3a, in which point A is the elastic limit state, and point B is the ultimate limit state. The above procedure completely investigates the behavior of the beam from its elastic to plastic state and is termed elastoplastic analysis. For design purposes, one may skip the elastoplastic analysis, jump to the plastic ultimate state B to calculate the collapse load factor. Since the simply-supported beam is a statically determinate structure, the moment diagram shown in Figure 9.1c is independent of the elastoplastic process. This implies that Equation (9.5) is independent of the elastoplastic analysis. Such a method to determine the collapse loading from the plastic ultimate state directly is called ***limit analysis***.

The limit analysis for the statically determinate beam is based on the fact that its moment diagram is independent of the elastoplastic procedure. However, the moment distribution of a statically indeterminate structure varies with its elastoplastic evolution. To solve this problem, let's consider a clamped-roller beam shown in Figure 9.4(a), whose plastic moment is $M_P = 7.93 \times 10^5$ N · m.

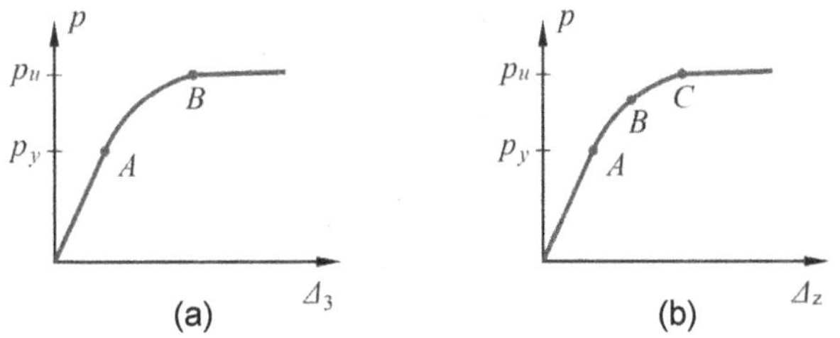

Figure 9.3. The $\lambda - \Delta$ relationship between the load factor and structural displacement. (a) Δ_3 is the vertical displacement of the structure, point A is the elastic limit state, and point B is the ultimate limit state. (b) Δ_z is the vertical displacement of the structure, point A is the elastic limit state, point B corresponds to the state during the elastoplastic process, and point C is the ultimate limit state.

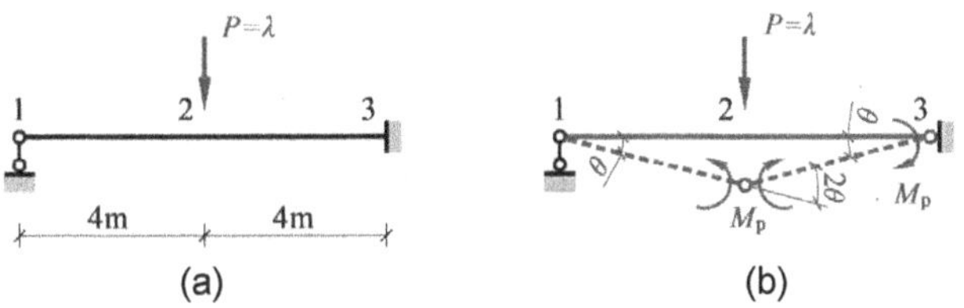

Figure 9.4. (a) A clamped-roller beam subjected to an increasing load at the middle point along the span of the beam. (b) The collapse mechanism of the beam.

Here, let's first skip the elastoplastic process, and determine its collapse load factor by limit analysis, that is, the collapse mechanism of the beam is first obtained, after which the collapse load factor is determined. Since this beam is statically indeterminate to the second degree, its collapse mechanism requires the formation of two plastic hinges. Furthermore, the two plastic hinges occur at sections 2 and 3, where the bending moments reach the local extreme values. Figure 9.4b shows the unique possible collapse mechanism of the beam.

Impose a virtual displacement on the collapse mechanism as shown in Figure 9.4b. Since the collapse mechanism is in equilibrium at the collapse limit state, all the forces on it satisfy the following virtual-displacement equation:

$$(-2M_{\rm p}\theta) + (-M_{\rm p}\theta) + \lambda_{\rm c} \cdot 4\theta = 0 \tag{9.6}$$

or

$$(4\lambda_{\rm c} - 3M_{\rm p})\theta = 0 \tag{9.7}$$

For Equation (9.7) to be valid for any arbitrary small θ, the coefficient of θ must vanish. This requirement gives the collapse load factor as follows:

$$\lambda_{\rm c} = \frac{3M_{\rm p}}{4} \tag{9.8}$$

Next, let's check this result by elastoplastic analysis. Solving the statically indeterminate beam, shown in Figure 9.4a, the moment diagram in the elastic stage is given in Figure 9.5a. This shows that the elastic-limit load factor is:

$$\lambda_{\rm y} = \frac{2M_{\rm y}}{3} \tag{9.9}$$

where $M_{\rm y}$ is the yield moment of the beam.

By further increasing the external load, the first plastic hinge at section 3 is formed. The beam will be converted to a statically determinate beam, shown in Figure 9.5b as long as the first plastic hinge forms at section 3. Since a plastic hinge can offer no additional resistance to deformation, keeping on increasing the load will

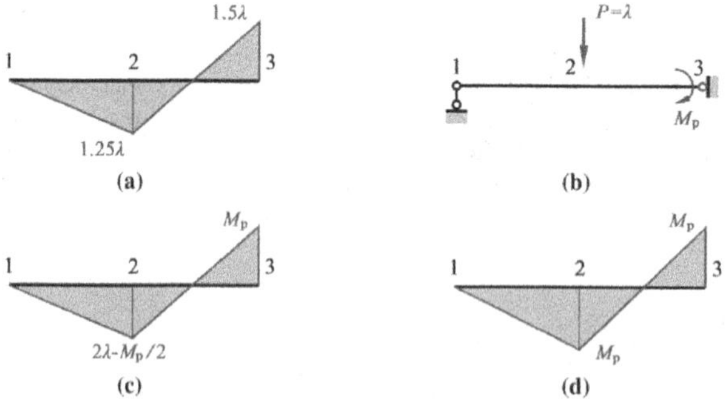

Figure 9.5. (a) The moment diagram of the beam in the elastic stage. (b) The beam has been converted to a statically determinate beam subjected to a force $P=\lambda$ at the beam center and a torque $M_{\rm p}$ at the right end due to the formation of the plastic hinge at section 3. (c) The moment diagram of the beam in the elastic-plastic stage. (d) The moment diagram at the collapse limit state.

lead to a moment redistribution, that is, the forces are redistributed to other parts of the structure. Solving the statically determinate beam provides the moment diagram (Figure 9.5c) in the elastic-plastic stage. One can further increase the external load until the moment at section 2 reaches the plastic moment, i.e.,

$$2\lambda - \frac{M_p}{2} = M_p \tag{9.10}$$

which yields,

$$\lambda_c = \frac{3M_p}{4} \tag{9.11}$$

At this instant, the beam becomes an unstable collapse mechanism due to the formation of the second plastic hinge at section 2.

9.2 Theorems of Plasticity

Definition 10.1 (Admissible Loading) ***statically-plastically admissible loads*** *or simply* ***admissible loads*** *of a structure are the proportional loads to which any static moment response of the structure is less than or equal to the corresponding plastic moment. The load factor of admissible loads is denoted by* λ_c^-.

The term "*static moment response*" indicates that the bending moment is calculated by *static structural-analysis approaches*, such as the force method, the displacement method and the matrix displacement method. This implies that the moment responses to admissible loading are not only statically admissible, but also plastically admissible.

Definition 10.2 (Possible collapse loading) *Possible collapse loads of a structure are the proportional loads determined by the equilibrium requirement of any possible collapse mechanism of the structure. The load factor of possible collapse loads is denoted by* λ_c^+.

Theorem 9.3 (Upper Bound Theorem) *The true collapse load factor* λ_c *is less than or equal to any possible-collapse-load factor, that is,*

$$\lambda_c \le \lambda_c^+ \tag{9.12}$$

Theorem 9.3 shows that for any possible-collapse-load factor there is an upper bound of the collapse load. The following theorem shows that for any admissible-load factor there is a lower bound of the collapse load.

Theorem 9.4 (Lower Bound Theorem) *The collapse load factor* λ_c *of a structure is larger than or equal to an admissible-load factor of the structure, i.e.,*

$$\lambda_c^- \le \lambda_c \tag{9.13}$$

Theorem 9.5 (Uniqueness Theorem) *The collapse load of a structure is unique.*

Theorem 9.6 (Necessary condition I for a plastic hinge) *If a plastic hinge occurs at a point in a uniform segment, then the moment at the point reaches an extreme value.*

In mathematics, a *normal vector of a curve* at a point is defined as a unit vector pointing towards the center of curvature. When the local region of a plastic hinge is enlarged, then it is differentiable. The normal vector at a plastic hinge is called a ***normal vector of the plastic hinge***.

Theorem 9.7 (Necessary condition II for a plastic hinge) If there is some loading on a plastic hinge, then the normal vector of the plastic hinge is in the opposite direction of the loading.

9.3 Applications of the Upper Bound Theorem

EXAMPLE 9.1

Consider a statically indeterminate uniform beam, shown in Figure 9.6, subjected to proportional loads P_2 and P_3. The plastic moment of the beam is M_P. Determine the collapse load factor by limit analysis.

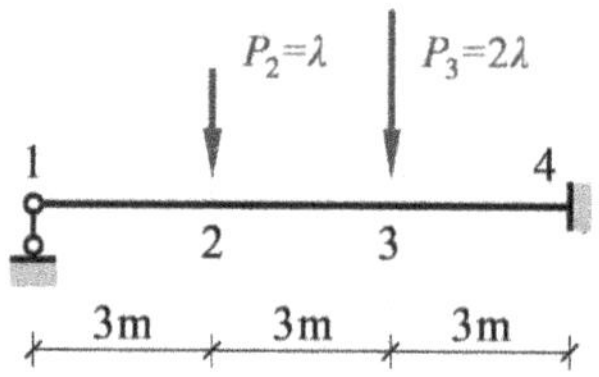

Figure 9.6. A clamped-roller beam subjected to increasing proportional loads.

***Solution*:**

Since the bending moments *may* reach local extreme values at sections 2, 3, and 4, Plastic hinges may occur at those locations according to Theorem 9.4. Figure 9.7 shows all the four possible collapse mechanisms/scenarios of the statically indeterminate beam. However, the cases shown in Figure 9.7a and Figure 9.7b can be excluded by Theorem 9.5. Therefore, only two possible collapse mechanisms of Figure 9.7c and Figure 9.7d remain to be checked.

Consider first the collapse mechanism shown in Figure 9.7c by imposing a virtual displacement on it. Since the collapse mechanism is in equilibrium at the

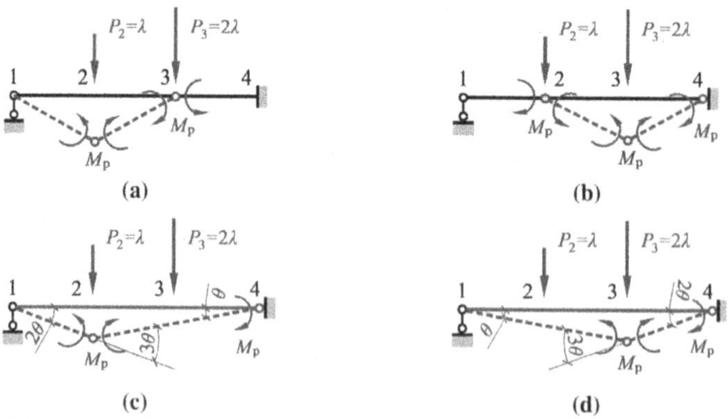

Figure 9.7. Four possible collapse mechanisms/scenarios of the statically indeterminate beam.

collapse limit state, all the forces on it satisfy the following virtual-displacement equation.

$$(-M_P \cdot 3\theta) + (-M_P \cdot \theta) + \lambda_c \cdot 2\theta \cdot 3 + 2\lambda_c \cdot \theta \cdot 3 = 0 \quad \text{(a)}$$

or

$$(12\lambda_c - 4M_P)\theta = 0 \quad \text{(b)}$$

For equation (a) to be valid for any arbitrary small θ, the coefficient of θ must be zero. This requirement gives:

$$\lambda_c^1 = \frac{M_p}{3} \quad \text{(c)}$$

Now, let's check the possible collapse mechanism shown in Figure 9.7(d). The virtual-displacement equation is,

$$(-M_P \cdot 3\theta) + (-M_P \cdot 2\theta) + \lambda_c \cdot 3\theta + 2\lambda_c \cdot 6\theta = 0 \quad \text{(d)}$$

which provides,

$$\lambda_c^2 = \frac{M_p}{3} \quad \text{(e)}$$

All this shows that plastic hinges will occur simultaneously at sections 2 and 3 as soon as the load factor reaches $\lambda_c = M_P/3$.

EXAMPLE 9.2

Consider a continuous beam shown in Figure 9.8.

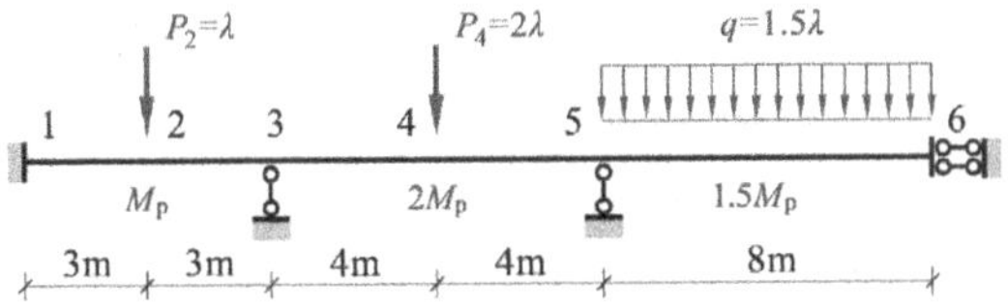

Figure 9.8. A continuous beam subjected to proportional loads.

***Solution*:**

The sufficient condition I for the plastic hinge, as per Theorem 9.4, shows that plastic hinges may form at sections 1, 2, …, and 6 because the moments *may* reach extreme values at those sections. All three possible collapse mechanisms are shown in Figure 9.9, where the combined collapse mechanisms have been excluded, such as that shown in Figure 9.10, according to the sufficient condition II for the plastic hinge, as per Theorem 9.4.

For the mechanism of Figure 9.9(a), the virtual displacement equation can be expressed as,

$$(-M_P \cdot \theta) + (-M_P \cdot 2\theta) + (-M_P \cdot \theta) + \lambda \cdot 3\theta = 0 \quad \text{(a)}$$

which provides,

$$\lambda_1 = \frac{4M_p}{3} \quad \text{(b)}$$

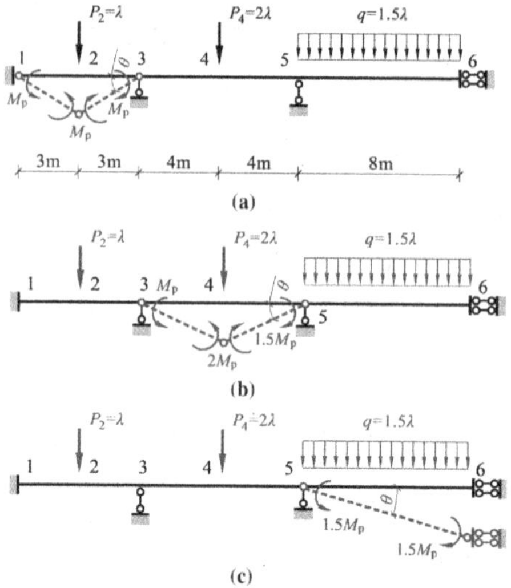

Figure 9.9. All the three possible collapse mechanisms of the continuous beam.

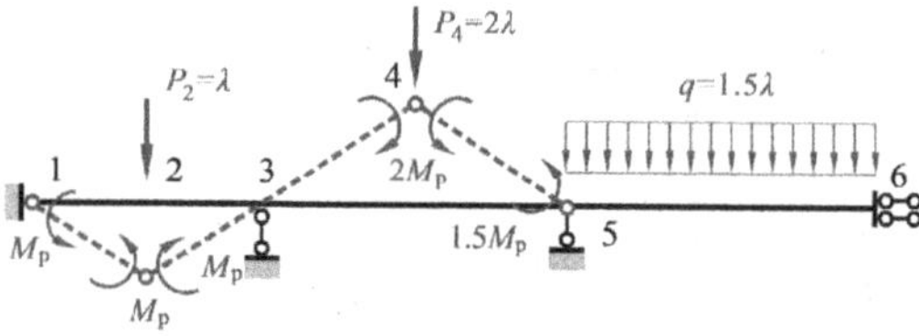

Figure 9.10. An example of combined mechanisms which have been excluded by Theorem 9.5 because the normal vector of the plastic hinge 4 is in the direction of load P_4.

The virtual displacement equation for the possible collapse mechanism of Figure 9.9b is:

$$(-M_P \cdot \theta) + (-2M_P \cdot 2\theta) + (-1.5M_P \cdot \theta) + 2\lambda \cdot \theta = 0 \tag{c}$$

It is noticed that, because the plastic hinges 3 and 5 occur at the left side of section 3 and the right side of section 5, respectively, their plastic moments are M_P and $1.5M_P$, but not $2M_P$. Solving Equation (a) for λ gives:

$$\lambda_2 = \frac{9M_p}{16} \tag{d}$$

Analogously, the virtual displacement equation for the collapse mechanism of Figure 9.9c can be written as:

$$\left(-1.5M_p \cdot \theta\right) + \left(-1.5M_p \cdot \theta\right) + \frac{1}{2}\theta 8^2 \cdot 1.5\lambda = 0 \tag{e}$$

where the last term on the left side of this equation is the virtual work done by the distributed load $q = 1.5\lambda$, and which gives:

$$\lambda_3 = \frac{M_p}{16} \tag{f}$$

Finally, the collapse load factor of the continuous beam is $\lambda_c = \lambda_3 = M_P/16$.

In general, the procedure for limit analysis of a continuous beams is summarized as the following three steps:

1) Identify possible collapse mechanisms in *each of the spans*.
2) Apply the virtual displacement principle to find the possible collapse load factors for each of the single-span collapse mechanisms.
3) Identify the critical span, whose collapse load factor is the smallest one, and take the smallest collapse load factor as the collapse load factor of the continuous beam.

EXAMPLE 9.3

As shown in Figure 9.11a, a clamped-roller uniform beam having plastic moment M_P is subjected to an increasing distributed load $q = \lambda$. Calculate its collapse load by limit analysis.

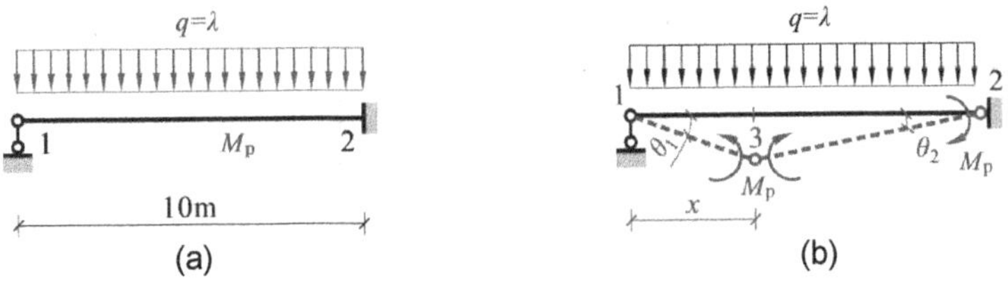

Figure 9.11. (a) A clamped-roller beam subjected to an increasing uniform distributed load. (b) The possible collapse mechanism of the beam.

***Solution*:**

The virtual displacement equation can be written as:

$$-M_p(\theta_1 + \theta_2) - M_p\theta_2 + \lambda \cdot \frac{1}{2} \cdot 10\Delta_C = 0 \tag{a}$$

Substituting,

$$\begin{cases} \theta_1 \approx \dfrac{\Delta_C}{x} \\ \theta_2 \approx \dfrac{\Delta_C}{10 - x} \end{cases}$$

into Equation (a) leads to,

$$\lambda = \frac{M_p}{5}\left(\frac{1}{x} + \frac{2}{10 - x}\right), \tag{b}$$

and its derivative is,

$$\frac{d\lambda}{dx} = \frac{1}{5}\left[\frac{2}{(10 - x)^2} - \frac{1}{x^2}\right] \tag{c}$$

λ has a minimum value when x satisfies:

$$x^2 + 20x - 100 = 0 \tag{d}$$

that is,

$$x_1 = -10\left(\sqrt{2}+1\right) \quad \text{and} \quad x_2 = 10\left(\sqrt{2}-1\right) \tag{e}$$

Inserting x_2 into Equation (b) gives the minimum value of λ:

$$\lambda_{\text{min}} = 0.1166 M_{\text{P}} \tag{f}$$

The upper bound theorem, Theorem 9.1, shows that the collapse load is $q_c = 0.1166M_P$.

EXAMPLE 9.4

As shown in Figure 9.12, a portal frame is subjected to a horizontal load of $P_2 = \lambda$ and vertical load of $P_3 = 2\lambda$, where λ is the common load factor of the proportional loads. The plastic moments of the columns and beam are $1.5M_P$ and M_P, respectively. Find the collapse load factor λ_c by limit analysis.

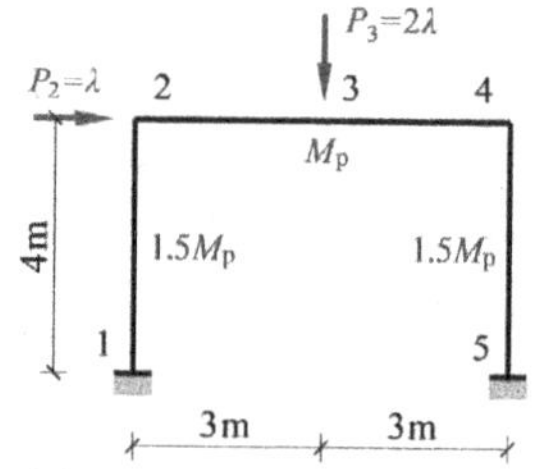

Figure 9.12. A portal frame subjected to increasing proportional loads.

Solution:

Since the moments of the frame subjected to the proportional loads may reach extreme values at sections 1, 2, …, and 5, the plastic hinges may form at these sections. Figure 9.13 shows all the three possible collapse mechanisms, which are called the *beam*, *sway*, and *combined mechanisms*. The virtual displacement equations of the three mechanisms are:

$$-4M_P\theta + 2\lambda \cdot 3\theta = 0 \tag{a}$$

$$-5M_P\theta + \lambda \cdot 4\theta = 0 \tag{b}$$

$$-6M_P\theta + \lambda \cdot 4\theta + 2\lambda \cdot 3\theta = 0 \tag{c}$$

which provide,

$$\lambda_1 = \frac{4M_P}{5}; \quad \lambda_2 = \frac{5M_P}{4}; \quad \lambda_3 = \frac{3M_P}{5} \tag{d}$$

Thus, the collapse load factor is $\lambda_c = 3M_P/5$.

9.4 Limit Analysis by Linear Programing

Again, consider the clamped-roller beam shown in Figure 9.6. Figure 9.14 shows the primary system for the force method, where X_1 is the redundant force. The moment

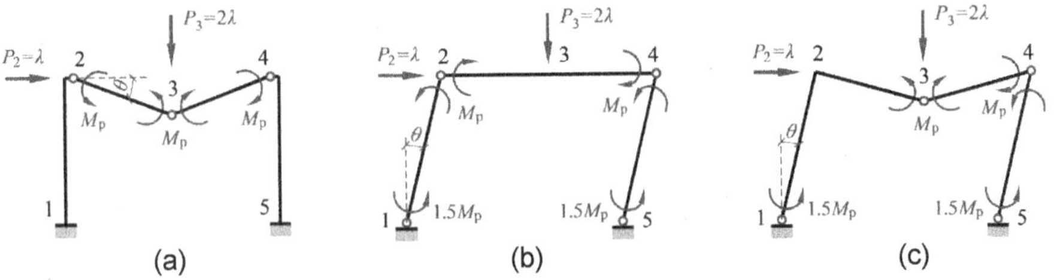

Figure 9.13. All the three possible collapse mechanisms: (a) Beam mechanism. (b) Sway mechanism. (c) Combined mechanism.

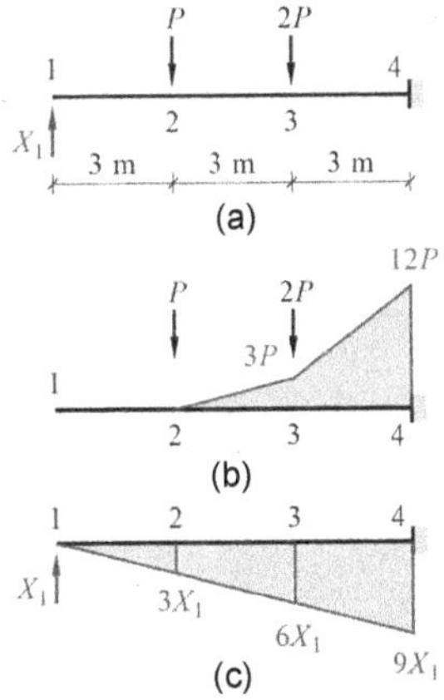

Figure 9.14. (a) The primary system for the force method. (b) The moment diagrams of the primary structure under the loading. (c) The moment diagrams of the primary structure under redundant X_1.

diagrams of the primary structure under the loading and redundant X_1 are illustrated in Figure 9.14b and c.

Let p be a load factor under which none of the moments in the clamper-roller beam exceeds the plastic moment M_P, namely,

$$\begin{cases} |12p - 9X_1| \le 7.93 \times 10^5 \\ |3p - 6X_1| \le 7.93 \times 10^5 \\ |3X_1| \le 7.93 \times 10^5 \end{cases} \tag{9.14}$$

The upper bound theorem shows that the collapse load factor p_u is the maximum p satisfying the constraints given in Equation (9.14) This implies that finding the collapse load factor of the clamped-roller beam is to find the optimal solution of the following linear programming problem:

$$\begin{aligned} &\text{maximize} \quad s \equiv \{f\}^{\mathrm{T}}\{x\} \\ &\text{subject to} \quad [A]\{x\} \le \{b\} \end{aligned} \tag{9.15}$$

or

$$\begin{aligned} &\text{minimize} \quad s \equiv -\{f\}^{\mathrm{T}}\{x\} \\ &\text{subject to} \quad [A]\{x\} \le \{b\} \end{aligned} \tag{9.16}$$

where,

$$\{x\} = [q \ X_1]^{\mathrm{T}} \tag{9.17}$$

$$\{f\}^{\mathrm{T}} = [-1 \ \ 0] \tag{9.18}$$

$$[A] = \begin{bmatrix} 12 & -9 \\ -12 & 9 \\ 3 & -6 \\ -3 & 6 \\ 3 & 0 \\ -3 & 0 \end{bmatrix} \tag{9.19}$$

$$\{b\} = 7.93 \times 10^5 \, [1 \ \ 1 \ \ 1 \ \ 1 \ \ 1 \ \ 1]^{\mathrm{T}} \tag{9.20}$$

The MATLAB codes to solve the linear programming problem (9.16) are:

```
% Limit Analysis of a clamped-roller beam
f=-[1 0 ];
A=[ 12 -9
     -12 9
     3 -6
     -3 6
     3 0
     -3 0
];
b = [7.93 e5 7.93e5 7.93e5 7.93e5 7.93e5 7.93e5];
linprog (f, A, b)
```

which lead to:

$$p_{\mathrm{u}} = 264.3 \text{ kN} \tag{9.21}$$

EXAMPLE 9.5

Find the collapse load factor p of the frame shown in Figure 9.15.

Solution: Construct the primary system shown in Figure 9.16.

```
f=[-1 0 0 ];
A=[ 22.5 0.0 12.0
    -22.5 0.0 -12.0
    16.5 -3.0 12.0
    -16.5 3.0 -12.0
    16.5 -6.0 12.0
    -16.5 6.0 -12.0
    9.0 -6.0 9.0
    -9.0 6.0 -9.0
    4.5 -6.0 6.0
    -4.5 6.0 -6.0
    0.0 6.0 0.0
```

```
    0.0 -6.0 0.0
    4.5 0.0 6.0
    -4.5 0.0 -6.0
    0.0 0.0 3.0
    0.0 0.0 -3.0
];
b=7.93e5*[1 1 1 1 1 1 1 1 1 1 1 1 1 1 1 1 1 ];
S= linprog (f, A, b )
M_1=[22.5 0.0 12.0]*S
M_2=[16.5 -3.0 12.0] *S
M_3=[16.5 -6.0 12.0] *S
M_4=[ 9.0 -6.0 9.0] *S
M_54=[ 4.5 -6.0 0.0] *S
M_56=[ 0.0 6.0 0.0] *S
M_57=[ 4.5 0.0 6.0] *S
M_7=[ 0.0 0.0 3.0] *S
S= 1.0e+05 *
1.7622
-0.7526
-2.6433
M_A=7.9300e+05
M_EG=-7.9300e+05
M_G=-7.9300e+05
```

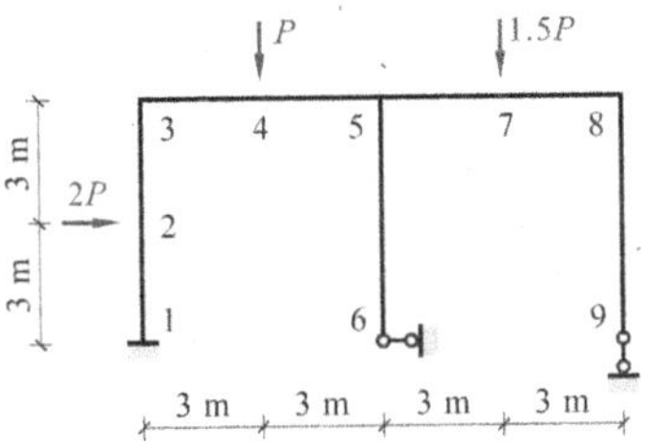

Figure 9.15. A frame system subjected to three external forces.

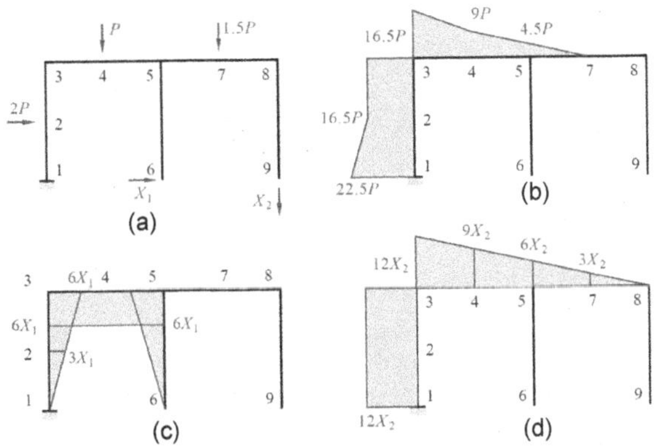

Figure 9.16. Construction of the primary system and corresponding moment diagrams.

EXERCISES

9.1 Two frames subjected to the proportional loads shown in Figure 9.17 have columns with a plastic moment $1.5M_P$ and beams with a plastic moment M_P. Determine the collapse load factors.

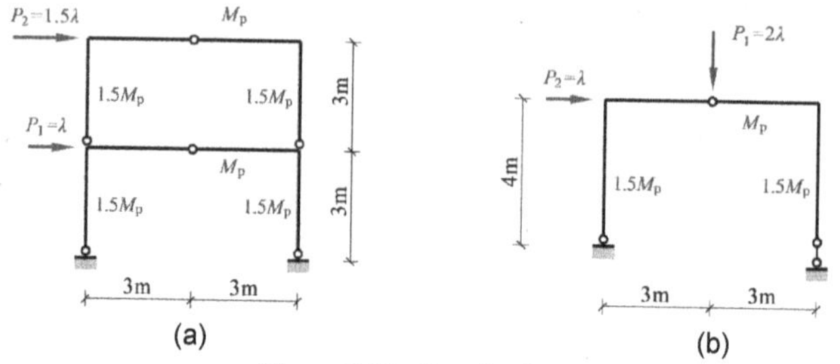

Figure 9.17. Exercise 9.1.

9.2 Determine the collapse load factors for the continuous beams shown in Figure 9.18. Plastic moments of the beams are indicated in the figures.

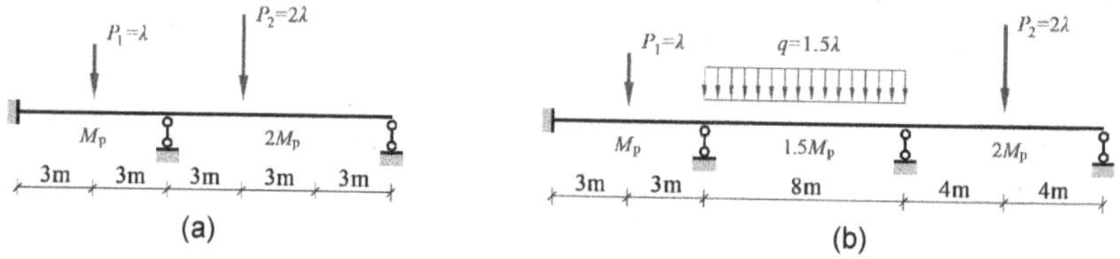

Figure 9.18. Exercise 9.2.

Appendix

Unit Conversions

Quantity	Unit	×	Unit	×	Unit
Length	in	25.4	mm	0.03937	In
	ft	0.3048	m	3.281	Ft
Area	in^2	645.16	mm^2	1550×10^{-3}	in^2
	ft^2	0.0929	m^2	10.76	ft^2
	ft^2	2.3×10^{-5}	Acre	43500	ft^2
Volume	cubic yard	21.7	Bushels	1.244	ft^3
	in^3	1.639×10^{-4}	mm^3	6.102×10^{-5}	in^3
Mass	slug	14.594	kg	2.205	lb_m
	slug	32.2	lb_m	0.454	Kg
Pressure	lb/in^2	6.895	kPa	0.1450	lb/in^2
	lb/ft^2	0.0479	kPa	20.89	lb/ft^2
	psi	6.895	kPa	0.1450	psi
	ksi	6.895	MPa	0.1450	ksi
	psi	6.895×10^{-6}	GPa	1.450×10^{-5}	psi
	ksi	6.895×10^{-3}	GPa	145	ksi
	lb_m/ft^3	16.02	kg/m^3	0.0624	lb_m/ft^3
Force	lb_f	4.448	N	0.2248	lb_f
Moment	lb·in	0.1130	N·m	8.851	lb·in
	lb·ft	1.356	N·m	0.7376	lb·ft
Energy	in·lb	0.113	Joule	8.85	in·lb
	kw-h	2655.18	klbs-ft	3.766×10^{-4}	kw-h
Velocity	m/s	2.237	mph	0.447	m/s
	mph	1.467	ft/s	0.682	mph
	mph	17.6	in/s	5.68×10^{-2}	mph
	knot	0.514	m/s	1.944	knot
Power	watt	0.73755	ft-lb/s	1.356	watt
	watt	1.341×10^{-3}	hp	745.7	watt

Acceleration of gravity (1 g) = 9.81 m/s^2 = 32.2 ft/s^2 (at latitude 45°)

Index

R

S

T

U

V

W

Y